FLORA & FAUNA HANDBOOK No. 11

INSECT AND SPIDER COLLECTIONS OF THE WORLD

(Second Edition)

Flora & Fauna Handbooks

This series of handbooks provides for the publication of book length working tools useful to systematics for the identification of specimens, as a source of ecological and life history information, and for information about the classification of plant and animal taxa. Each book is sequentially numbered, starting with Handbook No. 1, as a continuing series. The books are available on a standing order basis, or singly.

Each book treats a single biological group of organisms (*e.g.*, family, subfamily, single genus, etc.) or the ecology of certain organisms or certain regions. Catalogs and checklists of groups not covered in other series are included in this series.

The books are complete by themselves, not a continuation or supplement to an existing work, and do not require another work in order to use this one. The books are comprehensive, and therefore, of general interest.

The books in this series to date are:

Handbook No. 1, 1985. **THE SEDGE MOTHS,** by John B. Heppner

Handbook No. 2, 1992, 2nd printing, slightly revised. **INSECTS AND PLANTS: Parallel Evolution and Adaptation,** by Pierre Jolivet

Handbook No. 3, 1988. **THE POTATO BEETLES,** by Richard L. Jacques, Jr.

Handbook No. 4, 1988. **THE PREDACEOUS MIDGES OF THE WORLD,** by Willis W. Wirth and William L. Grogan, Jr.

Handbook No. 5, 1990. **A CATALOG OF THE NEOTROPICAL COLLEMBOLA,** by José A. Mari Mutt and Peter F. Bellinger

Handbook No. 6, 1990. **THE ENDANGERED ANIMALS OF THAILAND,** by Stephen R. Humphries and James R. Bain

Handbook No. 7, 1990. **A REVIEW OF THE GENERA OF NEW WORLD MYMARIDAE,** by Carl M. Yoshimoto

Handbook No. 8, 1990. **MAYFLIES OF THE WORLD: A Catalog of the Family and Genus Group Taxa,** by Michael D. Hubbard

Handbook No. 9, 1993. **CATALOG OF THE SOFT SCALE INSECTS OF THE WORLD,** by Yair Ben Dov

Handbook No. 10, 1993. **NORTH AMERICAN PSOCOPTERA,** by Edward L. Mockford

Handbook No. 11, (2nd Edition, 1993). **INSECT AND SPIDER COLLECTIONS OF THE WORLD,** by Ross H. Arnett, Jr., G. Allan Samuelson, and Gordon M. Nishida

Handbook No. 12, 1993. **SPIDERS OF PANAMA,** by W. Nentwig

THE INSECT AND SPIDER COLLECTIONS OF THE WORLD

By

Ross H. Arnett, Jr.
Florida State Collection of Arthropods
Gainesville, Florida

G. Allan Samuelson
Bishop Museum
Honolulu, Hawaii

and

Gordon M. Nishida
Bishop Museum
Honolulu, Hawaii

SECOND EDITION

CRC Press
Taylor & Francis Group
Boca Raton London New York

CRC Press is an imprint of the
Taylor & Francis Group, an **informa** business

CRC Press
Taylor & Francis Group
6000 Broken Sound Parkway NW, Suite 300
Boca Raton, FL 33487-2742

Reissued 2019 by CRC Press

A Library of Congress record exists under LC control number:

Publisher's Note
The publisher has gone to great lengths to ensure the quality of this reprint but points out that some imperfections in the original copies may be apparent.

ISBN 13: 978-0-367-25069-0 (hbk)
ISBN 13: 978-0-429-28586-8 (ebk)

Visit the Taylor & Francis Web site at http://www.taylorandfrancis.com and the CRC Press Web site at http://www.crcpress.com

Contents

INTRODUCTION

The countries of the world follow in alphabetical sequence. These countries have been taken from several sources and include necessary changes due to the recent reorganization of various political regions. A very thorough list of countries and islands, compiled by Gordon Nishida, has been incorporated along with the study of two new atlases. Place names in parentheses ( ) are cross-referenced to the countries to which they belong or to the current name now in use. Still, it is difficult to determine exactly what should be treated as a separate country. If they belong to a "commonwealth" but function independently, we have given them country status. At the present writing, we have listed 209 countries by our definition. During the course of the preparation of this edition, we have written to each of these government bodies and as a result have added many previously unlisted collections, or we have determined that no collection exists in these countries, and have so indicated. It is difficult to be certain if a "no-reply" means there is no collection, or simply that the letter was ignored. In the list that follows we have given the following: **Zoogeographic Realm, the capital city of each** (for possible future correspondence), **the population** (to give an idea of the usefulness of a collection in that region) and the, **the size in square miles** (to compare with the population). When we did not know of an existing collection we sent the questionnaire to the "Minister of Information" with good results. Otherwise we have written directly to the curators of known collections.

The **zoogeographic Realm** is given for each country, but of course, some include more than one Realm. Oceania is treated as a Realm. The older names "Ethiopian" and "Oriental" are changed to Afrotropical and Indomalayan respectively.

Unique **letter codens** are used to abbreviate the name and address of each collection for use in scientific papers and monographs to cite collection data. Codens are sometimes called "acronyms" but this term is inappropriate because usually an acronym is an abbreviation that is pronounceable. Acronyms, of course, are sets of initials, and so are codens. We have chosen to use the term "coden," which is defined as a code classification assigned to documents and consists *typically of four **capital** letters* (see Webster, 3rd ed.). We used this system for the Coleoptera collections (Arnett & Samuelson, 1969), and in the first edition of this work (Arnett, *et al.* 1986). These works have been widely cited in the literature and adopted by other directories and catalogs. We are asked why we insist on four letters: we don't; we have just followed the common useage for the sake of uniformity. We could use three, two, or even one letter, or going the other way, five, six, and so on. Almost everyone who has compiled a list of collection codes has used four letters. We think it is convenient. However, in all known cases, we have listed the three letter codens in the list and referred them to the four letter coden. We started that way, and it has been accepted by the majority of authors. That is satisfying.

This new edition. As the returns came in from the curators throughout the World, two changes became apparent. First, most collections have undergone considerable expansion. Perhaps this is the result of increased interest in cataloging the Earth's biota. Second, the staff and support for these collections have been greatly reduced. In many cases this is a critical situation. If it were not for the help of "overtime" work by teaching staff, students, and "amateurs," collections might have decreased instead of increased. This "renewed" interest is exciting, but not new. Perhaps some of the many millions of dollars that are spent on hysteric shouting and sign carrying could be diverted to true scientific cataloging of the Earth's biota, this book would show even greater progress.

Obviously, many collections were omitted from the first edition due to our lack of knowledge about collections, and many curators had not return the completed questionnaires at the time we went to press. To compile the first edition, we mailed, in March of 1983, the original questionnaire to all known museums. We used first class mail for mail to museum in U.S.A., Canada, and Mexico, and air mail to mail to the other countries. After a lapse of six months we mailed another questionnaire, also by first class or air mail, to all of those who did not respond to the original mailing. We also sent announcements to editors of many journals. These were published, giving us worldwide circulation. All of this was expensive and time consuming, and accounts, in part, for the long delay in the publication of the 1986 directory. Although several large museums failed to respond even after repeated letters, we decided to go to press without their input. Apparently we didn't reach the right person at these institutions. Since then, after the first edition was published, most of the large museums returned the forms. We have learned by experience that addresses and telephone numbers change rapidly even though the collections do not move! We have also learned that questionnaires are usually not welcomed and may be found deep in the pile of unanswered mail. Also, it is apparent that we have written to many places that do not have a collection of insects, or at least, only a few on display. Many of these institutions have not answered. We have removed several of these after carrying them for nearly a quarter of a century (1967 when we started, to 1992 when this edition was basically completed). Other institutions, known to have collections, an d have not replied, are listed, this time, with editorial notations, all constructive. After four or more attempts, we must conclude that for some reason, these institutions do not see the value of this project. We are dealing with a science that depends heavily on voucher material. The search for this material is time consuming for any researcher. We believe our efforts are time saving for these scientists.

All museums listed in the first edition were again contacted in June of 1992 for update and corrections. The results are included in this definitive edition. We call this a definitive edition because were are treating all known public collections of insects and hope we have produced a lasting volume. In most cases private collections are merged into the public collection with which they are affiliated. This adds to the permanent nature of this edition. We hope that the usefulness of this kind of directory will be apparent to those in other disciplines and that others will produce similar directories, using these codens, or adapting these to serve as a uniform system for all of systematic biology.

A copy of our single **questionnaire**, useable by both public and private collections is included at the end of the book. Please copy this form and return to the publisher for possible supplements.

PART I of this book is a list and description of the public collections known to us by the cut-off date of October 1, 1992. They are listed alphabetically by country. In some cases the curator did not return the proof of their entry in the 1986 edition sent with the questionnaire. We assume that there was very little change, so we used the entry as reported in the 1986 edition, and at the end of the entry, we inserted the date in brackets, i.e., [1986]. In all other cases we entered the date [1992], meaning the date we received the completed new questionnaire. In a very few cases in the U.S.A. we have never received information about a collection we know to exist, or we have only the information given in the 1969 book. We have made a note of this, and have written for the fourth time to each of these.

The designation "Director" means the curator or person in charge of the collection, not necessarily the director of the museum. This has given some concern. We have used "curator" wherever we have been asked to do so.

We have, rarely, abstracted the information provided by the curators, but usually we have included all information received, sometimes slightly edited for consistency. We continue to list most foreign collections even though they have not returned the forms, but we do not assert that the addresses are correct and we are not always certain that the collection does include insects other than those on display.

We have eliminated the separate section of private collections, but have included the description of these collections arranged alphabetically by the name of the owner, under the public collection with which they are affiliated. This list is far from complete and consists mainly of collections in the USA and Canada. In almost every case the private collections listed are registered with a public collection (see end of part I for those not registered). This registration procedure assures, in most cases, that the collection can be found even though the address of the owner changes or the collection is donated to a public institution. The private collection owner has selected the museum with which to register and has made arrangements for this with an appropriate curator.

We encountered one problem when we published a description of the private collections of Coleoptera (1969) found in North America, started to revise and expand it for the first edition of this work. Over 50% of the collections listed in the first volume, have been moved, donated, or their owners cannot be found, even though these collections contain voucher material and sometimes even primary types. In the current editiion of this directory this problem has been solved by requiring each private collection to be registered with a public collection. Data for this part have been compiled from forms that designate the collection registration. It should be made clear that this designation does not mean that the collection has been deposited in that museum, or that it will be in the future.

It does indicate, however, that if one is trying to locate a particular private collection, and mail is returned, one is able to contact a curator at the designated public collection for information on the current location of the private collection, although even this is not certain. We expect that this will help eliminate at least part of the previous problem.

In no case have we omitted collections not registered. We hope that all collectors will register their collections in the future, thus enabling them to be found after eventual change, which will help assure the scientific value of the collection. As you will see many private collections previously listed in the Coleoptera list (1969) are now "missing" (indicated by *address unknown* or *?*). Please send us any information you have on their whereabouts. All codens listed in the earlier work are repeated here because many of them have been used in the Coleoptera literature.

We must add a **disclaimer**. We cannot be held responsible either for the condition of the collections listed here or for any exchanges or loans negotiated with their owners. We do expect that the collections included are held by responsible individuals.

Part II is an alphabetical **list of 4-letter codens** with name of collection or owner, country, state/province, city and name of the collection. This cross-reference comprises the last part of this book. When a coden is used in the literature and the user wishes to find out what collection it refers to, use this section to find the collection in the text, details about the collection, who to write to, and where to write, call, or fax.

All known codens are given in this list. References to other published directories are cited at the end.

The **index to taxa** and the **index to personnel** complete this book.

Acknowledgments. We are grateful to the Florida State Collection of Arthropods, Division of Plant Industry, Florida Department of Agriculture, Gainesville, for help with the postage necessary for sending the original (1986) questionnaires for the first edition. Many persons have helped by sending corrections, addresses of collections, and in other ways. We particularly thank Dr. George E. Ball, University of Alberta, Dr. Philip S. Ward, University of California, Davis, Dr. Christopher K. Starr, University of the West Indies, Dr. Neal L. Evenhuis and Dr. Scott E. Miller, Bishop Museum, Honolulu, and Dr. Yair Ben-Dov, The Volcani Center, Israel. Without their aid and that of the hundreds of curators and private collectors in completing the questionnaires, this book would not be possible.

THE USE OF THE INFORMATION CONTAINED IN THE DIRECTORY

The main purposes of this compilation are to inform those doing systematic research of the availability of stored data in the form of specimens and associated information, and to help them arrange loans for the study of these specimens, or if loans cannot be arranged, to help them plan an itinerary of visits to collections. It is expected, of course, that those borrowing the material are responsible research scientists and that they will use these rules as guides for the proper treatment and use of specimens.

In the text of this directory, we use the phrases: "material may be borrowed for study; usual loan rules apply." The following are believed to be generally accepted rules for this procedure. In some cases curators of some collections have their own special rules. In other listings, no mention of borrowing specimens is made. **In each instance, however, individual arrangements must be made.** The following rules are only guides.

1. Specimens are shipped on loan to an established person whose interests in the proper handling of specimens has been demonstrated. Generally, specimens requested by students are sent to their major professors and the professors are responsible for their safe return. Shipment to the borrower is generally, but not always, at the expense of the shipper; material returned is sent at the expense of the borrower unless other arrangements have been made.

2. In general, unidentified material is sent out on loan. If requested by species, part of a series of identified specimens may be lent. Only rarely is an entire collection sent.

3. When special arrangements have been made, certain institutions will lend type specimens to established revisers. As a rule, primary types are not sent out to students. Part of a series of secondary types may be lent.

4. The sending of pin, vial, or slide data transcribed from the specimens should NOT be requested unless arrangements are made to pay for the labor involved.

5. All identified borrowed material, even if incorrectly identified when borrowed, must be returned unless specific and detailed arrangements have been made for the exchange of specimens or the gift of the material.

6. It is generally assumed that the primary type and at least half of the secondary type series of all new species will be returned to the collection from which it is borrowed. In some cases, collections (particularly

small private collections) do not keep types. Then it is the duty of the borrower to make specific arrangements for the deposit of the types in a suitable collection. The lender usually is the one to designate the depository, but the specialist may suggest a suitable collection. Types should be deposited in a public collection.

7. Part of the series of unidentified specimens sent on loan and identified by the specialist may be kept, but never without specific arrangements. However, usually up to one-half of the series may be kept. Single specimens from a locality are normally returned. Usually it is necessary to be certain that both sexes are represented from any single locality in the series of specimens to be returned.

8. If dissections of the genitalia or other parts are to be made, advanced arrangements and permission must be obtained. At the same time, agreement should be reached as to the method to be used for storing the dissected material.

9. It is the custom for the borrower to add an identification label to all specimens he identifies, even if the material has been previously identified.

10. **Curators are advised that it is improper to hold back part of a series and to add identifications to the held-back specimens after the specialist returns the remainder of the series.** It is not possible to assume that the series are all the same species. If the curator does not wish to lend all of the material of a group at one time (or as is usual, to hold back specimens from being retained by the identifier) the borrower should be notified of this and be given the opportunity to see the remainder of the series after the first part is returned. To do otherwise is unscientific and can easily lead to misidentification of specimens.

11. Labels rarely should be permanently removed from specimens. New pin labels are added at the bottom. Previous identification labels may be folded (if the paper will allow) and the new label added so it may be seen readily from above.

12. The borrower should keep a record of all borrowed specimens studied, indicating locality, number of specimens of each sex (where such may be readily determined), the collector's name, the date collected, and the name of the collection where the specimens of each species are deposited. Often photocopies of these data are returned with the borrowed material.

Though listed here, the user of this directory should not presume that permission has been granted by any institution or collection for the use of any specimens or any facilities described in these entries. Advanced arrangements with the curators of each institution or individual collector should be made on a personal basis.

When requesting the loan of specimens, be specific. Explain the nature of the research project, the exact group involved, to species if necessary, and the exact geographical areas to be covered. In families where the taxa may be difficult to recognize and sort, and specimens are abundant, the researcher should offer to visit the institution and sort the material to be borrowed. If this is not immediately possible, suggest that portions of the unidentified material be sent for sorting and immediate return. It is not uncommon for the borrower to send sketches and guides for the recognition of the taxa as a help to the curator in the sorting of the material. The desired length of time requested for the loan should be given as accurately as possible. If the deadlines for the return of material cannot be met, the lender should be informed and a loan extension requested.

Visits to collections are always arranged in advance with a request for work space and to reach an understanding as to what facilities will be available. In many places microscopes are available, but if possible, the researcher should take a microscope and "tool kit" containing those items needed to make dissections, mounting materials, and labels. Notes, record books, even literature, and other necessary materials should not be forgotten.

RULES FOR SUBMITTING SPECIMENS FOR IDENTIFICATION

Collectors and curators may not wait for a specialist to request material for study. Instead they often write to systematists and ask them to identify specimens. As with borrowing specimens, certain rules should be followed. The following are suggested procedures for this purpose.

1. Do not assume that specialists desire more specimens for their own sake; usually they are interested only in establishing new records or in additional species. The specialists are doing you a favor by spending valuable time to make identifications. At consultant wages, this could be expensive. Therefore, a request for identification is equivalent to asking for a substantial donation of time either from the individuals involved, or from their employing institution.

2. Never ship specimens for identification without making prior, detailed arrangements with the specialist, including:
   a. return shipping costs
   b. time that specimens are to be returned
   c. specimens to be retained as identification fee
   d. disposition of type specimens, if any.

3. All specimens submitted for identification must be:
   a. properly prepared and preserved
   b. provided with exact locality data
   c. sorted to group.

4. Never send bulk collections or unsorted collections with a request that the material be picked over to find things of interest.

5. Remember that a refusal to make identifications of large masses of material does not necessarily indicate a lack of interest, but sometimes just a lack of time or facilities.

6. Under certain circumstances, a specialist may request a fee for providing an identification. This practice has become nearly universal for court cases, commercial activities such as pest control operations, or environmental impact studies. Make sure that both you and the specialist agree on a fee, if any, before the identifications are made.

Note: We have included telephone numbers when supplied, but FAX numbers were not requested on the forms sent, so few are listed. However, now a new directory is available. "*1993 International FAX Directory for Biologists*" compiled by Neal L. Evenhuis may be obtained from: Pacific Science Association, P. O. Box 17801, Honolulu, HI 96817 USA.

PART I

LIST OF INSECT AND SPIDER COLLECTIONS ARRANGED ALPHABETICALLY BY COUNTRY

[19__] =Date received reply.

(Abu Dhabi, see United Arab Emirates.)

(Aden, see Yemen.)

(Afars and Issas, see Djibouti.)

1. AFGHANISTAN, Democratic Republic of

[Palearctic. Kabul. **Population:** 14,480,863; **Size:** 251,773 sq. mi. *Letter sent to Embassy, Washington, D. C., no reply. No known insect collection.*]

(Ajman, see United Arab Emirates.)

(Åland, see Finland)

2. ALBANIA, People's Socialists Republic of

[Palearctic. Tirana. **Population:** 3,147,352. **Size:** 11,100 sq. mi.]

INSECT COLLECTION, MUSEUM OF NATURAL HISTORY, STATE UNIVERSITY, TIRANA. [*Letter returned marked "Refused."*]. [MSUT] [1986]

(Aleutian Islands, territory of USA)

(Aldabra Island, see Seychelles.)

3. ALGERIA, Democratic and Popular Republic of

[Palearctic. Algiers. **Population:** 24,194,777. **Size:** 919,595 sq. mi.]

MUSÉE DE BENI ABBES, CENTRE NATIONAL DE RECHERCHE SUR LES ZONES ARIDES, BENI ABBES, W. BECHAR. [BAAC] [1986]

Director: Dr. N. Bounaga. Phone 649283. No professional staff. The collection consists of about 65 boxes of insects, half of which are arranged to family and otherwise identified. Several boxes of Coleoptera are from the Fezzan Mission of 1944. [1992]

4. AMERICAN SAMOA

[Oceanian (Polynesia). Pago Pago. **Population:** 39,254. **Size:** 75.2 sq. mi. Unincorporated territory of U.S.A.; *wrote to Pago Pago, 96799. No reply. An agricultural insect collection exists.*]

(Amirante Islands, see Seychelles.)

(Amsterdam Island, see New Amsterdam Island.)

(Andaman Islands, see India.)

5. ANDORRA, Valleys of

[Palearctic. Andorra la Vella. **Population:** 49,422. **Size:** 180 sq. mi. *No reply. No known insect collection.*]

6. ANGOLA, People's Republic of

[Afrotropical. Includes exclave of Cabinda. **Population:** 8,236,461. **Size:** 481,354 sq. mi.]

MUSEU DO DUNDO, RUA DE NOSSA SENHORA DE MUXIMA, LUANDA. [MDLA] [*No reply.*]

7. ANGUILLA

[Neotropical. British dependent territory: technically part of the British associated state of St. Christopher-Nevis-Anguilla, but administered as a British dependent territory. The Valley. **Population:** 6,875. **Size:** 36 sq. mi. No reply. *No known insect collection.*]

(Antarctica. Antarctic. The U.S.A. does not recognize sovereignty, but territorial limits may be noted by listing the unclaimed sector and the claimed sectors: Argentina, Australia, Chile, France, New Zealand, Norway, and the United Kingdom. *No known insect collections.*)

8. ANTIGUA and BARBUDA, State of

[Neotropical. St. John's. **Population:** 70,925. **Size:** 171 sq. mi. British Commonwealth: includes Barbuda and Redonda. *No reply. No known insect collection.*]

(Antipodes Islands, see New Zealand.)

(Archipelago de Colon, see Galapagos Islands, Ecuador.)

9. ARGENTINA, Argentine Republic

[Neotropical. Buenos Aires. **Population:** 31,532,538. **Size:** 1,068,302 sq. mi.]

DIVISION ENTOMOLOGIA, MUSEO ARGENTINO DE CIENCIAS NATURALES, AV. ANGEL GALLARDO 470 (C.C. 220, SUC. 5), 1405 BUENOS AIRES. [MACN]

Director: Dr. Axel O. Bachmann. Phone 982-7083. Nine curators. The collection consists of about one million specimens, half of them mounted (mainly pinned); of these about 200,000 are arachnids preserved in alcohol. The insect collection is housed mainly in Deyrolle boxes, some in drawers; the unpinned material is dry-preserved on cotton sheets in paper envelopes. Special collections include some or all of the following: H. Burmeister; C. Berg; C. Bruch; J. Brethes; A. Frers, G. Pellerano; A. Breyer; E. Dallas; E. Giacomelli; R. N. Orfila; A. Stevenin; A. Piran; S. Ruscheweg; S. Nosswitz; P. Mendoca; Lynch-Arribalzaga; S. Schajovskoy; A. O. Bachmann; E. E. Blanchard. [1992]

DEPARTAMENTO DE PATOLOGIA VEGETAL, INTA. C. C. No. 25, CASTELAR, BUENOS AIRES. [DPBA]. [*No reply.*]

MUSEO DE LA PLATA, DIVISION ENTOMOLOGIA, UNIVERSIDAD NACIONAL DE LA PLATA, PASEO DEL BOSQUE, 1900 LA PLATA. [MLPA]

Director: Dr. Richardo A. Ronderos. Phone: 021-218805. Professional staff: 18 curators. The collection has excellent coverage of Argentina and neighboring countries (Bolivia, southern Brazil, Chile, Paraguay, Peru, and Uruguay). All orders are represented but Collembola, Orthoptera, Hemiptera, Homoptera, Mallophaga, Anoplura, Coleoptera, Lepidoptera, Hymenoptera, and Thysanoptera are the most important. Nearly 3,500,000 specimens and 3500 types, and an equal amount of unpinned specimens are housed in 4,000 drawers, boxes, and on slides. A special Acarina collection is also housed in the Entomology Division. Limited material is available for exchange; loans for study to specialists is possible by official application. Special collections: Breyer; Berg; Bosq; Denier; Jorgensen; Ogloblin; Pertovsky; Richter; Tremoleres. Publication sponsored: "Revista del Museo de La Plata (Nueva Serie) Seccion Zoologia." [1986]

COLECCION ENTOMOLOGICA, CATEDRA DE ZOOLOGIA AGRICOLA, FACULTAD DE CIENCIAS AGRARICA, UNIVERSIDAD NACIONAL DE CUYO, ALTE BROWN 500, C. P. 5505, CHACRAS DE CORIA-LUJAN DE CUYO, MENDOZA. [CZAA]

Director: None listed. Phone: 960 004. Professional Staff: Ing. Agr. Guido S. Macola; Ing. Agr. Jose G. Garcia Saez; Ing. Agr. Silvio J. Lanati. Essentially this is an entomological collection of species from the province of Mendoza, with a few from other countries. The collection is housed in about 550 Deyrolle boxes. [1986]

INSTITUTO INVESTIGACIONES ENTOMOLOGICAS SALTA "INESALT," 9 DE JULIO 14, ROSARIO DE LERMA, SALTA 4405. [IIES]

Director: Mr. Manfredo A. Fritz. Phone: (087) 93-1023. Professional staff: Antonio Martínez (Coleoptera), and Manuel J. Viana (Coleoptera).

The collection contains about 50,000 specimens, mainly Coleoptera, Lepidoptera, and Hymenoptera; includes 150 primary types. [1992]

INSTITUTO PATAGONICO DE CIENCIAS NATURALES DE SAN MARTIN DE LOS ANDES, CASILLA DE CORREO #7, 8370 SAN MARTIN DE LOS ANDES, NEUQUEN. [IPCN]

Director: Sr. Mario O. Gentili. Phone: 0944-7208. Professional staff: Lic. Patricia Gentili. The collection consists of 60,000 pinned specimens, 70% of which are Lepidoptera, the remainder other groups; no spiders. Most of the collection is from Patagonia, south of the Rio Colorado, from the cold temperate forests and arid areas surrounding it (steppes and high Andes above the tree line). There is a special collection by the Fundacion Bariloche consisting of 6,500 pinned specimens from the Isla Victoria, Lago Nahuel Huapi, Neuquen. [1986]

FUNDACION E INSTITUTO MIGUEL LILLO, UNIVERSIDAD NACIONAL DE TUCUMAN, MIGUEL LILLO 251, 4000 TUCUMAN. [IMLA] [=FML]

Director: Lic. Zine D. A. de Toledo. Phone: 230056. Professional staff of 25. About 500,000 insects pinned and 2,300,000 in cotton beds. Housed in Deyrolle size boxes in metal cabinets (now being arranged in unit trays). Some specimens in alcohol including about 5,000 spiders. Publications sponsored: "Acta Zoologica Lilloana" (37 vols.), and "Opera Lilloana" (monographs, irregular). [1986]

10. ARUBA

[Part of Netherlands Antilles. Neotropical. Oranjestad. **Population:** 62,322. **Size:** 74.5 sq. mi. *Received reply. No insect or spider collections.*]

(Ascension Island, see St. Helena.)

(Ashmore and Cartier Islands, see Australia.)

(Auckland Islands, see New Zealand.)

(Austral Islands (=Tubuai), see French Polynesia).)

11. AUSTRALIA, Commonwealth of

[Includes Ashmore and Cartier Islands, Coral Sea Islands Territory (Cato Island and Diamond, Coringa and Willis Islets), Australian Antarctic Territory, and Macquarie Island (Australian). Canberra. **Population:** 16,260,436. **Size:** 2,966,151 sq. mi.]

AUSTRALIAN CAPITAL TERRITORY

Canberra

AUSTRALIAN NATIONAL INSECT COLLECTION, DIVISION OF ENTOMOLOGY, CSIRO, P.O. BOX 1700, CANBERRA CITY, A.C.T. 2601. [ANIC]

Director: Dr. Ebbe S. Nielson. Phone: (616) 246-4258. Fifteen professional staff. This is the largest and most comprehensive assemblage of specimens of Australian insects and related arthropods (mites and spiders) in the World. The collection currently comprises about ten million specimens of which about 15,000 are primary types. Most specimens are dry mounted and pinned, in over 12,000 cabinet drawers. Other specimens are either stored in liquid preservative or on microscope slides. [1992]

NEW SOUTH WALES

Beecroft

FOREST COMMISSION OF N. S. W., P. O. BOX 100, BEECROFT, N. S. W.. [FCNI]

Director: Mr. Robert Eldridge. Phone: 61-2-872-0111. Professional staff: Ms. Chris Ann Urquart, Ms. Deborah Kent. The Commission has one of the largest specialist collections of forest and timber insects in Australia and dates back to 1923, when W. W. Froggatt was appointed as New South Wales' first Forest Entomologist.

Currently, the pinned collection is housed in 400 drawers (estimated at 50,000 specimens). Most orders are represented but approximately 40% of the specimens are Coleoptera (Scarabaeoidea 5%; Bostrichoidea 4%; Chrysomeloidea 10%; Curculionoidea 6%). Lepidoptera form another 20% and a further 10% consists of voucher specimens. The alcohol stored collection consists of approximately 3,000 series of larval stages, 5,000 series of Isoptera, and 20,000 series of soil and litter fauna awaiting sorting. The slide collection is composed mainly of Psyllidae.

Although there are some paratypes in the collection, all holotypes are lodged with the Australian Museum, Sydney, or the Australian National Insect Collection, Canberra. The collection is used principally for reference and research; all material is available for loan to approved institutions. [1992]

Sydney

AUSTRALIAN MUSEUM, P. O. BOX A285, SYDNEY SOUTH N.S.W. 2000. [AMSA]

Director: Mr. M. S. Moulds. Phone: (02) 339-8221. Professional staff: Dr. D. Bickel, Mr. B. Day, Dr. D. McAlpine, Mr. M. Moulds. The collection contains about 4 million specimens, including approximately 3,500 primary and 6,000 secondary types. The majority of specimens are Australian but there is strong non-Australian representation for Psocoptera, Coleoptera, higher Diptera, and butterflies. Notable collections include the World's largest holdings of Psocoptera and Australian Acalyptrate Diptera, major collections of Neuroptera, Megaloptera, and Coleoptera, University of Sydney's School of Public Health and Tropical Medicine Acalyptrate Diptera, Hardy Diptera, Waterhouse butterflies, Phillips Fijian moths, Bock Diptera (Drosophilidae), Goldfinch moths, Evan's Homoptera, Elston Coleoptera, and McNamara New Guinea insects. The spider collection, the largest collection of Australian spiders, consists of 600,000 specimens, with 302 primary and 1092 secondary types. Special collections of spiders includes about 10,000 Australian cave spiders and the V. V. and J. L. Hickman collection of Tasmanian spiders.

The museum produces "Records of the Australian Museum; Technical Reports of the Australian Museum" and "Australian Natural History." [1992]

MACLEAY INSECT COLLECTION, MACLEAY MUSEUM, UNIVERSITY OF SYDNEY, SYDNEY, N.S.W. 2006. [MAMU]

Director: Dr. D. S. (Woody) Horning, Jr. Phone: (02) 692-3538. The Macleay Insect Collection is the oldest and most historic insect collection in Australia. There are more than 600,000 specimens in the collection dating from the 1750's and more than 60% are exotic, making it the largest collection of exotic insects in Australia. Over 9,000 types so far have been recognized in the collection but there are many yet to be discovered. The collection was started in Great Britain before 1800 by Alexander Macleay (1767-1848). He purchased and traded specimens with significant entomologists of the day. For instance, he purchased all of the non-British material from the Donovan collection, and a selection of material from Sir Ashton Lever's collection. These and other collections contain much type material. When he came to Australia as Colonial Secretary in 1826, he brought the finest and most extensive collection in the hands of a private individual at the time. His son, William Sharp Macleay, brought a large collection of Cuban and North American insects (including some from the Thomas Say collection) when he came to Australia in 1839. When William Sharp died in 1865, the collection was passed on to his nephew, Sir William John Macleay. He enlarged the collection greatly and presented it to the University of Sydney in 1888. George Masters was curator of the collection from 1872 to 1912. Unfortunately he relabelled many of the old insects, destroying the original labels. Thus it is difficult, if not impossible in some cases, to determine type material. The collection was neglected from his time to the early 1960's, hence it has suffered extensive damage. The collection is now housed in air-conditioned quarters and a full-time entomologist is curating it. The collection is stored in more than 1,400 drawers in wooden cabinets (some of them are Chippendale cabinets built in the 1780's), including 45 new 14 drawer cabinets. The insects are slowly being sorted and transferred to new wooden 14 drawer cabinets, using the unit tray system. The collection is almost totally unsorted but is accessible for searching. It is particularly strong in Coleoptera (especially Buprestidae, Cerambycidae, Scarabaeoidea, and Curculionidae) and Hymenoptera (especially Formicidae, Pompilidae, and Sphecoidea). Many historical collectors are represented including General Thomas Hardwicke, P. A. Latreille (a good friend of William Sharp Macleay), Sir Stamford Raffles, H. F. de Saussure, and A. R. Wallace. There are more than 2,000 vials of spirit material, most of which contains a world-wide spider collection. The collection is available for study; the usual loan rules apply. [1992]

Rydalmere

NEW SOUTH WALES AGRICULTURAL SCIENTIFIC COLLECTION TRUST, BIOLOGICAL AND CHEMICAL RESEARCH INSTITUTE, P. M. B., 10, RYDALMERE, N. S. W. 2116. [NSWA]

Director: Dr. M. J. Fletcher. Phone: (02) 683-9777. Professional staff: Dr. E. Schicha, and Mr. G. R. Brown. The collection is primarily an economic collection which serves as a reference for an identification service and as a basis for taxonomic research. It contains extensive records of economic insects and mites from New South Wales as well as significant holdings of non-economic groups from both N. S. W. and, to a lesser extent, other States of Australia. Very little non-Australian material is held. The estimated size of the collection is 300,000 pinned specimens, 17,500 alcohol specimens, and over 16,000 microscope slides (mostly mites). [1992]

NORTHERN TERRITORY

Darwin

NORTHERN TERRITORY MUSEUM of ARTS AND SCIENCES, P. O. BOX 4646, DARWIN, NORTHERN TERRITORY 0801. [MAGD]

Director: Dr. Alice Wells. Phone: (089) 824250. The collection consists of 250,000 insect specimens and 3,500 arachnids, including 27 holotypes and 362 paratypes. "The Beagle" is published by this institution. [1992]

QUEENSLAND

Saint Lucia

INSECT COLLECTION, DEPARTMENT OF ENTOMOLOGY, UNIVERSITY OF QUEENSLAND, SAINT LUCIA, QUEENSLAND. 4067. [UQIC]

Director: Miss Margaret A. Schneider (Curator). Phone: (07) 377-3656. This is a general collection of about 1,000,000 Australian insects and arachnids collected from all parts of Australia but especially the eastern states, with an emphasis on Queensland. There is also a small collection of foreign insects. The collection has representatives of most families and is particularly strong in Coleoptera (esp. Scarabaeoidea and Carabidae), Hymenoptera (Apoidea and Proctotrupoidea), and Diptera generally. The Diptera section includes a very large collection of Ceratopogonidae consisting of approximately 40,000 ethanol vials with a few to hundreds of specimens per vial. The mosquito collection is also large, named as completely as possible and representative of all parts of Australia. Holotypes described from the collection are deposited in the Queensland Museum, Brisbane. Material is normally not available for exchange but is available for loans to approved institutions and to individuals associated with such an institution; usual loan rules apply. There are no special collections; collections which have been acquired are incorporated into the general collection. The collection is housed in 420 metal cabinet drawers, 1,220 wooden cabinet drawers, 10 microscope slide cabinets, and 200 double-sided wooden storage boxes for excessive numbers of duplicate material. Soft-bodied insects and arachnids are stored in vials of ethanol contained in large jars of ethanol. No publications are sponsored. [1992]

Affiliated Collection:

Daniels, Greg, Department of Entomology, University of Queensland, Brisbane, Qld. 4072, Australia. [GDCB] This is a specialized collection of Diptera from Australia and the Pacific islands, with special emphasis on eastern Australia. There are 10,000 Asilidae, 8,000 Orthorrhapha of all families, excluding Empidoidea, and about 2,000 Syrphidae and Pipunculidae, housed in 100 drawers. The ecological data are limited. Material is available for loan and limited exchange; usual rules apply; specimens may be retained only by special arrangement, with the stipulation that holotypes be deposited in either the Australian Museum or the Queensland Museum [1992]. (Registered with UQBA.)

South Brisbane

QUEENSLAND MUSEUM, P. O. BOX 3300, SOUTH BRISBANE, QLD. 4101. [QMBA]

Director: Dr. Robert J. Raven. Phone: 61-7-840-7698. Professional staff: Mr. Philip Lawless. Chelicerata from Australia, New Caledonia, Fiji, New Guinea, about 80,000 specimens. Publish "Memoirs of the Queensland Museum."[1992]

Indooroopilly

QUEENSLAND DEPARTMENT OF PRIMARY INDUSTRIES, DIVISION OF PLANT PROTECTION, MEIERS ROAD, INDOOROOPILLY, QUEENSLAND 4068. [QDPC]

Director: Mr. J. F. Donaldson. Phone: (07) 877-9419. Professional staff: Miss J. F. Grimshaw. The collection has a bias toward pests of Queensland agriculture, and their predators and parasites. A large number of specimens have host data in addition to locality and other data. The records system is currently being computerised. The dry collection consists of about 275,000 pinned insect specimens, stored in trays in 1,334 drawers. There is a large collection of scale insects with 3,000 vials and 14,800 slides (mainly Diaspididae and Pseudococcidae). Some other minute whole insects and various dissected parts are stored on 11,300 slides. An alcohol collection of 7,700 vials contains various larvae and soft bodied insects.

The collection of spiders is stored 1,500 vials in metal drawers and an additional 2,000 vials, sorted to family are stored in large jars of alcohol. A slide collection of about 4,880 slides and 340 vials comprise the mite collection.

The collection does not keep any primary type material. The policy is to deposit primary types designated from the material in our collection to the Queensland Museum. Paratypes are retained. Material may be borrowed for study; usual loan rules apply. [1992]

Mareeba

DEPARTMENT OF PRIMARY INDUSTRIES, P. O. BOX 1054, MAREEBA, QUEENSLAND 4880. [QPIM]

Director: Mr. Ross I. Storey. This collection contains over 50,000 specimens housed in approximately 300 drawers and 100 boxes. The Department has had a collection in northern Queensland for about 50 years, concentrating on economic species. In recent years, however, much non-economic material has been accumulated especially in Coleoptera and Hymenoptera, so that now in Coleoptera, at least, it is one of the better collections available for tropical Queensland species. Only Australian material is held. [1986]

SOUTH AUSTRALIA

Adelaide

SOUTH AUSTRALIAN MUSEUM, NORTH TERRACE, ADELAIDE, SOUTH AUSTRALIA 5000. [SAMA]

Director: Mr. Lester Russell. Phone: 207-7500. Professional staff: Dr. Eric G. Matthews (Senior Curator of Insects); Dr. D. C. Lee (Senior Curator of Arachnids), Ms. J. A. Forrest OAM (Collection Manager of Insects) Mr. D. Hirst (Collection Manager of Arachnids). The collection has Australia wide and Indo Pacific representation plus some World specimens for comparative purposes. There are about 800,000 pinned specimens, 350,000 specimens in alcohol, 20,000 slides. There are 8,440 holotypes, of which 5,000 are from the A. M. Lea beetle collection. The collection includes insects and myriapods only. Arachnids are in a separate department. Special collections include: A. M. Lea and T. Blackburn (Coleoptera, minus holotypes which are in the Museum of Natural History, London); O. B. Lower, F. Angel, and N. McFarland (Lepidoptera); A. A. Girault and A. P. Dodd (Microhymenoptera). All collections are integrated. The dry insect collection is housed in 99 cabinets each with 50 drawers, specimens are in unit trays. Alcohol collection is housed in vials and bottles in wooden cabinets; slides in 900 slide boxes on shelves.

The arachnid collection is divided into two main parts: Acarina, and other arachnids. The Acarina collection comprises 50,000 slide mounted specimens, largely from Australasia and Indo-Pacific, and 4,000 vials of which much is partly sorted Berlesiate from Australia. It houses the collections of H. Womersley and R. V. Southcott. There are 600 holotypes and 2,000 secondary types. The collection of other arachnids has a good Australasian representation comprising 45,000 specimens in alcohol (mostly Araneae) and 350 slide mounts (mostly Pseudoscorpions). The W.

J. Rainbow spider types from Lord Howe and Norfolk Islands are among the 120 holotypes and 260 secondary types. [1992]

DUNCAN SWAN INSECT COLLECTION, DEPARTMENT OF ENTOMOLOGY, WAITE AGRICULTURAL RESEARCH INSTITUTE, UNIVERSITY OF ADELAIDE, GLEN OSMOND, 5064 SOUTH AUSTRALIA. [WARI]

Director: Dr. A. D. Austin (curator and senior lecturer in systematic entomology). Phone (08) 372-2265; 372-2273. One technical officer assistant. The collection has Australia-wide representation with significant material of parasitic Hymenoptera from S. E. Asia. Of particular comprehensive representation are the Coccoidea, parasitic Hymenoptera and Orthoptera of Australia; the remainder of the collection is essentially a synoptic collection of relevance to agriculture in southern Australia. The collection contains approximately 250,000 pinned specimens in 350 drawers fitted with unit trays; an alcohol preserved collection (10,000 vials) of arachnids and immature stages of insects, and a slide collection of 8,000 slides (40% of which are Coccoidea), complete the collection. The

SOUTH AUSTRALIAN DEPARTMENT OF AGRICULTURE [SADA], at ADELAIDE will merge with the Duncan Swan collection, above, in 1994. [1992]

TASMANIA

Hobart

TASMANIAN DEPARTMENT OF AGRICULTURE, HOBART, TASMANIA. [TDAH]

Curator: T. D. Semmens. Phone: (002) 784338. Professional staff: Dr. P. B. McQuillan, Dr. J. E. Ireson, Mrs. M. A. Williams. The collection contains about 145,000 specimens, including 600 types. Further details may be obtained from the curator. [1992]

TASMANIAN MUSEUM & ART GALLERY, GPO BOX 1164 M, HOBART, TASMANIA. [TMAG]

Curator: Mr. Roger Buttermore, Acting Curator of Invertebrate Zoology. Phone: (002) 231422. The collection contains about 20,000 specimens of insects, including 128 type specimens, and about 4,500 spiders, including 15 types. [1992]

TASMANIAN FOREST INSECT COLLECTION, FORESTRY COMMISSION, GPO BOX 207 B, HOBART, TASMANIA 7001. [FCTH]

Director: Mr. R. Bashford. Phone: (002) 338153. The collection consists of 160 drawers holding approximately 11,500 pinned specimens, especially Cerambycidae and Buprestidae of Tasmanian forest trees. The insects of eucalypts are well represented with Chrysomelidae being the major family. Wet collection of immature stages consists of 30 tube drawers containing approximately 1,500 insects and other invertebrates including spiders and snails. [1992]

VICTORIA

Burney

VICTORIAN AGRICULTURAL INSECT COLLECTION, DEPARTMENT OF FOOD & AGRICULTURE, INSTITUTE OF PLANT SCIENCE, SWAN STREET, BURNLEY, VICTORIA 3121. [VDAM]

Director: Dr. M. Malipatil. Phone: 61-03-8101-511. Approximately 40,000 specimens (70% dry; 25% wet and 5% on slides, ost from Victoria. About 75 cotypes of Australian Coleoptera described by A. M. Lea; the oldest specimen in the collection was collected in 1863. Computer database of collection in progress. [1992]

Melbourne

DEPARTMENT OF ENTOMOLOGY, MUSEUM OF VICTORIA, 71 VICTORIA CRESCENT, ABBOTSFORD, VIC-TORIA 3067. [MVMA]

Curator: Dr. A. Neboiss (retired); Ass't Curator: Mr. K. L. Walker. Total holdings of insect and arachnid collection (dry and in alcohol) in excess of 2.5 million specimens. Approx. 10,000 primary insect types and 500 arachnid types. Collection emphasis is on Australian fauna, with representation varying from group to group. Recently there has been a strong development of Trichoptera, Plecoptera, Diptera: Chironomidae, and Hymenoptera: Colletidae. Special collections include J. Curtis collection (mainly British insects, before 1860, with numerous types); Francis Walker (all insect orders, including specimens collected by Wallace); Compte de Castelnau (Coleoptera, worldwide, including types); Godeffroy collection (part of, mostly Australian and S.W. Pacific, also arachnids); A. W. Howitt (Australian Coleoptera); and G. Lydell (Australian Coleoptera). The collection is housed in standard insect cabinet drawers, partly arranged in unit trays; the alcohol collection storage consists of vials grouped systematically in glass jars. Publication: "Memoirs of the Museum of Victoria." [1986]

WESTERN AUSTRALIA

Como

DEPARTMENT OF CONSERVATION & LAND MANAGEMENT, P. O. BOX 104, COMO, WESTERN AUSTRALIA 6152 [CLMP]

Director: Mr. T. E. Burbridge. Phone (09) 3670305. Curators: Como collection: Dr. I. Abbott, Mr. A. J. Wills; Curator Monjimup collection: Dr. J. D. Farr, Mr. S. G. Dick. The collection is located at two sites: Manjimup (sub collection) and Como (main collection). The Manjimup collection was recently established (1988), with 710 specimens pinned and in alcohol. The Como collection has 6,540 pinned specimens, representing approximately 200 species, and 3,000 specimens in alcohol, representing about 800 species. Voucher specimens: 17 pinned, 12 in alcohol. [1992]

Perth

DEPARTMENT OF ZOOLOGY, UNIVERSITY OF WESTERN AUSTRALIA, NEDLANDS, PERTH, WESTERN AUSTRALIA 6009. [USAC]

No information received about the main collection. The Barbara York Collection [BYMC], pertains only to the spider collection, which consists of about 7,000 specimens from Australia and New Guinea. These are stored in alcohol vials and arranged taxonomically. The vials are stored in large glass bottles. [1986]

SPIDER AND INSECT COLLECTION, WESTERN AUSTRALIAN MUSEUM, FRANCIS ST., PERTH, WESTERN AUSTRALIA 6000. [WAMP]

Acting Director: Mr. M. V. Robinson. Phone (09) 328-4411. FAX (09) 328-8686. Curator: Dr. T. F. Houston. The insect collection consists chiefly of Western Australia species although there is an assortment of small collections from elsewhere in Australia and the world at large. There are approximately 200,000 specimens on pins and 5,000 lots in alcohol. The accession rate is currently about 5,000 specimens annually. Over 400 primary types are held, a list of which was published in 1980. Ecological data are few and are limited to recent material. Specimens are available for systematic studies subject to the usual rules. The collection is stored in unit trays in glass-topped steel drawers, stored in sealing steel cabinets (50 each). The museum publishes: "Records of the Western Australian Museum."

The collection of Arachnids and Myriapods (Dr. Mark S. Harvey, Curator; phone 328-4411) contains about 150,000 specimens stored in alcohol in bottles and jars, with about 500 primary types. The collection is mainly of spiders from Western Australia. Material may be borrowed for study; usual loan rules apply. Library facilities limited. A list of primary types is published in the "Records of the Western Australian Museum." [1992]

Affiliated Collection:

Baker, Dr. F. H. U., Western Australian Museum, Francis St., Perth, Western Australia 6000, Australia. [FHUB] This collection contains about 3,000 Australian Curculionoidea with over 900 named species. (Registered with WAMP.)

12. AUSTRIA, Republic of

[Palearctic. **Population:** 7,577,072. **Size:** 32,377 sq. mi.]

Admont

NATURHISTORISCHES MUSEUM DER BENEDIKTINER-ABTEI ADMONT, 8911 ADMONT, FOUNDED 1866. [NMBA] [*No reply; probably no insect collection.*]

Dornbirn

VORARLBERGER NATURSCHAU, MARKESTRASSE 33, 6850 DORNBIRN, A-6850. [VNGA]

Director: Mag. Herbert Waldegger. Phone: 05572-23235. The complete collection includes only animals from Vorarlberg: Coleoptera, *ca.* 20,000 species; Lepidoptera, *ca.* 8,000 species; other insects, *ca.* 6,000 species. There is no collection of spiders in the museum. (Founded 1960.) [1992]

Eisenstadt

BURGENLANDISCHES LANDESMUSEUM, MUSEUMGASSE 5, 7000 EISENSTADT, (FOUNDED 1926). [BLGA] [*No reply; probably no insect collection.*]

Graz

STEIERMÄRKISCHES LANDESMUSEUM JOANNEUM, ABTEILUNG FÜR ZOOLOGIE, RAUBERGASSE 10, A-8010 GRAZ, (FOUNDED 1811). [SLJG]

Head curator: Dr. Karl Adlbauer. Phone: 011-43-316-8772662. Professional staff: Dr. Karl Adlbauer, Dr. Ulrike Hausl-Hofstätter. This old museum (founded 1811) contains over 40 separate collections gathered by such people as W. Bernhauer (*Carabus*) and many others. A complete list is available from the curator. The total number of specimens exceeds 150,000. [1992]

Innsbruck

INSTITUT FÜR ZOOLOGIE DER UNIVERSITAT INNSBRUCK, UNIVERSITATSSTRASSE 4, 6020 INNSBRUCK. [IZUI] [*No reply; probably no insect collection.*]

TIROLER LANDSMUSEUM FERDINANDEUM, MUSEUMSTRASSE 15, 6020 INNSBRUCK, [TLMF]

Director: Prof. Dr. Gert Ammann; Head Curator: Dr. Gerhard Tarmann. Phone: 0512-572284. The insect collection at this museum (founded 1823) consists of a total of about 700,000 specimens, 400,000 Alpine Lepidoptera (including 20 holotypes), 250,000 Palearctic Coleoptera, with about 40 holotypes, 50,000 other orders of insects, mainly from the Alps. Publication: "Veröffentlichungen des Tiroler Landesmuseum Ferdinandeum." [1992]

Klagenfurt

LANDESMUSEUM FÜR KÄRNTEN, ZOOLOGISCHE ABTEILUNG, MUSEUMGASSE 2, A-9020 KLAGENFURT. [NHMK]

Director: Univ. Doz. Dr. Gernot Piccottini. Phone (0463) 536-30551. Professional staff: Dr. Paul Mildner. The collection includes mainly specimens of the local insect fauna. In 300 drawers there are represented the following groups, 90% of them collected in the federal state of Carinthia (Kärnten): Odonata, Saltatoria, Dermaptera, Blattodea, Hemiptera, Hymenoptera, Coleoptera, Neuroptera, Trichoptera, Lepidoptera, and Diptera. There are about 100,000 specimens. A special collection of Coleoptera (about 40,000 specimens) from Carinthia is housed in boxes. [1992]

Kremsmunster

STIFTSSAMMLUNGEN DES BENEDIKTINERSTIFTSKREMS-MUNSTER, 4550 KREMSMUNSTER, (FOUNDED 1700). [SBKA] [*No reply; probably no insect collection.*]

Salzburg

HAUS DER NATUR, MUSEUMSPLATZ 5, 5020 SALZBURG. [HNSA]
Director: Prof. Dr. Eberhardt Stüber. Phone: 0043-662-84-2653. Staff: Volunteers from the Entomological Group of the Haus der Natur: Fritz Mairhuber, Gernot Embacher, Elisabeth Gaiser. The collection consists of about 400,000 specimens housed in 1,230 drawers. It contains a local collection of Salzburg Lepidoptera and Coleoptera. Publication: "Haus der Natur: Jahresbericht." [1992]

Wien

AMT DER NIEDERÖSTERREICHISCHEN LANDSREGIERUNG, POSTFACH 6, 1014, WIEN. [ANLW] [*Replied: no research insect collection.*] [1992]

NATURHISTORISCHES MUSEUM WIEN, POSTFACH 417, BURGRING 7, 1040 WIEN. [NHMW]
Director: Hofrat Univ. Doz. Mag. Dr. Maximilian Fischer. Phone: 52177-316. Professional staff: 7 curators. The collection contains insects of all orders, from all zoogeographical regions, with about 6 million specimens stored in glass topped drawers. Publication sponsored: "Annalen des Naturhistorischen Museums Wien." [1992]

SAMMLUNG DER FORSTLICHEN BUNDESVERSUCHSANSTALT WIEN, A-1131 WIEN, SECKENDORFF GUDENTWEG 8. [FBWA]
Professional staff: Ing. Carolus Holzschuh. Phone: 0222-87838-131; FAX 8775907. The collection consists of 400 drawers of Central European Lepidoptera, Coleoptera, and Hymenoptera; no types. Publication: "Koleopterologische Rundschau" 1975-1989, vols. 52-65. [1992]

(Azores, Island Districts, see Portugal.)

13. BAHAMAS, Commonwealth of

[Neotropical. Excludes Turks and Caicos Islands, which is a British

Colony. Nassau. **Population:** 242,983; **Size:** 5,382 sq. mi. *No reply; no known insect collection.*]

14. BAHRAIN, State of

[Palearctic. Al-Manamah. **Population:** 480,383. **Size:** 268 sq. mi. *No reply; no known insect collection.*]

(Baker Island, see U.S.A.)

(Balearic Islands, Island Province of Spain, see Spain.)

(Ball's Pyramid Island, see Australia.)

(Banaba Island (=Ocean I.), see Kiribati.)

15. BANGLADESH, People's Republic of

[Indomalayan. Dhaka. **Population:** 109,963,551. **Size:** 55,598 sq. mi.]

ZOOLOGY MUSEUM, UNIVERSITY OF DHAKA, DHAKA-2. [ZMUD] [*No reply; may have an insect collection.*]

(Banks Islands, see Vanuatu.)

16. BARBADOS

[Neotropical. Bridgetown. **Population:** 256,784. **Size:** 166 sq. mi.]

BARBADOS MUSEUM AND HISTORICAL SOCIETY, ST. ANN'S GARRISON. [BMGB]

Director: Mr. David Devenish, B.A. Phone: 427-0201. Professional staff: 3 curators. Much of the collection is unclassified, but this should be corrected in the near future. Lepidoptera, mostly butterflies, 17 drawers; Coleoptera, 2 drawers; Hymenoptera, 1 drawer. The collection is stored in a wooden cabinet fitted with 20 drawers. Publication sponsored: The Journal of the Barbados Museum and Historical Society. [1986]

CARIBBEAN AGRICULTURAL RESEARCH INSTITUTE, IPC PROJECT COLLECTION, SUGAR TECHNOLOGY RESEARCH UNIT, EDGE-HILL, ST. THOMAS. [CARD]

Director: Dr. Ralph Phelps (Acting). Phone: 4250074. Professional staff: M. M. Alam, Entomologist-in-charge. The collection is stored in Cornell cabinets with drawers and in insect boxes. There is a total of 1,321 species of insects recorded from Barbados. A checklist published in 1985. [1986]

(Barbuda, see Antigua.)

(Bassas da India, see Mozambique.)

(Basutoland, see Lesotho.)

(Bear Island (Bjornoya), Island of Svalbard, Norway.)

(Bechuanaland, see Botswana.)

(Belau, see Palau.)

17. BELGIUM, Kingdom of

[Palearctic. Brussels. **Population:** 9,880,522. **Size:** 11,783 sq. mi.]

Brussels

COLLECTIONS NATIONALES BELGES D'INSECTES ET D'ARACHNIDES, INSTITUT ROYAL DES SCIENCES NATURELLES DE BELGIQUE, 29, RUE VAUTIER, B1040, BRUSSELS. [ISNB]

Director: Dr. Patrick GrootaertIng. Phone: (02) 648-0475. Professional staff: Dr. Leon Baert. Some 8.5 million insects and arachnids comprise this collection in separate collections for Belgium and the world; many types (some lists have been published); material available for study loans. There are also public exhibits. The library is very rich in entomological literature. The entomological part is managed by the entomology section which also contains the library of the "Societe royale belge d'Entomologie." The collection contains specimens from the following collections: Carpentier, Fauvel, D'Orchymont, Fagel, Roelofs, Candeze, Putzeys, brothers Oberthur (unidentified), Lheureux, Hoschek, Deuquet, J. Thomson, Gillet, Dejean, De Castelnau, Chapuis, Boppe, Laboissiere, Wesmael, Tosquinet, Bondroit, Predhomme de Borre, and Becker. The collection is housed in insect boxes, 12 alcohol jars and vials, and on microscope slides. The collection of the "Societe royale belge d'Entomologie" is housed here. Publication sponsored: "Bulletin de l'Institut royal des Sciences naturelles de Belgique, Entomologie." [1986, update by eds. 1992]

Gembloux

COLLECTIONS ZOOLOGIQUES, ZOOLOGIE GENERALE & APPLIQUÉE, FACULTE DES SCIENCES AGRONOMIQUES, GEMBLOUX, B 5030. [FSAG]

Director: Pr. C. Gaspar. Professional staff: Ir. Charles Verstraeten and Miss Camille Thirion. Large collection of insects and other Arthropoda from Europe (Belgium, France, Spain, Italy, etc.), Africa (Morocco, Central African Republic, Rwanda, Zaire, etc.); few specimens from other continents, except for Hymenoptera, Sphecoidea. About 450,000 specimens representing major orders. The types are principally Hymenoptera, Sphecidae (from Dr. J. Leclercq), Pompilidae (R. Wahis); Diptera, Tabanidae (Dr. M. Leclercq); Coleoptera, Elateridae (Dr. L. Laurent); Homoptera, Fulgoroidea (Dr. V. Lallemand). Special collection of Protozoa Thecamoeba with many types of Mr. D. Chardez. [1992]

Liege

MUSÉE DE ZOOLOGIE, UNIVERSITÉ DE LIEGE, INSTITUT DE ZOOLOGIE, QUAI VAN BENEDEN, 22, B-4020, LIEGE. [IEVB]

Director: Prof. Dr. Jean-Claude Ruwet. Phone: 19-32-41-66500.02. The collection contains the Edmond de Laeveyre collection of 400 boxes, and Léon Candèze collection of the Lepidoptera of the World stored in 14 cabinets. [1992]

Louvain-la-Neuve

COLLECTION DU LABORATOIRE D'ECOLOGIE, PLACE CROIX DU SUD NO. 5, LOUVAIN-LA-NEUVE 1348. [ECOL]

Director: Prof. Ph. Lebrun. The collection consists of world Papilionidae, Geometridae, and Acari, mainly Oribatids. [1986]

Tervuren

SECTION D'ENTOMOLOGIE, MUSEE ROYAL DE L'AFRIQUE CENTRALE, LEUVENSESLEENWEG 13, B-3040 TERVUREN. [MRAC]

Director: Dr. Rudy Jocqué. Phone (32) 2-769-5401. Professional staff: Mr. F. Puylaert. This is a general collection of insects and spiders of the Afrotropical region, especially from Zaire, Rwanda, Burundi, mountainous areas of East Africa and Ivory Coast. Contains the greatest part of the collections of the former "Institut des Parcs natio naux Congo." Includes a special collection of St. Helena. The 10 million insect specimens and 510,000 spiders (including primary types are housed in boxes, alcohol vials, or on microscope slides. Publications sponsored: "Journal of African Zoology," "Revue de Zoologie Africaine," and "Annales du Musee royal de l'Afrique centrale, Sciences zoologiques." [1992]

18. BELIZE

[Neotropical. Formerly British Honduras, **Population:** 171,735; **Size:** 22,965 sq. mi.]

NATIONAL INSECT COLLECTION, CENTRAL FARM, CAYO DISTRICT. [NICC]

Curator: John E. Link. Phone: (501) 092-2129; 092-2131; FAX (501) 092-2640. The collection contains about 950 species. [*No further information available; information received from Christopher Starr.*]

19. BENIN, People's Republic of

[=Dahomey (Afrotropical) Porto-Novo. **Population:** 4,497,150. **Size:** 43,484 sq. mi.

OPEN AIR MUSEUM OF ETHNOGRAPHY AND NATURAL SCIENCES, PARAKOU. [OAMB] [*No reply; probably no insect collection.*]

20. BERMUDA, Colony of

[Neotropical. British Colony. Hamilton. **Population:** 58,137. **Size:** 21 sq. mi.]

BERMUDA DEPARTMENT OF AGRICULTURE AND FISHERIES, BOTANICAL GARDENS, P.O. BOX HM 834, HAMILTON 5. [BMHP]

Director: Dr. I. W. Hughes. Phone (809) 296-4201. Professional staff: Mr. Kevin D. Monkman. This is a general collection covering most insect orders found in Bermuda. Specimens are housed in 58 Cornell drawers and 450 alcohol vials. Ecological data are limited. There is a special collection of Lepidoptera, mainly from New York and New Hampshire. Most of the 871 specimens were collected by A. Zerkowitz. It is housed in chipboard insect boxes. [1986]

21. BHUTAN, Kingdom of

[Indomalayan/Palearctic. Timphu. **Population:** 1,503,180. **Size:** 17,954 sq. mi.]

TADZONG MUSEUM, PARO. [TDMP] [*No reply. Probably does not have a collection of insects.*]

(Bioko, see Equatorial Guinea.)

(Bismarck Archipelago, see Papua New Guinea.)

(Bjornoya, see Bear Island.)

22. BOLIVIA, Republic of

[Neotropical. La Paz. **Population:** 6,448,297. **Size:** 424,164 sq. mi.]

COLECCION ENTOMOLOGICA, MUSEO NACIONAL DE HISTORIA NATURAL, CASILLA 5829, LA PAZ. [ANCB]

Director: Sr. Eduardo Forno. Phone: 795-364. The collection consists of about 3,750 specimens, mostly Lepidoptera. There is no spider collection. The collection is used for reference and is stored in boxes. [1986]

INSTITUTO DE ECOLOGIA, UNIVERSIDAD MAYOR DE SAN ANDRES, CASILLA 20127, LA PAZ. [UMSA] [*No reply.*]

COLECCION BOLIVIANA DE FAUNA, CASILLA 8706, LA PAZ. [CBFC]

Director: Dr. Jaime Sarmiento. Phone: 795364; FAX 591-2-797399. The Coleccion Boliviana de Fauna (CBF) unified the zoological collections which were separately implemented from the ANCB (entry above) and the Instituto de Ecologia. Presently the holdings are about 60,000 specimens of vertebrates and invertebrates, including 8,000 butterflies and 25,000 other insects and invertebrates. [1992]

MUSEO DE HISTORIA NATURAL "NOEL KEMPFF MERCADO" [UASC] [*No reply.*]

(Bonaire, see Netherlands Antilles.)

(Bonin Islands, Islands of Japan, see Ogasawara Arch.)

(Bophuthatswana, see South Africa.)

(Borneo, Island of East Indies, see Sabah, Sarawak, Brunei, Kalimantan.)

23. BOSNIA-HERZEGOVINA, Republic of

[Palearctic. Sarajevo. Formerly a part of Yugoslavia. (*At the time of this writing uncertain conditions prohibit details. Mail not delivered.*)]

ZEMALJSKI MUJSKI, SARAJEVO. [ZMSZ] [*No reply, 1986.*]

24. BOTSWANA, Republic of

[=Bechuanaland. Afrotropical. Gaborone. **Population:** 1.189,000. **Size:** 224,711 sq. mi.]

NATIONAL MUSEUM AND ART GALLERY, CHURCHILL ROAD, INDEPENDENCE AVE., P.O. BOX 131, GABERONE. [MAGB]

Director: A. C. Campbell. Phone: 53792. No entomological staff. Approximately 1,000 specimens, mainly Coleoptera of Botswana, represent a collection which was started this year. These are housed in boxes. The museum sponsors "Botswana Notes and Records," an annual journal. Volume 14 is current.

(Bougainville Island, part of Papua New Guinea.)

(Bounty Island, see New Zealand.)

(Bouvet Island, see Norway.)

(Bouvetoya, see Norway.)

25. BRAZIL, Federative Republic of

[=Brasil. Includes Atol das Rocas, Archipelago de Fernando de Noronha, Ilha da Trindade, Ilhas Martin Vaz, and Penedos de São Pedro e São Paulo. Neotropical. Brasília. **Population:** 150,685,145. **Size:**

3,286,488 sq. mi. *Although the response from many collections is provided, there are still the names of many other collections that have been supplied to us but no replies have been received after three attempts. It is likely that many of these do not have taxonomic collections.*]

M. ALVARENGA. Location of collection unknown. [MACB]

ALAGOAS

DEPARTAMENTO DE BIOLOGIA, SETOR DE ZOOLOGY, CENTRO DE CIENCIAS BIOLOGICAS, UNIVERSIDADE FEDERAL DE ALAGOAS, PRACA AFRANIO JORGE S/N, 57.000 MACEIO, ALAGOAS. [CCUF] [*No reply.*]

AMAPA

MUSEU TERRITORIAL DE HISTORIA NATURAL "ANGELO MOREIRA DA COSTA LIMA," SECRETARIA DE EDUACACAO E CULTURA, AV. FELICIANO COELHO, 1509, 68.900 MACAPA, AMAPA. [AMCL] [*No reply.*]

AMAZONAS

LABORATORIO DE ZOOLOGIA, DEPARTAMENTO DE CIENCIAS MORFOLOGICAS, INSTITUTO DE CIENCIAS BIOLOGICAS, FUNDACAO UNIVERSIDADE DO AMAZONAS, CAMPUS UNIVERSITARIO 69.000 MANAUS, AMAZONAS. [DCMB] [*No reply.*]

COLECÃO SISTEMÁTICA DA ENTOMOLOGIA (INPA), INSTITUTO NACIONAL DE PESQUISAS DA AMAZÔONIA, ESTRADA DO ALEIXO, 1756, C. P. 478, 69.011 MANAUS. [INPA]

Director: Dr. Seixas Lourenço. Phone: 642-3377, FAX 092-642-3340. Curatorial staff: Dr. Célio Magalhães (General Curator), BitNet: CELIOMAG@BRFUA.BITNET; Dr. José Albertino Rafael (Diptera); Dr. Cláudio Ruy Vasconcelos da Fonseca (Coleoptera); Dra. Ana Yoshi Harada (Hymenoptera). The systematic collection is located in the Collection Building on the INPA campus. This is one of the largest collections of Amazonian invertebrates in the world. Presently there are 930 primary and secondary types. The collection currently contains more than 500,000 specimens prepared and approximately two million preserved in alcohol awaiting specialists for sorting and identification. About 210,000 insects in this collection are pinned, labelled, and separated to family. These are stored in 42 insect cabinets in an air-conditioned room. About 3,000 specimens are slide mounted. The collection is almost exclusively an Amazonian collection, although there are a few specimens from other regions. Publication sponsored: "Acta Amazonica." [1992]

BAHIA

CENTRO DE PESQUISAS DO CACAU, CEPEC, CEPLAC, DIVISAO DE ZOOLOGIA AGRICOLA, C. P. 7, 45.600 ITABUNA, BAHIA. [CPDC] [*No reply.*]

CEARA

CENTRO DE CIENCIAS AGRARIAS, UNIVERSIDADE FEDERAL DO CEARA, CAMPUS DO P. I. C. I., C. P. 354, 60.000 FORTALEZA, CEARA. [CCAC] [*No reply.*]

DISTRITO FEDERAL [=BRASILIA]

[Prior to moving the capital to Brasilia, Distrito Federal was in Rio de Janeiro. During the earlier period many works treated the fauna of "Distrito Federal" in Rio de Janeiro. Thus, we now identify the district with Brasilia to avoid any confusion.- V. O. Becker]

CENTRO DE PESQUISAS AGROPECUARIAS DO CERRADO, EMBRAPA, C. P. 73.300 PLANTINA, D.F. [CPAC]

Contact: Dr. Vitor O. Becker. The collection contains over 30,000 specimens, mostly Lepidoptera, from central Brazil. Most of these have been reared and identified. [1992]

Affiliated Collection:

Becker, Dr. Vitor O. [VOBC]. This is a collection of over 200,000 moths from the Neotropical region. About half is Brazilian; the rest is Central American and Antillean. Over 20,000 species are represented, with a little over half identified. All material is spread, sorted, and housed in 1,000 Cornell drawers in a concrete room with humidity control. [1992] (Registered with CPAC.)

FACULDADE DE FILOSOFIA CIENCIAS E LETRAS, NOSSA SENHORA DO PATROCINIA, RUA MADRE M. BRASILIA, 965, 13.300 ITU, DF. [FFCL] [*No reply.*]

DEPARTAMENTO DE BIOLOGIA ANIMAL, INSTITUTO DE CIENCIAS BIOLOGICAS, UNIVERSIDADE DE BRASILIA, 70.910 BRASILIA, D. F. [DBAI] [*No reply.*]

MATO GROSSO

INSTITUTO DE BIOCIENCIAS, UNIVERSIDADE FEDERAL DE MATO GROSSO, AV. FERNANDO CORREA S/N, 78.098 CUIABÁ, Mato Grosso. [UFMI]

Director: Ms. Mirian Arabela da Silva Serrano, M.Sc. Phone: (+55) 65 315-8872. Professional staff: Ms. Helena Antonia Gusmao Pinheiro Duarte and Ms. Marinex Issac Marques. This is a collection of the fauna of Mato Grosso, central-east Brazil, which includes savannah and flooded grassland, swamps and flood plains. The collection includes about 15,000 mounted specimens, and more than twice this number of unmounted specimens, or about 80 insect drawers. Aquatic insects are specialities, with emphasis on Chironomidae. [1992]

MINAS GERAIS

MUSEUM OF ENTOMOLOGY, UNIVERSIDADE FEDERAL DE VIÇOSA, 36.750 VIÇOSA, MG. [UFVB]

The Museum is organized according to a model of international museums. All specimens are pinned or on points, with labels, arranged by family and stored in unit trays in glass topped museum drawers placed in wooden cabinets, which are carefully fumigated. The collection is a regional collection with more than 90% of its specimens from the state of Minas Gerais, mostly collected in Viçosa. Currently there is an estimated 74,000 specimens dry mounted and an additional 15,000 specimens stored in fluid. By orders there are 15,000 Coleoptera, 7,000 Heteroptera, 2,500 Homoptera, 8,200 Lepidoptera, 35,000 Hymenoptera, 3,600 Diptera, 1,300 Orthropteroids, and 1,200 miscellaneous orders. [1992]

DEPARTAMENTO DE FITOSSANIDADE, ESCOLA SUPERIOR DE AGRICULTURA, LAVRAS, C. P. 37, 37.200 LAVRAS, MG. [DFLC] [*No reply; probably no insect collection.*]

DEPARTAMENTO PARASITOLOGIA, INSTITUTO DE CIENCIAS BIOLOGICAS, AV. ANTONIO CARLOS 6627, C. P. 2486, 30.000 BELO HORIZONTE, MG. [DPIC] [*No reply.*]

MUSEU DE HISTORIA NATURAL, UNIVERSIDADE FEDERAL DE MINAS GERAIS, RUA GUSTAVODA SILVEIRA S/N, 30.000 BELO HORIZONTE, MG. [BHMH] [*No reply.*]

PARA

CENTRO DE PESQUISAS AGROPECUARIAS DO TROPICO UMIDO, EMBRAPA, TRAV. DR. ENEAS PINHEIRO S/N, C. P. 48, 66.000 BELEM, PA. [CPAP] [*No reply.*]

FACULDADE DE CIENCIAS AGRARIAS, UNIVERSIDADE FEDERAL DO PARA, AV. PERIMETRAL S/N, C. P. 917, 66.000 BELEM, PA. [FCAP] [*No reply.*]

DEPARTAMENTO DE ENTOMOLOGIA, MUSEU PARAENSE EMILIO GOELDI, C. P. 399, BELEM, PARA 66000. [MPEG]

Director: Dr. William L. Overal. Professional staff: five entomologists. The collection is a regional collection of Brazilian Amazon Basin. About 500,000 specimens are included among the pinned material which is housed in 1,120 glass-topped drawers. Most important among these collections is the Adolph Ducke Hymenoptera collection and the J. and B. Bechyne collection of Chrysomelidae. Important holdings also include collections of Vespidae (O. W. Richards), Papilionoidea (K. S. Brown, Jr.),

Dermaptera (Brindle), Blattoidea (I. Rocha), Collembola in alcohol and on slides (R. Arle), Isoptera in alcohol (A. G. Bandeira), Tabanidae (G. B. Fairchild), Scarabaeidae (B. Ratcliffe), and Formicidae (W. W. Kempf and K. Lenko). With the exception of termites and ants, the alcohol collections are small and include about 18,000 specimens, including about 1,000 spiders. A separate collection of Berlese samples is maintained. Publications sponsored: "Boletim do Museu Paraense Emilio Goeldi, Serie Zoologia." [1986]

PARAIBA

DEPARTAMENTO DE SISTEMATICA E ECOLOGIA, CENTRO DE CIENCIAS EXATAS E DA NATUREZA, UNIVERSIDADE FEDERAL DA PARAIBA, 58.000 JOAO PESSOA, PB. [DSEC] [*No reply.*]

PARANA

CENTRO NACIONAL DE PESQUISAS DA SOJA, EMBRAPA, RODOVIA CELSO GARCIA KM. 375, C. P. 1061, 86.100 LONDRINA, PR. [CNPS] [*No reply.*]

DEPARTAMENTO DE BIOLOGIA, FUNDACAO UNIVERSIDADE FEDERAL DE MARINGA, AV. COLOMBO 3690, C. P. 331, 87.100 MARINGA, PR. [DPUP] [*No reply.*]

DEPARTAMENTO DE CIENCIAS MORFOLOGICAS, UNIVERSIDADE FEDERAL DO PARANA, C.P. 756, 80.000 CURITIBA, PR. [DCMP] [*No reply.*]

MUSEU DE ENTOMOLOGIA PE. JESUS SANTIAGO MOURE, UNIVERSIDADE FEDERAL DO PARANÁ, DEPARTAMENTO DE ZOOLOGIA, C.P. 19020, 81531-970 CURITIBA, PR. [DZUP]

Curator: Ms. Maria Christina de Almeida. Phone: (041) 266-3633, Ramal 146; FAX (041) 266-2042. Professional staff: Twenty specialists; list and further details available on request. The collection contains about 3,000,000 specimens: Neotropical Lepidoptera (260,000 specimens, 244 primary types), with about 90% of the collection composed of Coleoptera, Hymenoptera (Vespidae with 3 primary and 103 secondary types), Diptera, Homoptera, and Heteroptera, with representation of the other orders. The private collection deposited are: Romualdo Ferreira D'Almeida, David Gifford, Richard Frey, Paulo Gagarin, Heinz Ebert, and Felipe Justus. The institute sponsors "Acta Biologica Paranaense." [1992]

MUSEU DE HISTORIA NATURAL CAPAO DA EMBUIA, RUA PROF. NIVALDO BRAGA 1225, 80.000 CURITIBA, PR. [MNCE] [*No reply.*]

RIO DE JANEIRO

DEPARTAMENTO DE BIOLOGIA ANIMAL, UNIVERSIDADE

SANTA URSULA, RUA FERNANDO FERRARI, 75, 22.231 RIO DE JANEIRO, RJ. [DBAU] [*No reply.*]

INSTITUTO DE BIOLOGIA, UNIVERSIDADE FEDERAL DO RIO DE JANEIRO, KM 47 ANTIGA RODOVIA RIO SÃO PAULO, 23.460 SEROPEDICA (ITAGUAI), RJ. [IBUS] [*No reply.*]

FUNDACAO INSTITUTO OSWALDO CRUZ, AV. BRASIL 4365, C.P. 926, 20.000 RIO DE JANEIRO, RJ. [FIOC] [=FOCB] [*No reply.*]

MUSEU NACIONAL, QUINTA DA BOA VISTA, SÃO CRISTOVAO, 20.942 RIO DE JANEIRO, RJ. [QBUM]

Director: Miguel A. Monné. Phone: (021) 264-8262. This is a collection 165,000 specimens of Cerambycidae (Coleoptera), including 668 primary types. There are 4,830 identified species. Nearly 750,000 specimens are not yet pinned. Publication sponsored: "Boletim Museu Nacional."

RIO GRANDE DO NORTE

ESCOLA SUPERIOR DE AGRICULTURA, AV. PRESIDENTE COSTA E SILVA S/N, C.P. 137, 59.600 MOSSORO, RN. [ESRN] [*No reply.*]

RIO GRANDE DO SUL

DEPARTAMENTO DE ZOOLOGIA, INSTITUTO DE BIOLOGIA, UNIVERSIDADE FEDERAL DO RIO GRANDE DO SUL, AV. PAULO GAMA, S/N, 90.000 PORTO ALEGRE, RS. [UFRG] [*No reply.*]

FACULDADE DE AGRONOMIA E VETERINARIA, UNIVERSIDADE FEDERAL DO RIO GRANDE DO SUL, AV. BENTO CONCALVES, 7712, C. P. 776, 90.000 PORTO ALEGRE, RS. [FAVU] [*No reply.*]

MUSEU ANCHIETA, AV. NILO PECANHA, 1521, C.P. 358, 90.000 PORTO ALEGRE, RS. [MGAP] [*No reply.*]

MUSEU DE CIENCIAS, PONTIFICIA UNIVERSIDADE CATOLICA DO RIO GRANDE DO SUL, AV. IPIRANGA, 6681 PREDIO 10, C.P. 1429, 90.000 PORTO ALEGRE, RS. [MCPU] [*No reply.*]

MUSEU DE CIENCIAS NATURAIS DA FUNDAÇAO ZOO-BOTÂNICA DO RIO GRANDE DO SUL, AV. DR. SALVADOR FRANCA, 1427 C. P. 1188, 90.000 PORTO ALEGRE, RS. [MCNZ] [=MCN]

Director: Ms. Christina T. Guimarães Gresele. Phone: (051) 336-1079, FAX (051) 336-1778. Professional staff: Mr. Claudio José Becker, Ms. Hilda Alice de Oliveira Gastal, Dra. Magali Hoffman, Maria Elizabeth Lanzer de Souza, Dra. Maria Helena Mainieri Galileo, Ms. Erica

Helena Buckup, and Maria Aparecida de Leão Marques. The insect collection is composed of 67,207 specimens, the great majority of which came from Rio Grande do Sul, but there are specimens from other states of Brazil and also from other countries of South America. All major orders are respresented. There are holotypes and paratypes of Hemiptera (mainly Pentatomidae), Coleoptera (mainly Cerambycidae), Hymenoptera, and Mantodea. The spider collection contains about 22,000 viles with an estimated 40,000 specimens from the same regions as the insects. There are 344 holotypes and paratypes, representing 102 species in eight families. [1992]

SANTA CATARINA

DEPARTAMENTO DE BIOLOGIA, CENTRO CIENCIAS BIOLOGIA UNIVERSIDADE FEDERAL DE SANTA CATARINA, CAMPUS UNIVERSIDADE TRINDADE, 88.000 FLORIANOPOLIS, SC. [SCUF] [*No reply.*]

SÃO PAULO

DEPARTAMENTO DE BIOLOGIA, FACULDADE DE FILO-SOFIA CIENCIAS E LETRAS DE RIBEIRÃO PRETO, UNIVERSIDADE DE SÃO PAULO, AV. DOS BANDEIRANTES S/N, 14.000 RIBEIRÃO PRETO, SP. [RPSP] [*Collection exists; no reply.*]

DEPARTAMENTO DE CIENCIAS BIOLOGICAS, UNIVERSIDADE FEDERAL DE SÃO CARLOS, RODOVIA WASH-INGTON LUIZ KM 235, C.P. 676, 13.560 SÃO CARLOS, SP. [DCBU] [*No reply.*]

DEPARTAMENTO DE DEFESA FITOSSANITARISTA, FACULDADE CIENCIAS AGRARIA E VETERINARIA DE JABOTICABAL, 14.870 JABOTICABAL, SP. [DDFF] [*No reply.*]

DEPARTAMENTO DE ENTOMOLOGIA, ESCOLA SUPERIOR DE AGRICULTURA LUIZ DE QUEIROZ, UNIVERSIDADE DE SÃO PAULO, AV. DR. ARNALDO 01255 SP. [DEES] [*Collection exists; no reply.*]

DEPARTAMENTO DE EPIDEMIOLOGIA, FACULDADE DE SAÚDE PÚUBLICA, UNIVERSIDADE DE SÃO PAULO, AV. DR. ARNALDO, 715, 01.255 SÃO PAULO, SP. [DEFS] [*Collection exists; no reply.*]

DEPARTAMENTO DE ZOOLOGIA, UNIVERSIDADE ESTADUAL DE CAMPINAS, C.P. 1170, 13.100 CAMPINAS, SP. [DZIB] [*No reply.*]

INSTITUTO AGRONÔMICO DE CAMPINAS, SEÇÃO DE ENTOMOLOGIA, CAIXA POSTAL 28, CAMPINAS, SP. 13001. [IACC]
Director: André Luiz Lourenção (M. Sc.). Phone: (0192) 41-5188.

Professional staff: Cesar Pagoto Stein, M.S. The collection has as its priority the collection of, cataloging, and identification of phytophagous insects, host plants and their natural enemies. Each insect added to the collection receives two labels, the collecting data and host plant information, and an accession number which refers to the series of single species collected at this locality. The collection has a card index system, a card for each number, which is used to correlated the data on each species, with information on the nature of damage, presence or absence of natural enemies, necessity of control, and other observations. Only specimens with host plant data are included in the collection. The collection has good series of *Dysdercus* (Hemiptera: Pyrrhocoridae) of all areas, arranged and identified by the late Dr. Luiz Otavio Teixeira Mendes, specialist on the group. The collection was started in 1936 by Dr. L. O. T. Mendes and at present has 7140 entries. [1992]

INSTITUTO DO ACUCAR E DO ALCOOL, PLANALSUCAR, C.P. 158, 13.600 ARARAS, SP. [IAAA] [*No reply.*]

INSTITUTO BIOLOGICO, SECRETARIA DA AGRICULTURA, AV. CONSELHEIRO RODRIGUES ALVES, 1252, 04.604 SÃO PAULO, SP. [IBSP] [*No reply.*]

INSTITUTO BUTANTÃ, AV. VITAL BRASIL, 1500, C.P. 65, SÃO PAULO, SP. [IBUT] [*No reply.*]

INSTITUTO FLORESTAL, SECRETARIA DA AGRICULTURA, HORTO FLORESTAL, C. P. 1322, 01.000 SÃO PAULO, SP. [IFSA] [*No reply.*]

INSTITUTO DE PESQUISAS TECNOLOGICAS, C.P. 7141, 01.000 SÃO PAULO, SP. [IPTB] [*No reply.*]

MUSEU DE ZOOLOGIA DA UNIVERSIDADE DE SÃO PAULO, BIBLIOTECA, 7172, 01.051 SãO PAULO, SP. [MZSP]

Director: Dr. Paulo Emilio Vanzolini. Phone: 274-3455, or 274-3690. Professional staff: Dr. Eliana M. Cancello, Dr. Sonia Casari Chen, Dr. Nelson Papavero, Dr. Carlos Roberto Brandãao, Dr. Francisca Carolina do Val, Dr. Cleide Costa, Dr. Ubirajara R. Martins, and Dr. Mirian David Marques. Over 2 million specimens, stored in drawers housed in steel cabinets, in alcohol vials, and on slides. Publications sponsored: "Arquivos de Zoologia" and "Papeis Avulsos de Zoologia." [1992]

SERGIPE

DEPARTAMENTO DE BIOLOGIA, UNIVERSIDADE FEDERALDE SERGIPE, RUA VILA CRISTINA, 1061, 49.000 ARACAJU, SE. [DBSE] [*No reply.*]

(Britain, see United Kingdom.)

(British Antarctic Territory, U. K. Dependent Territory.)

(British Guiana, see Guyana.)

(British Indian Ocean Territory, U. K. Dependent Territory, including Chagos Archipelago. *Uninhabited.*]

26. BRITISH VIRGIN ISLANDS, Colony of the Virgin Islands

[Neotropical. U.K. Dependent Territory: includes Anegada, Jost Van Dyke, Tortola, and Virgin Gorda. Road Town. **Population:** 12,075. **Size:** 59 sq. mi. *No known insect collection.*]

27. BRUNEI, Commonwealth State

[Indomalayan. British protected State (=Borneo). Bandar Seri Begawan. **Population:** 316,565. **Size:** 2,226 sq. mi.]

BRUNEI MUSEUM, KOTA BARU. [BMKB] [*No reply.*]

28. BULGARIA

[Palearctic. Sofia. **Population:** 8,966,927. **Size:** 42,823 sq. mi.]

INSECT COLLECTION, NATIONAL MUSEUM OF NATURAL HISTORY, BULGARIAN ACADEMY OF SCIENCES, BOULV. TSAR OSVOBODITAL, BG-1000 SOFIA. [SOFM]

Director of Museum: Assoc.-Prof. Dr. Krassimir. Phone: +3592-88 28 94. Scientific Secretary of the Museum and Curator of the Insect Collection: Assoc.-Prof. Dr. Alexi Popov. Phone: +3592-88-51-15 (714). Professional staff (Three Entomologists): Dr. Kumanski, Dr. Popov, Mrs. Vassila Jordanova, and one technician. About 480,000 specimens housed in boxes and alcohol vials. Journal sponsored: Journal for the Museum (including insects): "Historia naturalis bulgarica" (started in 1989). [1992]

Affiliated Collection:

Ganev, Julius, 84, Rakovski Str., BG-1000 Sofia, Bulgaria. [JMFC] A large collection of 30,000 specimens of Lepidoptera and 4,000 specimens of Coleoptera, this collection is representative of the fauna of Bulgaria. It includes 300 specimens of tropical Lepidoptera and a special collection of Crambidae from Palearctic. The collection is stored in boxes. (Registered with SOFM.) [1992]

INSECT COLLECTION, INSTITUT OF ZOOLOGY, BULGARIAN ACADEMY OF SCIENCES, BOULV. TZAR 1, 1000 SOFIA. [ZISB]

Director: Prof. Vasil Golemanski. Phone: 88 47 08. FAX 88 51 15. Professional staff: Dr. Venelin Beschovski, Prof. Dr. Michail Josifov, Dr. Vassil Gueorguiev, Dr. Z. Hubenov. Ms. E. Vassileva. The collection, stored in 300 boxes, each with about 20,000 specimens, including approximately 150 types, from Europe, Asia Minor, Korea, Vietnam (undetermined), and South Africa (undetermined). Dr. Vladimir Sakalian also has 200 species of Buprestidae from Europe and Asia. Publications sponsored: "Acta Zoologica Bulgarica" and "Hydrobiology." [1992]

29. BURKINA FASO

[(=Upper Volta.) Afrotropical. Ouagoudougou. **Population:** 8,485,737. **Size:** 105,870 sq. mi. *No known insect collection.*]

30. BURMA, Socialist Republic of the Union of

[=Myanmar. Indomalayan. Rangoon. **Population:** 39,632,183. **Size:** 261,218 sq. mi.]

NATURAL HISTORY MUSEUM, FORMER BOAT CLUB, ROYAL LAKE, RANGOON. [NHMC] [*No reply.*]

31. BURUNDI, Republic of

[Afrotropical. Bujumbura. **Population:** 5,155,665. **Size:** 10,747 sq. mi. *No known insect collection.*]

(Caicos Island, see Turks and Caicos Islands.)

32. CAMBODIA

[=Democratic Kampuchea, formerly, Khmer Republic. Indomalayan. Phnom Penh. **Population:** 6,685,592. **Size:** 69,898 sq. mi. *No known insect collection.*]

33. CAMEROON, United Republic of

[=French Cameroon. Afrotropical. Yaoundé. **Population:** 10,531,954. **Size:** 183,569 sq. mi.]

DOUALA MUSEUM, P.O. BOX 1271, DOUALA. [DMDC] [*No reply.*]

(Campbell Island, see New Zealand.)

34. CANADA

[Nearctic. Ottawa. **Population:** 26,087,536. **Size:** 3,535,303 sq. mi.]

RIVEREDGE FOUNDATION. [RFAC]. Collection transferred to Provincial Museum of Alberta.

ALBERTA

BIOLOGY DEPARTMENT, UNIVERSITY OF CALGARY, CALGARY T2N IN4. [BDUC] [*No reply; probably a teaching collection.*]

DEPARTMENT OF BIOLOGY, CAMROSE LUTHERAN COLLEGE, CAMROSE T4V 2R3. [CLCC] [*No reply; probably a teaching collection.*]

DEPARTMENT OF FISHERIES AND FORESTRY COLLECTION, Calgary. [*Transferred to Forest Research Laboratory, Edmonton.*] [DFFA]

ALBERTA AGRICULTURE, PLANT INDUSTRY LABORATORY, 9TH FLOOR, O. S. LONGMAN BLDG., 6909 116 ST., EDMONTON T6H 4P2. [AAPI] [*No reply; probably no research collection.*]

NORTHERN FOREST RESEARCH CENTRE, FORESTRY CANADA, NORTHWEST REGION, 5320 122 ST., EDMONTON T6H 3S5. [NFRC]

Director: Mr. A. D. Kiil. Phone: (403) 435-7210. Professional staff: Dr. D. W. Langor (curator), Dr. W. J. A. Volney, Dr. H. F. Cerezke, and 4 technicians. The collection was started in 1939 and now consists of over 120,000 specimens of adult and immature insects and arachnids, mostly from the Canadian Prairies. This collection was originally formed by an amalgamation of collections based in Indian Head, Saskatchewan; Calgary, Alberta; and Winnipeg, Manitoba. Taxonomists associated with this collection were the late G. R. Hopping (Scolytidae), G. N. Lanier (Scolytidae), Dr. W. C. McGuffin (Geometridae), and Dr. H. R. Wong (Tenthredinidae). The special collections consist of forest and shade tree insects of the Canadian Prairies. Adult insects are stored in drawers housed in aluminum and wooden cabinets. Immatures, soft-bodied insects, and spiders are stored in ethanol in glass vials. Many mites have been mounted on microscope slides. [1992]

STRICKLAND ENTOMOLOGICAL MUSEUM, DEPARTMENT OF ENTOMOLOGY COLLECTION, UNIVERSITY OF ALBERTA, EDMONTON T6G 2E3. [UASM]

Director: Dr. George E. Ball. Phone: (403) 492-2084. FAX: (403) 492-1767. Professional staff: Mr. Danny Shepley. The collection contains between 500,000 and 1,000,000 specimens pinned insects, with small specimens preserved in alcohol, or on slides. For most orders, holdings consist principally of material from western Canada, and series of individual species are small. Holdings of Coleoptera are continent-wide for North America. Carabidae are especially well represented: the collection is world-wide, but mostly North American and Mexican. Ecological data are limited and spotty. Material is available for loan, under usual rules. No holotypes are kept in the collection. The collection contains the F. S. Carr collection of Coleoptera and the Kenneth Bowman collection of Lepidoptera of Alberta. The collection is housed in USNM type drawers which are stored in wood or metal cabinets. Slides are stored in metal cabinets and alcohol vials are in clear plastic racks. Visitors who wish to study material are welcome. [1992]

PROVINCIAL MUSEUM OF ALBERTA, 12845 102 AVENUE, EDMONTON T5N OM6. [PMAE]

Director: Dr. P. H. R. Stepney. Phone: (403) 453-9100. Professional staff: Dr. A. T. Finnamore and T. W. Thormin. The collection consists of about 213,000 specimens (80% Hymenoptera) of world-wide insects representing the major orders. The collection is housed in Cornell drawers and cabinets. [1992]

LETHBRIDGE RESEARCH STATION, CROP SCIENCES SECTION, AGRICULTURE CANADA, LETHBRIDGE TIJ 4BI. [AGRL]

Director: B. H. Sonntag. Phone: (403) 327-4561; FAX (403) 382-3156. Section Head: Dr. K. W. Richards. The collection consists of over 200 drawers in cabinets representing all major orders as a regional reference collection; no types. [1992]

OLDS COLLEGE, PLANT SCIENCE DEPARTMENT, OLDS, AB TOM IPO. [OCOA]

Mr. Ernest Mengersen. [*No further information available about this collection.*]

PLANT SCIENCES, ALBERTA ENVIRONMENTAL CENTRE, VEGREVILLE TOB 4LO. [PSAE]

Director: Dr. Sherman Weaver. Phone (403) 632-6761. Professional staff: Mr. H. G. Philip, Dr. M. Liu, Mrs. M. Y. Steiner. Several thousand specimens of many orders of insects and other arthropods, primarily pest species received by the diagnostic laboratory since 1969. A special collection of approximately 80 species of Alberta leafhoppers is maintained. Pinned specimens are kept in boxes; alcohol preserved specimens in vials. [1986]

BRITISH COLUMBIA

AGRICULTURE CANADA RESEARCH STATION, SUMMERLAND VOH IZO. [ACBC] [*No reply.*]

THE APHIDS OF BRITISH COLUMBIA, AGRICULTURE CANADA RESEARCH STATION, 6660 N.W. MARINE DRIVE, VANCOUVER V6T 1X2. [ACBV]

Director: Dr. A. R. Forbes. Phone: 224-4355. Professional staff: C. K. Chan, technician. The collection consists of 405 species of aphids mounted on slides collected from 1178 different host plants. [1992]

SPENCER MUSEUM, DEPARTMENT OF ZOOLOGY, UNIVERSITY OF BRITISH COLUMBIA, VANCOUVER V6T 1Z4. [SMDV]

Director: Dr. G. G. E. Scudder. Phone: (604) 822-3379 or 3682. Professional staff: Mr. S. G. Cannings, Curator. The Spencer Entomological Museum is the largest collection of insects in the province, and is probably the largest collection of British Columbia insects in the world. It currently contains approximately 600,000 specimens of which about 500,000 are pinned and housed in about 1,200 Cornell-type drawers;

75,000 are stored in alcohol in vials, and about 25,000 are mounted on microscope slides. Included in the collection are the Whitehouse, Bucknell, and Cannings and Stuart Odonata, the Bucknell and Spencer Orthoptera, the Dones and Scudder Hemiptera, the Spencer Anoplura, Mallophaga, and Siphonaptera, the Blackmore, Llewellyn-Jones, and Kimmich Lepidoptera, the Stace-Smith and Guppy Coleoptera, and the Foxlee Diptera and Hymenoptera. [1992]

PACIFIC FOREST RESEARCH STATION, ENVIRONMENT CANADA, 506 W. BURNSIDE ROAD, VICTORIA V8Z IM5. [PFRS]
[*No reply.*]

ROYAL BRITISH COLUMBIA MUSEUM, 675 BELLEVILLE ST., VICTORIA V8U IX4. [BCPM]
Executive Director: William D. Barkley. Phone: (604) 387-3685. Professional staff: Mr. Robert A. Cannings, Mr. Crispin Guppy. The collection is predominantly British Columbia material, 182,544 dry specimens of insects, and approximately 40,000 wet and unsorted specimens in alcohol. Stored in USNM, Cornell, and Cornell modified drawers. Six holotypes, cotypes for three species and paratypes for 72 species. Publications sponsored: BC Provincial Museum publishes handbooks on the fauna and flora of BC, RBCM Memoirs, Heritage Record, Discouvery Magazine. [1992]

MANITOBA

CRIDDLE COLLECTION, AGRICULTURE CANADA RESEARCH STATION, 195 DAFOE ROAD, WINNIPEG R3T 2M9. [ACRM]
Director: Dr. T. G. Atkinson. Phone 269-2100. This is primarily a reference collection of local insects. Some historical significance with core collection by N. Criddle dating from late 1800's. There are over 300 cabinet drawers. These is a special collection of 105 species of stored product insects. [1992]

DEPARTMENT OF FISHERIES AND OCEANS, FRESHWATER INSTITUTE, 501 UNIVERSITY CRESCENT, WINNIPEG, MB R3T 2N6. [FIEC]
Director General: Mr. P. Sutherland. Phone: (204) 938-5117. Professional staff: Dr. David M. Rosenberg, Mr. Allan P. Wiens, and Mr. Bohdan Bilyj. The majority of the collection consists of Chironomidae (Diptera) determined to species, collected generally from Manitoba, with minor collections taken from other provinces. Specimens are either slide mounted (approx. 50,000) or preserved in alcohol (800,000). Many paratypes (slide-mounted) are included, with the corresponding holotypes sent to the Canadian National Collection in Ottawa. Minor collections of the following groups are included (aquatic): Chaoboridae (approx. 2,000); Culicidae (approx. 700); Ceratopogonidae (approx. 300). There are also species reference collections used for various environmental impact studies on bogs, lakes, resevoirs, and streams. The slide mounts of this collection are stored on trays in cabinets. [1992]

MANITOBA MUSEUM OF MAN AND NATURE, 190 RUPERT AVENUE, WINNIPEG R3B ON2. [MMMN]

Director: Dr. [initials not given] Hempill. Phone: (204) 956-2830. General entomological collection of approximately 50,000 specimens of insects and 500 specimens of spiders. The collection consists of 15,000 to 20,000 specimens of Lepidoptera of which 8,000 are computerized. The collection is housed in drawers in cabinets. [1986]

ENTOMOLOGY DEPARTMENT, UNIVERSITY OF MANITOBA, WINNIPEG R3T 2N2. [EDUM] [*No reply.*]

NEW BRUNSWICK

AGRICULTURE CANADA RESEARCH STATION, P. O. BOX 20280, FREDERICTON E3B 4ZT. [ACNB]

Director: Dr. D. K. McBeath. Phone: (506) 452-3260. Professional staff: Dr. G. Boiteau and Dr. Y. Pelletier. The collection consists of 25 drawers of Lepidoptera and 20 drawers of other orders. It includes the MacGillivary collection of 32,000 slides (approximately 478 species) of Aphididae. [1992]

FOREST INSECT AND DISEASE SURVEY REFERENCE COLLECTION, FORESTRY CANADA - MARITIMES REGION, P.O. BOX 4000, FREDERICTON E3B 5P7. [FRLC]

Director: Dr. L. P. Magasi. Phone: (506) 452-3516. Curator: Ms. Georgette Smith. Phone: (506) 452-3569; FAX (506) 452-3525. The reference collection is a regional working collection for the Forest Insect and Disease Survey unit of Forestry Canada in the Maritime Provinces. It is specialized with emphasis on forest and shade tree species of Lepidoptera, Coleoptera, Hymenoptera, and approximately 300 genera of their associated parasitic Diptera and Hymenoptera found in New Brunswick, Nova Scotia, and Prince Edward Island. An associated microscope slide collection houses 2,000 slides of 300 species, mostly forest Homoptera, Heteroptera, and Acari. There is a color photographic slide collection which incorporates about 7,000 slides of insects, their associated damage, predators, and parasites. The pinned and pointed collection is housed in 500 USNM drawers and numbers more than 80,000 specimens. More than 4,000 four dram vials are in the liquid collection. The collection incorporates approximately 100,000 specimens representing 1,900 species of Lepidoptera, nearly 1,000 species each of Hymenoptera and Coleoptera with smaller numbers of other orders. Material is regularly submitted to the Biosystematics Research Centre in Ottawa so Maritime material is represented in the Canadian National Collection. Material is available for loan and on-site research. Field work is limited to the Maritime provinces. [1992]

DEPARTMENT OF FOREST RESOURCES, UNIVERSITY OF NEW BRUNSWICK, P. O. BOX 4400, FREDERICTON E3B 5A3. [DFRU]

Director: Dr. C. J. Sanders. Phone: (506) 453-4501. The collection is composed largely of pinned insects, identified to family only, obtained from student collections, and used for class work in insect taxonomy. Reference collection of important forest insects of n.e. NA, and a few special collections identified to species, exist within the general collection (e.g., n.e. Lepidoptera, Scolytidae, and Cerambycidae). Total number of pinned specimens not known exactly, but probably about 15,000-20,000. A small collection of larval forms determined to family is also present. The collection is housed mostly in drawers. Some student collections and laboratory specimens for class work are in boxes. [1986]

Affiliated Collection:

Brown, Prof. N. Rae and Brown, Sandra L., 100 Burpee St., Fredericton, NB E3A IL9, Canada. [NRBC] Phone: (506) 472-6570. This collection is com posed of about 2,000 specimens (identified to species) of Siphonaptera, Mallophaga, and Anoplura, plus duplicates in alcohol. In addition, there is a collection of over 1,000 pinned Lepidoptera from New Brunswick and other areas of North America, Mexico, and Central America. (Registered with DFRU.) [1986]

ATLANTIC REFERENCE CENTRE, HUNTSMAN MARINE SCIENCE CENTRE, ST. ANDREWS, NEW BRUNSWICK EOG 2XO. [ARCM]

Director: Dr. Kenneth Sulak. Phone: (506) 529-3945. Professional staff: Dr. Gerhard Pohle, and Mr. William Hogans. The collection is predominately composed of aquatic insects collected during ecological surveys of rivers in our region. The insect collection is a very small portion of the Centre's marine fish and invertebrate collection. Insects and arachnids are not priority items, although this is the best collection of these in the region. There are approximately 450 lots of specimens representing 90 families of insects. [1992]

NEW BRUNSWICK MUSEUM, 277 DOUGLAS AVE., ST. JOHN E2K 2E5. [NBMB]

Director: Dr. Frank Milligan. Phone: (506) 658-1842. Professional staff: Donald F. McAlpine. Approximately 50,000 pinned insects are stored in Cornell trays in steel cabinets, principally Atlantic Canadian Lepidoptera; estimated 5,000 samples of aquatic insects from New Brunswick in vials of alcohol. The museum sponsors: "New Brunswick Publications in Natural Science" and "New Brunswick Museum Monograph Series (Natural Science)." [1992]

NEWFOUNDLAND AND LABRADOR

NEWFOUNDLAND INSECTARIUM, P. O. BOX 476, TRANS CANADA HIGHWAY, DEER LAKE, NF AOK 2EO. [NINF]

Director: Lloyd H. Hollett. Phone: (709) 635-3861. Head curator: Mr. Gary A. Holloway. This collection is quickly becoming one of the most comprehensive collections in Newfoundland. The local portion of the collection has over 4,000 species of native insects. These represent all major orders with emphasis on Coleoptera, Lepidoptera, Heteroptera, Odonata, and Diptera. The collection is housed in 72 wooden drawers as well as many boxes. In addition to the Newfoundland material there is a large number of species (2,000-3,000) from over fifty countries throughout the World. Material is available for exchange and may be borrowed for study. [1992]

INSECT REFERENCE COLLECTION, NEWFOUNDLAND DEPARTMENT OF FORESTRY, FOREST PROTECTION DIVISION, P. O. BOX 2006, HERALD BUILDING, CORNER BROOK A2H 6J8. [NDFC]

Director: Mr. Lloyd H. Hollett. Phone: (709) 637-2413. This collection, started in 1978, now contains approximately 30,000 specimens, with over 3,500 identified species, nearly all pinned or pointed, with a small number of insect larvae and spiders in vials of alcohol. All major orders are represented, with emphasis on Lepidoptera, Coleoptera, Odonata, Diptera, and Hymenoptera. The collection is housed in 32 drawers in steel and wooden cabinets. Material is available for exchange and may be borrowed for study. [1992]

INSECT REFERENCE COLLECTION, FOREST INSECT AND DISEASE SURVEY, NEWFOUNDLAND FOREST RESEARCH CENTRE, ENVIRONMENT CANADA, P. O. BOX 6028, ST. JOHN'S AIC 5X8. [NFRN]

Director: Mr. K. E. Pardy. Phone: 772-4823. The insect museum has a total of 1,951 forest insects of which 1,355 have been collected from Newfoundland and Laborador. Adult specimens are usually pinned or pointed but some are preserved in vials containing alcohol. Most of the immature stages are in alcohol. The Forest and Disease Survey monitors the distribution and intensity of forest insects and tree diseases. A comprehensive reference collection is maintained to aid in the identification of all species collected. Some of the major orders represented in the museum are Lepidoptera, Coleoptera, Hymenoptera, Diptera, and Hemiptera. The orders and scientific names of the species have been systematically cataloged according to the Forest Insect Numerical Code developed by the Forest Insect and Disease Survey. The collection is housed in USNM drawers in metal cabinets. Vials are stored in racks in metal cabinets. [1986]

INSECT COLLECTION, BIOLOGY DEPARTMENT, MEMORIAL UNIVERSITY OF NEWFOUNDLAND, ST. JOHN'S AIC 5S7. [MUNC]

Director of collection: Dr. David J. Larson. Phone: (709) 737-7534. This collection consists of 200 Cornell drawers of pinned specimens and 7,000 vials of fluid preserved specimens (housed in trays in drawers). Emphasis is placed on the aquatic insects of Newfoundland and Labrador. [1986]

NORTHWEST TERRITORIES

[No known insect collections.]

NOVA SCOTIA

INSECTRY, NOVA SCOTIA DEPARTMENT OF NATURAL RESOURCES, R. R. # 1, BELMONT, NS BOM ICO. [ESRC]

Director: Mr. Eric Georgeson. Phone (902) 893-5749 or 662-3390. Professional staff: Mr. Mike LeBlanc, Ms. Jacqui, and Norma Collett. This is a working reference collection of those insects found in Nova Scotia woodlands. The main part of the collection is 304 Cornell drawers which are kept in 14 cabinets. The collection consists of the following specimens: Coleoptera, 3,085; Diptera, 5,794; Lepidoptera, 18,315; Hymenoptera, 894, and other orders, 520, making a total of 28,608 pinned specimens in the collection. A separate wet mount collection has over 165 vials of specimens. [1992]

INSECT COLLECTION, NOVA SCOTIA MUSEUM, 1747 SUMMER ST., HALIFAX B3H 3A6. [NSPM]

Director: Candace Stevenson. Phone: (902) 424-7370. Barry Wright, Curator. The collection contains 225,000 insects, mainly Lepidoptera from Nova Scotia (60%), USA (30%), and Palearctic, etc. (10%). Originally this collection was curated by J. H. McDunnough and D. C. Ferguson. It contains many paratypes of Microlepidoptera. There is a special collection of Coleophoridae (Lepidoptera) which are being revised. The collection is housed in Cornell drawers in steel cabinets. Publications sponsored: "Proceedings of Nova Scotian Institute of Science." [1992]

AGRICULTURE CANADA RESEARCH STATION, KENTVILLE B4N IJ5. [ACSK]

Director: Dr. P. W. Johnson. Phone: (902) 679-5700. Professional staff: Dr. R. F. Smith, and Dr. J. M. Hardman. There are 90 drawers containing samples of local insects mostly of agricultural interest. Phytoseiid mites are mounted on 3,000 slides. No spiders. A collection of 1,000 local ticks complete the collection. Some material is housed in boxes. [1992]

A. D. PICKET ENTOMOLOGICAL MUSEUM AND RESEARCH LABORATORY, DEPARTMENT OF BIOLOGY, NOVA SCOTIA AGRICULTURAL COLLEGE, P. O. BOX 550, TURO NS B2N 5E3. [NSAC]

Curator: Dr. Jean-Pierre R. Le Blanc, P. Agr. Phone: (902) 893-6606; FAX (902) 895-4547; Internet: JPLB_B@AC.NSAC. NS.CA. Professional staff: Ms. Anna K. Fitzgerald, Dipl. T. This is a collection covering the major orders. Most specimens were collected in Nova Scotia and Atlantic Canada. There are 276 Cornell drawers in wood and steel cabinets; 400 slide specimens, and 500 vials, mostly immatures, a total of 12,000 specimens. The Museum also houses a teaching collection comprised of 60 additional drawers containing specimens that illustrate important families of the major orders. Since 1991, the entire collection is computerized on the College mainframe computer (Datatrieve on Vax). [1992]

ONTARIO

DEPARTMENT ENVIRONMENTAL BIOLOGY, UNIVERSITY OF GUELPH, GUELPH NIG 2WI. [DEBU]

Director: Dr. S. A. Marshall. Phone: (416) 843-4120, ext. 2720. Collection includes the original Entomological Society of Ontario collection, primarily of local material. Roughly 2,000 Cornell type drawers of pinned material plus vials of specimens in alcohol are now present, although many groups have not been curated in the past 70 or 80 years since the collection has no funds and no curator other than the faculty curator. There are extensive holdings of Muscomorpha, especially acalypterates, which includes 150 drawers of world Sphaeroceridae. [1986]

BIOLOGY DEPARTMENT, MCMASTER UNIVERSITY, 1280 MAIN ST. W., HAMILTON L8S 4KI. [BDMU]

Director: Dr. Douglas M. Davies, Prof. Emeritus. Phone: (416) 525-9140, ext. 3554. Professional staff: Mrs. Elen Gyorkos. The collection consists of more than 200,000 specimens, mainly Diptera, family Simuliidae, some Tabanidae, and other families. Most other orders represented in lesser numbers; also some slides of lice and fleas from Ontario wild mammals and birds. Some slides and alcohol vials of mites and ticks, mainly from Ontario wildlife. The Simuliidae are mainly from northeastern Canada, western Europe, Sri Lanka, West Malaysia, West Java, and Australia. The Tabanidae are mainly from Ontario and Norway and some from western Europe. Much of the Simuliidae and Tabanidae collection is scheduled to go to the Canadian National Collection. [1986]

LONDON RESEARCH CENTRE, AGRICULTURE CANADA, 1400 Western ROAD, LONDON N6G 2V4. [DACL]

Director: Dr. Frank Marks. Phone: (519) 645-4452; FAX 645-5476. Professional staff: Dr. Alan D. Tomlin, Dr. Glenn McLeod, and Dr. Jeff Tomlin. Most specimens are from southern Ontario agricultural and wooded areas represented by approx. 5,000 pinned general collection, 6,000 fluid preserved insects and several hundred spiders, approx. 2,500 slide mounted insects, mostly Collembola, and some other apterygotes, and about 7,000 soil mites. There are about 1,000 voucher specimens of mites and insects from the Hudson Bay Lowlands and about 1,500 pinned noctuid moths. The pinned insects are in USNM drawers. [1992]

DEPARTMENT OF ZOOLOGY COLLECTION, UNIVERSITY OF WESTERN ONTARIO, LONDON N6A 5B7. [UWOC]

Director: Dr. W. W. Judd. Phone: (519) 679-6171. About 25,000 pinned specimens from Ontario, mainly southern, and other parts of Canada, stored in USNM type drawers in 32 cabinets. [1986]

CANADIAN NATIONAL COLLECTION OF INSECTS, CENTRE FOR LAND AND BIOLOGICAL RESOURCES RESEARCH, BIOLOGICAL RESEARCH DIVISION AGRICULTURE CANADA, OTTAWA, ON KIA OC6. [CNCI]

Director: Mr. G. A. Mulligan. Phone: (613) 996-1665; FAX (613) 995-

1823. Professional Staff: Thirty-two professional scientists, and 26 technicians and artists (write for directory).

The Centre for Land and Biological Resources Research (CLBRR) Biological Research Division (BRD) of Agriculture Canada maintains and develops the Canadian National Collection of Insects and Arachnids, the largest of its kind in Canada. It provides government departments and the Canadian public with a unique centre of systematic expertise, conducting research on all aspects of the biosystematics of organisms of importance to Canadians and providing a National Identification Service used by clients throughout Canada. The collection and research facilities are housed at the Central Experimental Farm of Agriculture Canada in Ottawa.

The collection of insects and arachnids currently contains approximately 10.5 million specimens of which 8.3 million are prepared, labelled, identified, and available for study. Remaining specimens are stored in envelopes or alcohol and are curated as studies on a particular group are completed.

The CNCI is the major repository for type material of insects and arachnids in Canada with many primary types represented in the collection. A published list of Coleoptera types is available, and type data for the whole collection are currently being prepared for input into a computerized system. In addition, type books, containing data on all types in the CNCI, can be consulted. The collection is also the primary repository for specimens of value, *e.g.*, voucher specimens, which should be preserved.

The CNCI holdings are primarily Nearctic with especially good representation from Canada, Arctic Canada, and areas of the U.S.A. adjacent to Canada. Holdings of South American and Old World material are extensive in some groups and collections in groups on which staff scientists are conducting broad taxonomic studies have and are being developed on a hemispherical or worldwide basis.

In addition to providing the research material for staff scientists and research associates, the collection is used by hundreds of Canadian and foreign scientists each year, both visiting the CNCI and requesting specimens on loan. Students from many universities studying taxonomic problems work at the CNCI during the year. The CanaColl Foundation, an independent non-profit organization, promotes taxonomic research on the CNCI by granting funds to visiting specialists for research on poorly known segments of the collection. The collections are excellent for research purposes because of holdings of well documented material in long series, and collected by specialized methods.

The Diptera Unit is staffed by 6 research scientists, 5 technicians and an artist. The collection consists of 1.8 million pinned specimens housed in 5,300 drawers, 40,000 slides of Chironomidae and Ceratopogonidae and 40,000 vials of adult and larval Chironomidae, Ceratopogonidae, and Simuliidae. In addition to the major Nearctic holdings, Diptera from other regions, primarily European and Neotropical are well represented.

The Coleoptera Unit is staffed by 5 research scientists, 4 technicians, and an artist. The collection, with 1.6 million specimens housed in 5,300 drawers and 800,000 specimens in alcohol awaiting preparation, is the second largest in North America. The collection of larval Coleoptera contains about 60,000 specimens, and is housed in vials. Holdings from the Nearctic, particularly the Arctic, are especially strong. Scolytidae, Elateridae, and Alleculidae are well represented worldwide and Staphylinidae are well represented from Neotropical and Palaearctic regions.

The Hymenoptera Unit is staffed by 6 research scientists and 5 technicians. The collection consists of 1,606,000 curated specimens in 9650 drawers; 508,000 specimens of various groups are stored in alcohol and a large number of Neotropical Ichneumonidae are stored dry in envelopes. The core of the collection is Nearctic, however, other regions, particularly the Neotropical, are also represented. The Ichneumonidae collection is the largest in the New World with over 500,000 specimens. The Proctotrupoidea collection, with 120,000 specimens, is the world's largest with almost all genera represented from all geographical regions.

The Lepidoptera-Trichoptera Unit is staffed by 6 research scientists and 4 technicians. The collection consists of 1,275,000 prepared Lepidoptera housed in 10,000 drawers and on 60,000 slides and 150,000 Trichoptera specimens in 625 drawers. In addition, 134,000 adult Lepidoptera await preparation in envelopes and 75,000 larvae are stored in vials and there are 250 vials of adult and larval Trichoptera. Besides the strong North American holdings there are moderate Lepidoptera holdings from the Mexican, Neotropical, and Palaearctic regions and small holdings from Ethiopian, Oriental, and Australian regions. The collections of Noctuidae, Geometridae, and Pyralidae are extensive. Himalayan Trichoptera are particularly well represented in the CNCI.

The Hemiptera Unit is staffed by 2 research scientists and 1 technician. The collection, consisting of 204,000 prepared Homoptera in 900 drawers and 100,000 aphids mounted on slides, is particularly strong from North America. Approximately 10,000 specimens of small Homopteran groups are also represented, in addition, 34,000 unsorted Hemiptera are stored in alcohol.

Small holdings of other insect orders are also maintained and developed by one research scientist. These holdings total one million prepared specimens, consisting of 425 species in 18 orders housed in 500 drawers and on 25,000 slides. Specimens are primarily North American.

The Arachnida Unit is staffed by 4 research scientists, 3 working on mites, 1 on spiders, and 3 technicians. The collection consists of 115,000 spiders in 925 vials and approximately 500,000 mites of which 200,000 are slide mounted, the remainder in alcohol. Holdings are primarily from North America with extensive collections from eastern Canada and the Arctic. Special spider collections include the Renault Collection of conifer-forest spiders from New Brunswick and the Dondale Collection from Nova Scotia apple orchards and Belleville meadows.

At present no series is published, the staff generally uses the "Canadian Entomologist," although many other publication outlets, e.g., "Journal of Arachnology," are also used. Large works are generally published as "Memoirs of the Entomological Society of Canada." A Canadian faunal series, "The Insects and Arachnids of Canada," covering different groups, is prepared by B.R.I. staff and published by Agriculture Canada.

The facilities of the Entomology Library are extensive and excellent. The holdings include many rare and out of print entomological works. Collecting expeditions to enrich the CNCI are sponsored primarily to parts of Canada and adjacent U.S.A. In addition, scientists are eager to acquire material from other parts of the world to expand the holdings of the CNCI. [1986]

Affiliated Collections:

Freitag, Dr. Richard, Department of Biology, Lakehead University, Thunder Bay, ON P7B 5EI, Canada. [RFCC] Phone: (807) 345-2121; 577-3990. This is a collection of over 4,650 specimens of Cicindelidae housed in 30 drawers in cabinets, and 224 vials of male and female genitalia. (Registered with CNCI.) [1986]

Kiteley, Eric J., 16 13th St., Roxoboro, PQ H8Y IL4, Canada. [EJKC] Phone: (514) 684-5215. The collection consists of about 50,000 specimens of Coleoptera, all families represented except Staphylinidae, Hydrophilidae, Anthicidae, and Scolytidae. The collection is housed in Schmitt type boxes. (Registered with CNCI.) [1986]

Matthews, John V., Jr., Geological Survey of Canada, 601 Booth Street, Ottawa, ON KIA OE8, Canada. [JVMJ] Phone: (613) 996-6371; 226-8781. This collection contains over 2,000 Coleoptera from various North American localities, but mostly from Arctic and Subarctic Canada, housed in boxes and drawers and in alcohol. The collection now includes the private collection of Richard E. Morlan, Ottawa, Canada. (Registered with CNCI.) [1986]

ENTOMOLOGY DIVISION, CANADA MUSEUM OF NATURE, P. O. 3443 STATION D, OTTAWA ON KIP 6P4. [CMNC]

Director: Dr. Robert S. Anderson. Phone: (613) 954-2581. Collection assistant: Mr. François Génier. This collection was initiated in 1984. At the present time there are over 315,000 specimens consisting entirely of Coleoptera collected predominantly in the New World, South Africa, and Australia. The following families are especially well represented: Carabidae (Cicindelinae), Leiodidae, Histeridae, Lucanidae, Scarabaeidae, Buprestidae, Cleridae, Nitidulidae, Tenebrionidae, Cerambycidae, Chrysomelidae, and Curculionidae. There are 78 primary types and over 1,200 secondary types (a list is currently in prepraration). The collection is housed in 1,200 Cornell drawers in steel cabinets. [1992]

Affiliated Collections:

Anderson, Robert S., Entomology Division, Canada Museum of Nature, P. O. Box 3443, Station D, Ottawa, ON KIP 6P4, Canada. [RSAN] Phone: (613) 954-2581. The collection consists of 12,300 specimens of Coleoptera, mainly Nearctic Silphidae (300), Nearctic and Neotropical Buprestidae (500), Nearctic and Neotropical Cerambycidae (1,500), and 10,000 specimens of Curculionidae from Nearctic, Neotropical, New Zealand, and Republic of South Africa, including some paratypes. The collection is housed in drawers, some in alcohol vials. (Registered with CMNC.) [1986]

Howden, Henry F. and Anne T., Biology Department, Carleton University, Colonel By Drive, Ottawa, ON KIS 5B6, Canada. [HAHC] Phone: (613) 788-2600, ext. 3872. This is an extensive collection of Coleoptera specializing in the Scarabaeoidea and Curculionidae of the world with large collections also in Cerambycidae, Buprestidae, Nitidulidae, and Cicindelidae. It is particularly strong in New World and Australian regions. All primary types are deposited on loan in the CNCI, which consist, at present, of 227 primary types comprising many families of Coleoptera. The collection is housed in about 592 drawers. (Registered with CMNC.) [1992]

Génier, François, 35 Pine St., Aylmer, PQ J9H 3Y8, Canada. [FGIC] Phone: (613) 954-1067. This collection consists of about 8,000 identified Scarabaeidae (Coleoptera). It is stored in 90 Cornell drawers. (Registered with CMNC.) [1992]

POINT PELEE NATIONAL PARK, CANADIAN PARK SERVICE, R.R. 1, LEAMINGTON, ON N8H 3V4. [PPNP]

Director: Chief Park Warden. Phone: (519) 322-2365, ext. 203. Professional staff: Natural Resource Conservation (Park Wardens) and Park Interpretive staff (total of 10 staff members). The collection consists of about 3,000 pinned specimens in 30 drawers. [1992]

GREAT LAKES FOREST RESEARCH CENTRE, P. O. BOX 490, SAULT STE. MARIE P6A 5M7. [GLFR] [*No reply.*]

BIOLOGY DEPARTMENT, LAURENTIAN UNIVERSITY, RAMSAY LAKE ROAD, SUDBURY P3E 2C6. [BDLU] [*No reply.*]

FOREST INSECT AND DISEASE SURVEY, GREAT LAKES FOREST RESEARCH LABORATORY, BOX 490, SAULT STE. MARIE, ON P6A 5M7. [FIDS]

Director: Mr. J. H. Cayford. Phone: (705) 949-9461. Professional staff: Dr. G. M. Howse (Unit Head), Dr. P. D. Syme (Taxonomist), Mrs. Kathryn Nystrom (Insect Identification Officer), Miss Celine Handfield (Curator/technician). The collection consists of approximately 136,000 specimens of pinned adults, larvae, and soft-bodied insects in liquid, some small specimens on slides, and an extensive color slide collection. The collection concentrates on forest insects; with Lepidoptera, Hymenoptera, and Coleoptera our strongest holdings. A special collection of blown or freeze-dried larval specimens of major defoliator species is available. Pinned adults are stored in glass topped or wooden topped drawers. Larval liquid collections are stored in metal drawers in custom made racks. All specimens are listed on inventory cards. [1986]

ONTARIO HYDRO, BIOLOGICAL RESEARCH SECTION, 800 KIPLING AVENUE, KD 118, TORONTO, M8Z 5S4. [OHBR]

Director: Dr. G. L. Vascotto. Phone: (416) 231-4111, ext. 6865. The collection represents insects and crustaceans, mostly in the larval or nymphal form, found in the Great Lakes and smaller lakes of northern Ontario. Specimens are either slide mounted (e.g., Diptera: Chironomidae), or preserved in alcohol. The Chironomidae are well represented (62 genera), with hundreds of larvae mounted. The orders Coleoptera, Ephemeroptera, Hemiptera, Odonata, Plecoptera, and Trichoptera are commonly encountered. The collection is stored in slide boxes and jars. [1986]

DEPARTMENT OF ENTOMOLOGY, ROYAL ONTARIO MUSEUM, 100 QUEEN'S PARK, CRESCENT, TORONTO M5S 2C6. [ROME]

Department Head: Dr. D. Christopher Darling, Associate Curator in Charge. Phone (416) 586-5531. Professional Staff: Dr. G. B. Wiggins, Curator Emeritus, Dr. Douglas C. Currie, Assistant Curator, Ms. Pat

Schefter, Curator Assistant, Ms. Dael Morris, Technician; Mr. Brad Hubley, Collection Manager. The collection consists of approximately 1 million specimens of insects, arachnids, and myriapods, of which some 70% have been identified. The collection has pinned and fluid preserved specimens, including about 15,000 papered Odonata and 10,000 slide mounts (2,000 of parasitic Hymenoptera and 8,000 of water mites). It is particularly strong in aquatic groups of both insects and mites, and in larval stages of most groups. The Trichoptera and Hymenoptera components are the most active in terms of new accessions. Geographic representation is primarily strong for North American, with strong northern, western, and Beringian components and there is also a bulk collection of eight hundred 125 ml bottles of unsorted insects from Indonesia, Guyana, Costa Rica, Peru, Philippines, India, Canada, and the United States, collected by Malaise traps, screen sweeping, and pan traps, which are kept in ultra-cold storage facilities.

Associated with the collection is an extensive library of reprints, books, and photographic slides. There are facilities for rearing insects under controlled environmental conditions. Working facilities for visiting scientists are available. Among the special collections are: Trichoptera (developed by G. B. Wiggans); aquatic Acarina (developed by D. W. Barr); Araneae (developed by T. B. Kurata); Canadian Cretaceous amber; Odonata (E. M. Walker collection); Orthoptera (E. M. Walker and F. A. Urquhart collections); Lepidoptera (developed by J. C. E. Riotte and others). Publications sponsored: Royal Ontario Museum Life Science Contributions; Occasional Papers, and Miscellaneous Publications. [1992]

BIOLOGY DEPARTMENT, UNIVERSITY OF WATERLOO, WATERLOO ON, N2L 3GI. [BDUW] [*No reply.*]

DEPARTMENT OF BIOLOGICAL SCIENCES, UNIVERSITY OF WINDSOR, WINDSOR, ON N9B 3P4. [BDWC]

Curator: Dr. Jan J. H. Ciborowski. Phone: (519) 253-4232, ext. 2725. This is a general pinned collection of local terrestrial insects with emphasis on Lepidoptera; extensive collections from 1930's, including 18 drawers of Lepidoptera, 31 drawers of other orders. A pinned collection of purchased tropical Lepidoptera and Coleoptera (South America; dates unknown) are stored in 54 drawers. A fluid-preserved collection of aquatic insects from Ontario, Alberta, and northern Canada is housed in eight drawers of vials. Cataloging is incomplete so full description of contents and number is not possible at present. [1992]

PRINCE EDWARD ISLAND

AGRICULTURE CANADA RESEARCH STATION, P. O. BOX 1210, CHARLOTTETOWN CIA 7M8. [APEI]

Director: Dr. C. B. Willis. Phone: (902) 566-6800. Professional staff: Dr. J. G. Stewart. Size of collection: About 9,000 specimens. Housed in drawers. [1992]

QUEBEC

INSECT COLLECTION, NATURAL HISTORY MUSEUM, BISHOP'S UNIVERSITY, LENNOXVILLE JIM IZ7. [ICBU]

Director: Dr. Donald F. J. Hilton. Phone: (819) 822-9600. This collection contains 125 USNM drawers with a total of about 12,000 specimens. These represent a general collection of most orders and families that occur in eastern Canada, although other parts of Canada and the U.S.A. are represented. Except for the butterflies, Tabanidae, and Odonata, specimens are not identified to species. This insect collection is primarily for teaching in entomology courses. A special effort is made to develop the collection of Odonata, now containing 7,000 specimens in cellophane envelopes arranged in file boxes. The collection is world-wide and specimens are available for exchange. [1992]

COLLECTION ENTOMOLOGIQUE OUELLET-ROBERT, DEPARTMENT OF BIOLOGICAL SCIENCES, UNIVERSITY OF MONTREAL, C.P. 6128, MONTREAL H3C 3J7. [QMOR]

Director: P. P. Harper. Phone: (514) 343-6790. Professional staff: Mrs. Monique Coulloudon. The collection contains about 20,000 specimens, half of which are from Quebec, best represented by Coleoptera, Odonata, Diptera, and some aquatics (Plecoptera, Trichoptera, and Ephemeroptera). This collection is based on the private research collections of Quebec entomologists Joseph Ouellet and Adrien Robert. Material is available for exchange and for study loan. Library and laboratory facilities are available through the department. [1986]

INSECT COLLECTION, DEPARTMENT OF BIOLOGY, UNIVERSITY OF LAVAL, QUEBEC, PQ GIK 7P4. [ULQC]

Director: Dr. J. M. Perron. Phone: (418) 656-2497). The collection contains a large representation of families found in Quebec, with Coleoptera, Lepidoptera, and Odonata well represented. It is housed in 600 USNM type drawers in cabinets. Material is available for exchange and loan for study. It has on loan the insect collection of the "Seminaire de Quebec" housed in 160 drawers, containing representation of families found in Quebec, exotic material, and part of the William Couper insect collection. [1992]

LYMAN ENTOMOLOGICAL MUSEUM AND RESEARCH LABORATORY, MCDONALD COLLEGE, MCGILL UNIVERSITY, ST. ANNE DE BELLEVUE, H9X 3M1. [LEMQ]

Director: Dr. D. J. Lewis. Phone: (514) 398-7911; FAX (514) 398-7900. Professional staff: Dr. C.-C. (George) Hsiung; Curator: Dr. V. R. Vickery; Emeritus Curator: Dr. S. B. Hill, Dr. E. G. Munroe, and Dr. J. A. Downes. The collection consists of about 2.5 million labelled specimens representing more than 45,000 species. The world Orthopteroids, which are stored in 1,500 drawers, Coleoptera (860 drawers), Lepidoptera (1,500 drawers) have received the greatest attention in modern times. The collection of Hymenoptera (520 drawers) and Heteroptera (440 drawers) have recently been expanded considerably. Other small orders (mainly Nearctic) include Diptera (240 drawers), Neuroptera, Odonata, Trichoptera, Mecoptera, and Plecoptera (220 drawers). The orthopteroid orders are well represented and include the late Dr. D. K. McE. Kevan's important collections and type specimens, especially of the family Pyrgomorphidae. The collection of Lepidoptera represents 3,818 Nearctic and 3,488 exotic species which include the founder Mr. H. Lyman's collection and the recent acquisition of the collection of Mr. A. C. Sheppard. The slide collection, mainly Collembola and Acari (soil mites), with smaller representation of Siphonaptera and other orders, contains 67,000 slides. Alcohol stored material is preserved in about 800 jars (65 ml.) and 53,856 vials (15 ml.). The major orders in alcohol are Acarina, Thysanoptera, Collembola, Ephemeroptera, Plecoptera, aquatic Diptera and orthopteroid larvae. The Museum reprint collection contains 80,000 titles and the combined library holdings include about 5,000 volumes plus complete sets of nearly all of the important entomological journals. The Museum is also a publishing institution, issuing larger works "Memoir of Lyman Entomological Museum and Research Laboratory" series of which there are now 17, and shorter papers, "Notes from the Lyman Entomological Museum and Research Laboratory" at irregular, but fairly frequent intervals. [1992]

SASKATCHEWAN

PFRA TREE NURSERY, INDIAN HEAD SOG 2KO. [PFRA]

Director: Dr. Gordon Howe. Phone: (306) 695-2284. The collection consists of more than 5,000 specimens representing 1,200 species in 850 genera. Most specimens were collected from trees and shrubs of the prairie provinces of Canada. It is housed in 35 drawers. [1992]

AGRICULTURE CANADA, REGINA RESEARCH STATION, 5000 WASCANA PARKWAY, P. O. BOX 440, REGINA S4P 3A2. [RARS]

Director: Dr. Raj Grover. Phone: (306) 780-7400. The collection consists of 2,063 species represented by 20,000 specimens. Most of the insects were taken from weed species as a part of biological control of weeds project. It is housed in 98 drawers in 8 cabinets. [1992]

SASKATCHEWAN MUSEUM OF NATURAL HISTORY, DEPARTMENT OF COMMUNITY SERVICES, 2340 ALBERT ST., REGINA, SK S4P 3V7. [SMNH]

Curator: Mr. Keith Roney. Phone: (306) 787-2801. The collection contains about 50,000 specimens of Saskatchewan insects and spiders. There are 18,000 N.A. Lepidoptera in the J. B. Wallis collection. About 1,000 spiders are unidentified. The collection is stored in drawers and vials. Publications sponsored: "Natural History Contribution Series." [1992]

Affiliated Collection:

Hooper, Ronald R., P. O. Box 205, Fort Qu'Appelle, SK SOG ISO Canada. [RRHC] Phone: (306) 332-4198. This is a collection of over 5,000 specimens, mostly Coleoptera of Saskatchewan, but also of butterflies, including Papilionidae of the world, housed in drawers and boxes. (Registered with SMNH.) [1992]

YUKON TERRITORY

[No known insect collection.]

(Canal Zone, territory in Panama no longer leased by U.S.A., see Panama for collections.)

(Canary Islands, see Spain.)

(Canton and Enderbury Islands, see Kiribati.)

35. CAPE VERDE, Republic of

[Includes Boa Vista, Brava, Fogo, Maio, Sal, Santo Antao, São Nicolau, São Tiago, and São Vicente. Palearctic. Cidade de Praia. **Population:** 353,885. **Size:** 1,557 sq. mi. *No reply; no known insect collection.*]

(Caroline Islands, see Pacific Islands, Trust Territory of the.)

(Cartier Island, see Australia.)

36. CAYMAN ISLANDS

[British dependency: includes Grand Cayman, Cayman Brac, and Little Cayman. Neotropical. Georgetown. **Population:** 23,037. **Size:** 100 sq. mi. *Reply; no insect collections maintained on the islands; extensive collecting, but these specimens deposited in various collections.*]

(Celebes [=Sulawesi], Island Provinces of Indonesia.)

37. CENTRAL AFRICAN REPUBLIC

[Formerly Central African Empire and Ubangi-Shari. Afrotropical. Bangui. **Population:** 2,736,478. **Size:** 240,535 sq. mi. *No reply.*]

STATION EXPERIMENTAL DE LA MABOKE PAR M'BAIKI. [SEMM]
[Contact: Dr. Roger Heim. Collection contains agriculturally important insects. Eds. 1992]

(Ceuta, see Spain.)

(Ceylon, see Sri Lanka.)

38. CHAD, Republic of

[Afrotropical/Palearctic. N'Djamena. **Population:** 4,777,963. **Size:** 486,180 sq. mi.]

CHAD NATIONAL MUSEUM, PLACE DE L'INDEPENDENCE, B. P. 503, FORT LAMY. [CCFL] [*No reply.*]

(Chagos Archipelago, see British Indian Ocean Territory.)

(Channel Islands, British Crown Dependency, including Guernsey and Jersey.)

(Chatham Islands, see New Zealand.)

(Chesterfield Islands, see New Caledonia.)

39. CHILE, Republic of

[Includes Easter Island, Islas Juan Fernandez, Islas San Felix, and Isla Sala y Gomez. Neotropical. Santiago. **Population:** 12,638,046. **Size:** 292.132 sq. mi.]

COLECCION ENTOMOLOGICA DEL INSTITUTO DE AGRONOMIA, UNIVERSIDAD DE TARAPACÁ, CASILLA 6 D, ARICA. [IDEA]

Principal Entomologist: Mauricio Jiménez R. Phone: 224157; FAX 058-224327. Professional Staff: Sr. Hector Vargas C., M. Sc., Sr. Dante Bobadilla G., and Pedro Callo D. The collection consists of nearly 60,000 pinned specimens from northern Chile (Coquimbo to Parinacota), and specimens in vials. About 1,000 microscope slides of miscellaneous insects in boxes. This is a local collection from the Atacama desert, including Arid Land from III and IV Regions of Chile. The collection includes Coleoptera, Coccinellidae and several other families from all of Chile, Lepidoptera, and Orthopteroidea; Hymenoptera, Ichneumonidae, Vespidae, Braconidae, Chalcidoidea. It is rich in Tachinidae, Simuliidae, Bombyliidae, Tabanidae, Calliphoridae, and Sarcophagidae (Diptera). A few types are deposited in the collection. Pinned specimens are stored in Cornell drawers in cabinets. Publication sponsored: "Idesia," a review of work at the Institute, formerly called CICA, Department of Agriculture. [1992]

MUSEO DE ZOOLOGIA, UNIVERSIDAD DE CONCEPCION, CASILLA 2407-10, CONCEPCION. [UCCC]

Curators: Mr. Thomás Cekalovic, Ms. Viviane R. Jerez. Phone: 56-41-240280, ext. 2152. Professional staff: Dr. Jorge N. Artigas (Diptera: Asilidae), Dr. Andrés O. Angulo (Lepidoptera: Noctuidae), Prof. Viviane R. Jerez (Coleoptera: Chrysomelidae); Prof. Luis E. Para (Lepidoptera: Microlepidoptera), Dra. María E. Casanueva (Arachnida, Acari), Mr. Tomás Cekalovic (Arachnida, Scorpionida, Coleoptera). The collection consists of 412,448 specimens representing 9,020 species, including 140 holotypes, 58 allotypes, and 52 neotypes, and 10,574 specimens, representing 325 species of Arachnida. Publications sponsored: "Gayana" and "Boletin de la Sociedad de Biologia de Concepción." [1992]

INIA SUBESTACION EXPERIMENTAL CONTROL BIOLOGICO LA

CRUZ, CASILLA 3, LA CRUZ, VA. REGION. [INLA]

Director: Dr. Renato Ripa S. Phone: 033-310666. Professional Staff: Sng. Sergio Rojas P., Dr. René R. Vargas, and Bil. Fernando Rodríguez A. This is a general collection of Chilean insects, particularly agricultural pests. It includes about 50,000 specimens stored in wooden boxes. A special collection of Chilean Tachinidae is included. [1992]

COLECCION NACIONAL DE INSECTOS, MUSEO NACIONAL DE HISTORIA NATURAL, CASILLA 787, SANTIAGO. [MNNC]

Curator of Entomological Section: Dr. Ariel Camousseight M. Phone: 6814095. FAX 381975. Professional staff: Dr. Ariel Camousseight M. (Phasmatoidea); M.Sc. Lic. Mario Elgueta D. (Curculionidae); Dra. Fresia Rojas A. (Hymenoptera, Apoidea and Trichoptera). The collection of insects is composed of about 140,000 mounted specimens stored in CASC drawers. The major groups represented are: Coleoptera (developed from the collection of R. A. Philippi and Ph. Germain) with 50,000 specimens; Hymenoptera 40,000 specimens, and Lepidoptera (collection of E. Ureta) with 12,000 specimens. The spider collection (collection of H. Zapfe), especially from Chile and nearby countries consists of 30,000 specimens most with ecological data. Included in the above collections are about 3,700 holotypes and paratypes, with a catalog of where the species names were published. The collection is housed in drawers in cabinets. Specimens may be borrowed by specialists for one year (renewable). Publications sponsored: "Boletin Museo Nacional de Historia Natural de Chile," "Noticiario Mensual Museo Nacional de Historia Natural de Chile," and "Publicacion Ocasional de Museo Nacional Historia Natural de Chile." [1992]

MUSEO ENTOMOLOGICO, FACULTAD DE AGRONOMIA, UNIVERSIDAD DE CHILE, CASILLA 1004, SANTIAGO. [MEUC] [*No reply. The museum includes collections from the University of California\University of Chile Arthropod Expeditions to Chile and Argentina.*] [Eds. 1992]

FACULTAD DE CIENCIAS FORESTALES, UNIVERSIDAD AUSTRAL DE CHILE, CASILLA 853, VALDIVIA, CHILE. [CFUA]

Director: Professor Luis A. Cerda. Phone: 3911-230. Collection contains mainly insects associated with forest trees. There are approximately 2,500 specimens housed in drawers and boxes. [1986]

UNIVERSIDAD CATOLICA DE VALPARAISO, CASILLA 6059, VALPARAISO, CHILE. [UCVC]

Director: Haroldo Toro. Phone: 251024. Professional staff: Luisa Ruz, Elizabeth Chiappa, J. C. Magunacelaya. The collection consists of about 35,000 specimens of Apoidea in 315 boxes. [1986]

40. CHINA, Peoples's Republic of

[Includes Hainan Island, Inner Mongolia, Sinkiang, and Tibet. Indomalayan/Palearctic. Beijing. **Population:** 1,088,169,192. **Size:**

3,695,500 sq. mi. *We have compiled a list of museums and government collections from various sources, including information given directly by Dr. Wu Yan Ru of IZAS at the 1992 meeting of the International Congress of Entomology. More collections have responded since the Congress but our list remains incomplete. Collections are arranged under their municipality or province. Province headings are not listed when there is no collection from that province. Note that surnames of staff are usually given first.*]

ANHUI

[No known insect collection.]

BEIJING

DEPARTMENT OF PLANT PROTECTION, BEIJING AGRICULTURAL UNIVERSITY, BEIJING 100094. [BAUC]

[Contact: Prof. Yang Ji-kun, taxonomist. Collection has approximately 560,000 specimens; strong in Neuroptera, Megaloptera, Strepsiptera, Rhaphidioptera, Psocoptera, Lepidoptera, and Hymenoptera. Eds. 1992]

INSECT COLLECTION, FOREST RESEARCH INSTITUTE, CHINESE ACADEMY OF FORESTRY, BEIJING 100091. [CFRB]

Director: Mr. Changlu Wang. Phone: (01) 258-2211-622. Professional staff: Jian Wu, Shuzhi, Honghin Wang, Gangrou Xiao. This collection consists of 135,000 pinned insect, of which 6,725 are identified. This includes 120 type specimens. Publication: "Forest Research." [1992]

INSECT COLLECTION, INSTITUTE OF ZOOLOGY, ACADEMIA SINICA, 19 ZHONGGUANCUN LU, HAIDIAN 100080, BEIJING. [IZAS]

Director: Dr. Wu Yan-ru. The collection contains about 3,500,000 specimens of which there are over 19,000 species, with over 3,050 types, housed in over 21,000 boxes. Collection strengths include Coleoptera, Lepidoptera, Diptera, Hymenoptera, Homoptera, Orthoptera, and Isoptera. [1986, updated by Eds. 1992]

BEIJING NATURAL HISTORY MUSEUM, 126 TIEN CHAIO STREET, BEIJING. [CNHP]

[Contact: Dr Wang Wenli. The collection includes amber types of Chinese insects. Eds. 1992]

FUJIAN

BIOLOGICAL CONTROL RESEARCH INSTITUTE, FUJIAN AGRICULTURAL COLLEGE, SHAXIAN, FUZHOU, FUJIAN 350002. [FACS]

Director: Prof. Dr. Zhao Xiufu (=Hsiu-fu Chao). Phone: 710713, ext. 537. Professional staff: Dr. Zhao, Dr. Lin Naiquan, Dr. Yuqing Tang.

This collection contains about 150,000 pinned insects, mainly parasitic Hymenoptera and Odonata. Trichogrammatidae and Mymaridae on slides number 30,000 specimens; Odonata in paper envelopes, about 70,000. There are about 200 type specimens, mainly Ichneumonidae, Braconidae, Agriotypidae, Trichogammatidae, Mymaridae, Aphelinidae, and Odonata. Publications: "Wuyi Science Journal" and special publications of the Biological Control Research Institute. [1992]

GANSU

[No known insect collection.]

GUANGDONG

CHINA ENTOMOLOGICAL RESEARCH INSTITUTE COLLECTION, 105 XINGANG WEST ROAD, GUANGZHOU, GUANGDONG PROVINCE. [CGEC]

Director: Ping Zheng-ming, Termite Taxonomist. Phone: 48651. The collection was started in 1958 and now consists of about 170,000 specimens, mostly from South China, including about 200 types. Particularly outstanding are the Isoptera, Rutelinae (Coleoptera), Collembola, and Acarina, Phytoseidae, etc. Most complete regional collections have come from the Ding Hu Mountains where collecting has been done for many years. The collection is housed in drawers, boxes, vials, and on slides. [1986]

INSECT COLLECTION, RESEARCH INSTITUTE OF ENTOMOLOGY, ZHONGHAN (SUN YAT-SEN) UNIVERSITY, GUANGZHOU. [ICRI]

Director: Prof. Dr. Liang Ge-Qiu. Prone: 4446300-6697. Professional staff: Chen Zhen-yao, Jia Feng-Long, Huang Zhihe, Zeng Hong, Wu Wu, Pu Zhe-Long, Hua Lizhong, and others. There are about 310,000 specimens, particularly Orthoptera (Acrididae, Tetrigidae, Tettigoniidae), Pentatomidae, Cerambycidae, and Hydrophilidae, and Dytiscidae housed in drawers. [1992]

GUANGDONG ENTOMOLOGY INSTITUTE, INSECT TAXONOMY DIVISION, ACADEMIA SINICA, GUANGZHOU, GUANGDONG PROVINCE. [GEIC] [*No reply.*]

INSTITUTE OF ENTOMOLOGY, GUANGZHOU (CANTON). [ICRG] [*Probably not a complete address; no reply.*]

SOUTH CHINA AGRICULTURAL COLLEGE, PLANT PROTECTION DEPARTMENT, GUANGZHOU, GUANGONG PROVINCE. [SCAC] [*No reply.*]

GUANGKI-ZHUANG, GUIZHOU, HEBEI

[No knwon insect collections.]

HEILONGJIANG

HEILONGJIANG NATURAL HISTORY MUSEUM, HARBIN CITY. [HNHH]
[Contact: unknown. However, a collection is present. Eds. 1992]

HENAN

NORTH-WEST COLLEGE OF AGRICULTURE, WUKUNG, SHENSI, HENAN PROVINCE. [NCAW] [*No reply.*]

HUNAN, JIANGSU, JIANGXI, and JILIN

[No known insect collections.]

JIANGSU

DEPARTMENT OF PLANT PROTECTION, NANJING AGRICULTURAL UNIVERSITY, NANJING 210014. [NAUJ]
Director: Dr. Tian Lixin, Professor. Phone: 432111-2241. Number of specimens in the collection is about 50,000. Collection is strong in Trichoptera, with 800 specimens and 100 types. [1992]

LIAONING

SHENYANG AGRICULTURAL COLLEGE, SHENYANG, LIA-ONING PROVINCE. [SACS] [*No reply.*]

NEI MONGOL

[=Inner Mongolia; no known insect collections.]

NINGXIA HUI

[No known insect collection.]

QINGHAI

QUINHAI INSTITUTE OF BIOLOGY, XINING, QINGHAI. [QIBX] [*No reply.*]

SHAANXI

ENTOMOLOGICAL INSTITUTE, SHANXI AGRICULTURAL UNIVERSITY, TAIGU. [EISC] [*No reply.*]

DEPARTMENT OF PLANT PROTECTION, NORTH-WEST AGRICULTURAL UNIVERSITY, YANGLINGZHEN 712100. [NWAU]
[Contact: Dr. Zhou Yao, Professor. Collection contains approximately 500,000 specimens. Strengths include Homoptera, Mecoptera,

Rhaphidioptera, and Lepidoptera. Eds. 1992]

SHANDONG

[No known insect collections.]

SHANGHAI

LABORATORY OF TAXONOMY AND ECOLOGY, INSTITUTE OF ENTOMOLOGY, ACADEMIA SINICA, CHUNGKING ROAD (S.) 225, SHANGHAI 200025. [IEAS]

[Contact: Dr. Xia, Kai-ling, Professor. Collection is apparently the second largest in China, with 650,000 specimens. Strengths include Protura, Homoptera, Orthoptera, and Diptera. [Eds. 1992]

MUSEUM OF SHANGHAI INSTITUTE OF ENTOMOLOGY, ACADEMIA SINICA, 225 CHUNGKING ROAD (S.) SHANGHAI, 200025 SHANGHAI PROVINCE. [MSIE]

Curator: Zi-Yi Luo; Senior Techniccans: Xiao-Ju Chen, Hui-Nian Zhang, and Li Li Zhou. In addition the professional staff of 17 (including some now retired, but still working) constitute the specialists on varius group. The collection contains about 650,000 pinned specimens, 40,000 specimens in alcohol, and 20,000 slides, including 680 type specimens. The collection started as the "Musee Heude" at the Universite L'Aurore in the early 1930's, which sponsored the "Notes D'Entomologie Chinoise." The institute now publishes "Contributions of Shanghai Institute of Entomology." [1992]

MUSEUM OF NATURAL HISTORY, SHANGHAI. [CMNS] [*Probably not a complete address; no reply.*]

SECOND MILITARY MEDICAL COLLEGE, DEPARTMENT OF PARASITOLOGY, SHANGHAI. [SMMC] [*Probably not a complete address; no reply.*]

SICHUAN

DEPARTMENT OF PLANT PROTECTION, SOUTH-WEST AGRICULTURAL UNIVERSITY, CHONGQING 630716. [SACA]

[Contact: Dr. Jiang Shu-nan, Professor. Collection contains about 150,000 specimens. Coleoptera is well represented. Eds. 1992]

SICHUAN INSTITUTE OF AGRICULTURE, CHENGDU. [SIAC] [*Probably not a complete address; no reply.*]

TIANJIN

TIANJIN (TIENTSIN) MUSEUM OF NATURAL HISTORY, TIANJIN 300074. [TMNH] [*No reply.*]

INSECT COLLECTION OF NANKAI UNIVERSITY, DEPARTMENT OF BIOLOGY, 94 WEIJIN ROAD, TIANJIN. [NKUM]

Director: Prof. Le-yi Zheng. Phone: (86) (022) 344300, ext. 2200. Professional staff: Mr. Guo-qing Liu (assistant curator), Ms. Chen Chen (technician). A total of approximately 300,000 specimens of Chinease insects, including about 200,000 specimens of Chinease Hemiptera-Heteroptera and a number of primary type specimens of this latter group described by Chineae Heteropterists T. Y. Hsiao, L. Y. Zheng, S. Z. Ren, X. L. Jin, G. Q. Liu, P. P. Chen, W. J. Bu, X. Z. Li, *etc.* after 1960. [1992]

XIZANG

[=Tibet; no known insect collections.]

YUNAN

DEPARTMENT OF ENTOMOLOGY, UNIVERSITY OF KUNMING, P. O. BOX 51, KUNMING, YUNAN. [UKKY] [*No reply.*]

KUNMING INSTITUTE OF ZOOLOGY, ACADEMIA, KUNMING 650107. [ISAS]

[Contact: Dr. Gan, Yun-xing. The collection contains about 450,000 specimens, with strengths in all groups from Yunnan, and generally in Lepidoptera, Diptera, and Hymenoptera. Eds. 1992]

ZHEJIAN

DEPARTMENT OF PLANT PROTECTION, ZHEJIAN AGRICULTURAL UNIVERSITY, HANGZHOU 310029. [ZUAC]

[Contact: Dr. Tang Jue, Professor. The collection has about 350,000 specimens, and its strengths include Hymenoptera. Eds. 1992]

(China, Republic of, see Taiwan.)

(Christmas Island, Territory of, Australian dependency in the Indian Ocean; *no known insect collection.*)

(Clipperton Island, see France.)

(Coco Island, see Burma.)

Cocos (Keeling) Islands, Territory of, Australian dependency; *no known insect collection.*]

41. COLOMBIA, Republic of

[Includes Archipelago de San Andres y Providencia, Isla de Malpelo, Roncador Cay, and Serrana and Serranilla Banks. Neotropical. Bogotá. **Population:** 31,298,803. **Size:** 440,831 sq. mi.]

MUSEO DE HISTORIA NATURAL, INSTITUTO DE CIENCIAS NATURALES, UNIVERSIDAD NACIONAL DE COLOMBIA, APTO. 7495, SANTA FÉ DE BOGOTÁ. [UNCB] [*No reply.*]

COLECCION ENTOMOLOGICA "LUIS MARIA MURILLO", ICA, TIBAITATÁ, APTO. AÉREO 151123, ELDORADO, BOGOTÁ. [CELM]

Director: Dr. Ingeborg Zenner-Polania. Phone: (91) 2672710, 2861721, ext. 335, 338. The total number of identified insects and mite species is 5,490 with the total collection consisting of about 120,000 specimens. The collection emphasizes insects and mites of economic importance and their natural enemies. It includes immature forms, represented by 321 larvae of mainly Lepidoptera and Coleoptera and 43 nymphs of Heteroptera and Homoptera. Paratypes of 12 Bruchidae (Coleoptera) and 10 Membracidae (Homoptera) have been deposited in this collection. The collection is housed in Cornell drawers. Publications: "Notas y Noticias Entomologicas," bimonthly. [1992]

MUSEO DE ENTOMOLOGIA "FRANCISCO LUIS GALLEGO," DEPARTAMENTO DE BIOLOGIA, FACULTAD DE CIENCIAS, A.A. 3840, MEDELLIN. [UNCM].

Director: Dr. Raul Velez-Angel. Phone: (57) 230-0280, 230-1975; FAX 230-2029. Professional staff: Adolfo Molina-Pardo (Hymenoptera: Apoidea), Alejandro Madrigal (Hemiptera: Tingidae), and Gilberto Morales (Lepidoptera: Rhopalocera). The general taxonomic collection consists of about 600 Cornell drawers in cabinets, holding about 100,000 specimens, of which 6,000 specimens are identified. In addition there are separate economic insect collections of approximately 10,000 specimens, and an immature collection of about 600 specimens. Each species is recorded on cards with details about habitats, etc. A color slide collection supplements the collection. A catalog of the collection is in preparation. [1992]

MUSEO DE HISTORIA NATURAL, UNIVERSIDAD NACIONAL DE CALDAS, APARTADO AEREO 275, MANI-ZALES. [UNCC] [*No reply.*]

COLECCION ENTOMOLOGICA, FACULTAD DE AGRONOMIA, UNIVERSIDAD DE NARINO, PASTO. [FAUN] [*No reply.*]

MUSEO DE HISTORIA NATURAL, UNIVERSIDAD DEL CAUCA, POPAYAN. [UCPC] [*No reply.*]

COLECCION ENTOMOLOGICA, FACULTAD DE CIENCIAS AGROPECUARIAS, UNIVERSIDAD NACIONAL DE COLOMBIA, APARTADO AEREO 237, PALMIRA. [UNCP].

Director: Dr. Adalberto Figueroa. Phone: 28181, ext. 42. Professional staff: Ing. Agr. Jaime de la Cruz, Ing. Agr. Ivan Zuluaga, Ing. Agr. Gersain Olaya. This is a regional collection from the southeastern part of Colombia and to some extent part of the Departamento de Choco. Most of the specimens are unidentified, except for pest species. About 12,000 specimens are housed in Cornell drawers, Schmitt boxes, and vials. Papers are published in the Acta Agronomica. [1986]

42. COMOROS, Federal and Islamic Republic of the

[Includes Anjouan, Grand Comore, Moheli, and other smaller islands. Excludes the island of Mayotte, presently administered by France but claimed by the Comoros. Afrotropical. Moroni. **Population:** 429,479. **Size:** 719 sq. mi. *No reply; no known insect collection.*]

43. CONGO, People's Republic of the

[Afrotropical. Brazzaville. **Population:** 2,153,685. **Size:** 132,047 sq. mi. *No reply; no known insect collection.*]

44. COOK ISLANDS (COOK)

[Self-governing country in free association with New Zealand. Includes the atolls of Pukapuka (Danger), Manihiki, Penrhyn, and Rakahanga, which the U.S.A. claims. Oceanian (Polynesia). Avarua. **Population:** 17,995. **Size:** 91.5 sq. mi. *Reply: No insect collections kept on islands.*]

(Coral Sea Islands Territory, External Territory of Australia.)

(Corsica, see France.)

(Cosmoledo Group, see Seychelles.)

45. COSTA RICA, Republic of

[Includes Coco Island (not Cocos). Neotropical. San José. **Population:** 2,888,227. **Size:** 19,730 sq. mi.]

INSTITUTO NACIONAL DE BIODIVERSIDAD, APTO. 22-3100, SANTO DOMINGO DE HEREDIA, 3100, HEREDIA. [INBC]

Director: Dr. Rodrigo Gámez. Phone: 9506) 36-7690; FAX (506) 36-2816. Coordinator of Arthropods: Mr. Angel Solís. The institute has the mandate of sampling and recording the plant and animal life of Costa Rica, preparing identification guides, and training new taxonomists. Although it is a private organization, it works closely with the Ministry of Natural Resources, Energy, and Mines, and is effectively the national collection, at least in insects. At present the arthropod collection comprises about 1.5 million specimens, with an average monthly increment of about 60,000. The facilities include 650 12-drawer cabinets. Specimens data are being registered in a relational data-base, with the intention of interrelating this with the institutes other data-bases. The mode of operation is quite different from that of most collections and is thus being watched as a possible model for comparable national or regional efforts elsewhere. Much of the collecting and curating that would normally be handled by doctoral-level researchers is instead done by parataxonomists and apprentice curators. These are individuals of greatly varying background, trained specifically for work at the institute. About 40 parataxonomists are now employed full-time. International cooperation is a key part of the institute's operation. Several individuals at major foreign collections serve as counsellors to the organization, and provision is made for visiting specialists to work at the collection. (ex. C. Starr, in litt.) [1992]

MUSEO NACIONAL DE COSTA RICA, APARTADO 749-1000, SAN JOSE. [MNCR]

Director: Lorena San Ramon de Gallegos. Phone: 21-0295. Professional staff: Ms. Francisco Fallas, B.S., Isidro Chacon. This collection contains those of Pittier, Tonduz, Biolly, and other local collectors. Most of this material is in a bad state of preservation. Recent general collections have been made for classroom study. There are about 10,000 specimens with labels. Emphasis has been on Costa Rican butterflies, moths, Blattaria, and Coleoptera, housed in drawers and boxes. "Brenesia" is a quarterly publication of the Department of Natural History. [1986]

MUSEO DE INSECTOS, UNIVERSIDAD DE COSTA RICA, CUIDAD UNIVERSITARIA. [MUCR]

Director: Dr. Paul Hanson. Curator: Lic. Humberto Lezama. The collection consists of about 300,000 specimens, mainly from Costa Rica, taxonomically arranged. The museum's primary mandate is to provide a reference collection of economically important insects, but is strong in certain groups. (C. Starr, in litt.) [1992]

MUSEO DE ZOOLOGIA, ESCUELA DE BIOLOGIA, UNIVERSIDAD DE COSTA RICA, CIUDAD UNIVERSITARIA. [MZCR]

Director: Dr. Douglas C. Robinson. Phone: 25-5555. Professional staff: Dr. Carlos E. Valerio, spider curator. The spider collection consists of about 800 vials, with additional vials of scorpions, opilions, pseudoscorpions, amblypigids, and others. The specimens are from Costa Rica and Panama, including 36 holotypes. [No information was submitted about their insect collection.]

ORGANIZATION FOR TROPICAL STUDIES, INC., APTO. 16, UNIVERSIDAD DE COSTA RICA, SAN JOSE, COSTA RICA. [OTSC]. [No insect collection, but listed to prevent repeated questions about material to borrow.]

(Crozet Island, see France. (Antarctic).)

(Crete (Kriti), Island Province of Greece.)

46. CROATIA, Republic of

[Palearctic. Zagreb. *At time of printing details about the new countries forms from the old Yugoslavia were not available.*]

ENTOMOLOGY COLLECTION, HRVATSKI NARODNI ZOOLOSKI MUZEJ, DEMETROVA UL. 1, 4100 ZAGREB. [HZMZ]

Director: Ms. Marija Stosic. Phone: (041) 445-005. Professional staff: Franjo Perovic (Hymenoptera, Symphyta), Branko Djurasin (Carabidae). This collection has generally good coverage for western part of Yugoslavia. The groups particularly well represented are Coleoptera, Lepidoptera, Hymenoptera, Orthopteroidea, Diptera, and Hemipteroidea. Special collections included the following: P. Novak, I. and K. Igalffy, Weingartner, Korlevic, Koca, Redensek, Grund, Langhoffer, Kozulic, and Locke. The 350,000 specimens are housed in about 1,000 drawers; supporting ecological data are limited except for new collections made after 1960. Exchanges are not normally made for old collections but newly collected fauna can be exchanged. Usual loan rules apply on demonstration of competence and/or association with a university, college, or museum. The library facilities cover the determined fauna. The museum collection also has about 2,500 specimens of spiders. [1986]

47. CUBA, Republic of

[Neotropical. Habana. **Population:** 10,353,932. **Size:** 42,803 sq. mi.]

INSTITUTO DE ZOOLOGIA, ACADEMIA DE CIENCIAS DE CUBA, CAPITOLIO NACIONAL, LA HABANA 2, CUIDAD DE LA HABANA 10200. [IZAC]

Curator: Lic. Luis F. de Armas C. The collection includes more than 600,000 specimens. About half of the specimens were collected by P. Alayo. It also contains the Gundlach collection of more than 20,000 specimens, and the Franganillo collection of arachnids. Only Hymenoptera and the Gundlach collection have been cataloged. (C. Starr, in litt.)

MUSEO NACIONAL DE HISTORIA NATURAL, CAPITOLIO NACIONAL, LA HABANA 2, CIUDAD DE LA HABANA 10200. [MNHC]

Curator: Luis R. Hernández. Phone: (537) 63-2589; FAX (537) 62-5604 or 62-5605. This collection, with a national mandate, was founded in 1989. It presently contains about 10,000 curated specimens and may grow fairly rapidly. Areas of strength include Cuban arachnids and Dictyoptera and Caribbean myriapods. [May include the collection of the late Ing. Fernando de Zayas.] (C. Starr, in litt.) [1992]

[*Note: there are several other collections noted by Starr, particularly several private collections, but details have not been received at the time this book went to press.*]

(Curacao, see Netherland Antilles.)

48. CYPRUS, Republic of

[Palearctic. Nicosia. **Population:** 691,966. **Size:** 3,572 sq. mi.]

NATURAL HISTORY MUSEUM, ZOOLOGICAL GARDENS, LIMASSOL. [ZGLC] [*No reply.*]

49. CZECH, Federal Republic of

[=part of Czechoslovakia, which is now two federal republics, but with one president (see also Slovakia) Palearctic. Prague. **Population:** 9,800,000. **Size:** Part of total area of 49,384 sq. mi.]

CZECH ACADEMY OF SCIENCE, INSTITUTE OF ENTOMOLOGY, NA SADKACH 702, 37005 CESKE BUDEJOVICE. [IECA] [*No reply.*]

ENTOMOLOGY, MORAVSKE MUZEUM, PRESLOVA UL. 659 37, BRNO. [MMBC] [*No reply.*]

KRAJSKE MUZEUM VYCHODNICH CECH, HUSOVO NAMESTI 124, HRADEC KRALOVE 500 02. [KMVC].

Director: Ms. Jana Starkova. Phone: 23-416. Professional staff: B. Mocek (Curator), M. Mikat. The collection of beetles contains about 250,000 specimens of all families from the Palearctic region, with a special collection of 20,000 Scarabaeidae of the Palearctic and Oriental region from Dr. Z. Tesar; Lepidoptera, 20,000 specimens from Czechoslovakia only; Diptera, 11,000 specimens; Hymenoptera, 10,000; Orthoptera, 5,000; Heteroptera, 5,000, housed in 2,000 boxes. The spider collection includes about 4,000 specimens stored in vials. The museum sponsors "Acta musei reginaehradecensis S. A." [1986]

ENTOMOLOGICAL COLLECTIONS, SLEZSKÉ MUZEUM OPAVA, DEPARTMENT OF NATURAL HISTORY, MASARYKOVA 35, 74646 OPAVA [SMOC].

Curator: Dr. Jindrich Rohacek. Phone: 0653-212870; Mrs. Anna Novakova, Assistant. The collection includes 750,000 specimens (Coleoptera 490,000; Lepidoptera, 128,000; Hymenoptera, 35,000; Diptera, 81,000, others 16,000). The material is mainly Palaearctic and particularly Central European; extra-Palaearctic material is limited (less than 30,000 specimens, mainly Lepidoptera and Coleoptera). There are 910 type specimens (mainly paratypes) of Coleoptera, Lepidoptera, and Diptera, with only a few Hymenoptera and Orthoptera. The collection is stored in drawers and boxes. The museum publishes "Casopis Slezskeho zemského muzea (series A)." [1992]

DEPARTMENT OF ENTOMOLOGY, NATIONAL MUSEUM (NATURAL HISTORY), 148 00 PRAHA 4, KUNRATICE 1. [NMPC].

Director: Dr. Joseph Jelínek (Coleoptera). Phone: 49-32-06. Professional staff: Dr. S. Bíly, Dr. I. Kovár (Coleoptera), Dr. F. Krampl (Lepidoptera), Dr. J. Jezek (Diptera), Dr. J. Macek (Hymenoptera); Dr. V.

Svihla (Heteroptera), Dr. P. Chvojka (Trichoptera). The collection of about 2,500,000, is world-wide, representing all groups, but mostly Coleoptera and Lepidoptera, with about 5,000 types. It is stored in boxes. The museum publishes: "Acta entomologica Musei nationalis Pragae" and "Acta faunistica entomologica Musei nationalis Pragae." [1992]

(Dahomey=Benin.)

(Daito Islands, see Japan.)

(Danzig, see Poland.)

50. DENMARK, Kingdom of

[Includes Kalaalitt Nunaat [=Greenland], Faroe Islands, and Bornholm Island. Palearctic. København. **Population:** 5,125, 676. **Size:** 16,638 sq. mi.]

NATURAL HISTORY MUSEUM, UNIVERSITETSPARKEN, 8,000 ÅRHUS, E. JUTLAND [NHMA].

Director: Dr. Peter Gjelstrup. Phone: 86129777. Professional staff: Frank Jensen, Toke Skytte. More than 3 million specimens representing most groups. Publications sponsored: "Natura Jutlandica."

DEPARTMENT OF ENTOMOLOGY, ZOOLOGICAL MUSEUM, UNIVERSITY OF COPENHAGEN, UNIVERSITETSPARKEN, DK-2100 KØBENHAVN Ø. [ZMUC]

Director: Dr. N. Møller Andersen. Phone: (+45) 3532 1000; FAX (+45) 3532 1010. Professional staff: Dr. Henrik Enghoff, Dr. N. P. Kristensen, Mr. William Buch, Mr. Ole Karsholt, Mr. Søren Langemark. Additional staff: Research Associates: Dr. Michael Hansen, Dr. Leif Lyneborg, Dr. Verner Michelsen, Dr. Thomas Pape, and Dr. Nikolaj Scharff. The collection consists of about 3 million dry insects (mainly pinned, some in envelopes); more than 3 million insects, arachnids, and myriapods in alcohol. The collections cover all geographical regions of the World. Particular strengths are all groups of terrestrial arthropods from Denmark, Faroe Islands, Iceland, Greenland. Most insect groups from the Philippines and the Bismarck Archipelago ("Noona Dan" Expedition); Palearctic Diplopoda Julida, Diplopoda from Africa and Thailand; Acari Oribatida worldwide (coll. M. Hammer); Araneae from East Africa; Protura worldwide (coll. S. L. Tuxen); Old World Odonata; Heteroptera Gerromorpha worldwide (coll. L. Cheng, R. Poisson, and others); Coleoptera larvae, particularly from Denmark; Palearctic and Nearctic Coleoptera, particularly Carabidae (coll. E. Suenson); primitive Lepidoptera worldwide, including material in fixative for anatomical studies; Lepidoptera from Madeira (coll. N. L. Wolff), austral South America (Mision Cientifica Danesa), Cameroun-Nigeria-Ethiopia (coll. J. S. R. Birket-Smith); Neotropical butterflies (coll. C. S. Larsen); Mecoptera and Neuropteroid orders worldwide (coll. P. Esben-Petersen); Trichoptera immatures from Northern Europe (coll. A. Nielsen); Trichoptera from

East Africa; Palearctic and Afrotropical Diptera, Therevidae; Palearctic higher Diptera; Hymenoptera from Mediterranea, East Africa, South India, and Thailand; Palearctic and Oriental Mutillidae; Neotropical Apidae (coll. P. Jørgensen). The collection includes about 10,500 primary types, including most of the insects described by J. C. Fabricius (1745-1808) and numerous others. The collection is housed in drawers in cabinets. The museum sponsors "Steenstrupia," edited by Dr. Henrik Enghoff, "Entomologica Scandinavica," and "Dyr i Natur og Museum." [1992]

(Diomedes Islands, see USA/USSR.)

51. DJIBOUTI

[Formerly known as the Afars & Issas and also French Somalia. Afrotropical. Dijbouti. **Population:** 320,444. **Size:** 8,958 sq. mi. *No reply; no known insect collection.*]

52. DOMINICA, State of

[British associated state. Neotropical. Roseau. **Population:** 97,763. **Size:** 290 sq. mi. *Reply; no insect collections on island.*]

53. DOMINICAN REPUBLIC

[Neotropical. Santo Domingo. **Population:** 7,136,748. **Size:** 18,680 sq. mi.]

MUSEO NACIONAL DE HISTORIA NATURAL, PLAZA DE CULTURA, SANTO DOMINGO. [MHND]

Curators: Kelvin Guerrero (insects) and Félix del Monte (arachnids). Phone: (809) 689-0106, ext. 09. The insect collection comprises about 40,000 specimens. Preservation is troubled by irregular electrical current. (C. Starr) [1992]

(Dutch Guiana, see Suriname.)

(Dubai, see United Arab Emirates.)

(East Germany, see Germany.)

(Easter Island (=Rapa Nui, Isla de Pascual), see Chile.)

54. ECUADOR, Republic of

[Includes Galapagos Islands (Archipelago de Colon). Neotropical. Quito. **Population:** 10,231,630. **Size:** 106,860.]

MUSEO ECUATORIANO DE CIENCIAS NATURALES, TA-MAYO 557 Y VEINTEMILLA, CASILLA 8976 SUC. 7, QUITO. [MECN] [*No reply.*]

ECUADORIAN INSTITUTE OF NATURAL SCIENCES, P. O. BOX 408, QUITO. [EINS] [*No reply.*]

INSECT COLLECTION, CHARLES DARWIN RESEARCH STATION, PUERTO AYORA, SANTA CRUZ, GALAPAGOS. [CDRS]

[Contact: Ms. Sandra Abedrabbo. These is a small collection of arthropods kept at the station; also a research library. Eds. 1992.]

QUITO CATHOLIC ZOOLOGY MUSEUM, DEPARTAMENTO DE BIOLOGIA, PONTIFICIA UNIVERSIDAD CATOLICA DEL ECUADOR, 12 DE OCTUBRE Y CARRION, APTO. 2184, QUITO. [QCAZ]

Director: Dr. Giovanni Onore. Phone: 593-2-529750, ext. 273. Professional staff: Dr. Laura Arcos (department Director). The collection of Ecuadorian insects is presently being reorganized and new specimens are being added. Approximately 150,000 specimens have been labeled and stored in locally made wooden glass-topped boxes. A good collection of tropical palm pests and stingless bees are being developed. Regular collecting trips are made both to the west coast and eastern jungle areas. Specimens are available for exchange and loan; usual loan rules apply. Types are kept. Library facilities are fair. Publication sponsored: "Revisita de la Universidad Catolica." [1992]

FACULTAD DE CIENCIAS NATURALES, UNIVERSIDAD DE GUAYAQUIL, APARTADO 471 GUAYAQUIL. [UGGE]

No director or professional staff given. The collection consists of over 25,000 specimens. [1986]

55. EGYPT, Arab Republic of

[Palearctic. El-Qahira (=Cairo). **Population:** 53,347,679. **Size:** 385,229 sq. mi. *The following institutions are known to have insects collections; even after three letters, each has failed to reply.*]

PLANT PROTECTION DEPARTMENT, MINISTRY OF AGRICULTURE, DOKKI, CAIRO. [PPDD] [*No reply.*]

INSECT COLLECTION, [C/O DR. NAGART, SYSTEMATIST AND CURATOR OF COLLECTION], DEPARTMENT OF ENTOMOLOGY, AIN SHAMS UNIVERSITY, AEBASLY, CAIRO. [ASUA] [*No reply.*]

INSECT COLLECTION, BIOLOGY DEPARTMENT, EL AZHAR UNIVERSITY, CAIRO. [AUCE] [*No reply.*]

INSECT COLLECTION, DEPARTMENT OF ENTOMOLOGY, FACULTY OF SCIENCE, CAIRO UNIVERSITY, GIZA, CAIRO. [CUGE] [*No reply.*]

(Erie, see Irish Republic.)

56. EL SALVADOR, Republic of

[Neotropical. San Salvador. **Population:** 5,388,644. **Size:** 8,260 sq. mi.]

MUSEO NACIONAL "DAVID J. GUZMAN," SAN SALVADOR [MNDG] [*No reply.*]

NATURAL HISTORY MUSEUM, PARQUE ZOOLOGICAL NACIONAL, "FINCA MODELO," SAN SALVADOR. [FMSS] [*No reply.*]

UNIVERSIDAD DE EL SALVADOR, SAN SALVADOR [UESS] [*No reply.*]

(Ellice Islands, see Tuvalu.)

(England, see United Kingdom.)

57. EQUATORIAL GUINEA, Republic of

[=Spanish Guinea. Comprised of Rio Muni and the islands of Macias Nguema Biyogo, Pagalu, Corisco, Bioko (=Fernando Póo) and Islas de Elobey. Afrotropical. Malabo. **Population:** 346,839. **Size:** 10,831 sq. mi. *No reply; probably no insect collection.*]

58. ESTONIA

[Palaearctic. Tallinn. **Population:** 1,556,000. **Size:** 17,400 sq. mi.]

INSTITUTE OF ZOOLOGY AND BOTANY, UL. VANEMUIZHE 43, TARTU, ESTONIAN SSR. [IZBE] [*No reply.*]

59. ETHIOPIA

[=Abyssinia. Afrotropical. Addis Ababa. **Population:** 48,264,570. **Size:** 483,123 sq. mi.]

NATURAL HISTORY MUSEUM, HAILE SELASSIE I UNIVERSITY, P. O. BOX 1176, ADDIS ABABA. [HSUE] [*No reply.*]

(Faeroe Islands, see Denmark.)

(Falkland Islands, colony and dependency of Britain, claimed by Argentina (as Islas Malvinas); includes West Falkland, East Falkland, South Georgia, and South Sandwich islands. Stanley. [*No known insect collection.*])

(Fernando de Noronha Island, see Brazil.)

(Fernando Póo Island, see Equatorial Guinea.)

60. FIJI, Dominion of

[Includes Viti Levu, Vanua Levu, and Rotuma islands. Oceanian (Melanesia). Suva. **Population:** 740,761. **Size:** 7,095 sq. mi.]

THE ENTOMOLOGIST, MINISTRY OF PRIMARY INDUSTRIES, KORONIVIA RESEARCH STATION, NAUSORI. [PIKN] [*No reply.*]

[Contact: Mr. Sada N. Lal. The collection has many endemic insects represented, even though the emphasis is on agricultural species. Eds. 1992.]

INSECT COLLECTION, DEPARTMENT OF BIOLOGY, SCHOOL OF NATURAL RESOURCES, THE UNIVERSITY OF THE SOUTH PACIFIC, P.O. BOX 1168, LAUCALA BAY, SUVA. [FIJI]

Director: Dr. W. Kenchington. Phone: 315610. The collection, primarily a teaching collection, consists of about 10,000 specimens from Fiji stored in boxes. There is a good representation of Macrolepidoptera and Odonata. It includes the Fiji Sugar Corporation collection (1920-1935) in poor condition. [1992]

61. FINLAND, Republic of

[Including Åland. Palearctic. Helsingfors. **Population:** 4,949,716. **Size:** 130,559 sq. mi.]

ZOOLOGICAL MUSEUM, FINNISH MUSEUM OF NATURAL HISTORY, UNIVERSITY OF HELSINKI, P. RAUTATIEK 13, SF-00100 HELSINKI. [MZHF] [=ZMH and UZMH]

Director: Prof. Olof Biström. Phone: 358-0-191 7430. The professional staff consists of 15 curators and research associates. Although a part of the University of Helsinki, the museum has the status of the national zoological museum in Finland. The collections include some 6-7 million processed samples, and nearly equal number of samples more or less unprocessed. This includes the special collection of Mannerheim (Coleoptera), Nylander (Hymenoptera), and Palmgren (Spiders), as well as those of Bergroth, Frey, Hellen, Lindberg, Lindqvist, Poppius, Reuter, Sahlberg, Stockmann, and Tengstrom. The collection is stored in unit trays in drawers, and alcohol material in vials. [1992]

ZOOLOGISKA MUSET, UNIVERSITETS OULU. SF-90570, OULU 57. [ZMUO]

Phone: (9) 81-345-411. This is mainly a local collection of northern European insects and spiders. About 300,000 identified specimens, pinned or preserved in alcohol, and many unidentified and unsorted samples. The groups best represented in the collection are the Lepidoptera (including immature stages), Coleoptera, Trichoptera, and Araneae. The Lepidoptera collection includes also some material from other regions. The collection is housed in 1,700 cabinet drawers and 500 insect storage boxes and in cabinets for samples preserved in alcohol. [1992]

(Formosa, see Taiwan, Republic of China.)

62. FRANCE, Republic of

[Includes the island of Mayotte presently administered by France but claimed by the Comoros. Palearctic. Paris. **Population:** 55,798,282. **Size:** 210,026 sq. mi.]

MUSEUM D'HISTOIRE NATURELLE D'AUTUN, 14 RUE ST-ANTOINE, F 71400, AUTUN. [MHNA]

[Contact: not known. Holdings include Geoffroy collections. Eds. 1992.]

SERVICE SCIENTIFIC CENTRAL D'ENTOMOLOGIE AGRICOLE. ORSTOM, 70-74, ROUTE D'AULNAY, BONDY, SEINE. [ORST] [*No reply.*]

NATIONAL COLLECTION OF INSECTS, MUSEUM NATIONAL D'HISTOIRE NATURELLE, 45, RUE BUFFON, PARIS 75005. [MNHN]

Director: Prof. Dr. Claude Caussanel. Phone: 407-93-409. Professional staff includes 24 professionals, some of whom are staff members of the CNRS and EPHE. The collection contains 100 million specimens from all over the world, which represents about 400,000 species and 200,000 type specimens. For the older collections housed here, see Horn & Kahle, 1935 and 1937. For recent accessions refer to: "Bull. Mus. natl. Hist. nat., annual supplement" (or "Miscellanea"). The collection is housed in 500,000 Schmitt type boxes and some drawers. [1992]

MUSEUM ZOOLOGIQUE, UNIVERSITY DE STRASBOURG, 29 BLVD. DE LA VICTOIRE, F-67 STRASBOURG. [MZSF] [*No reply.*]

MUSÉE D'HISTOIRE NATURELLE DE LYON, 28 BLVD. DES BELGES, 69006 LYON, RHONE. [MHNL]

Director: Dr. L. David. Phone: 78-93-22-33. Professional staff: J. Clary. The collection has a large representation of families (Coleoptera, Lepidoptera, Heteroptera, Hymenoptera) with a good coverage of local insects including the Audras, Roman, and Milliat collections (Coleoptera), Côte, Baraud, Riel, Rouast, those of Rey (Coleoptera and Heteroptera with many types, 250 boxes), Foudras, Dejean (Coccinelidae, with types), Onzel (Lepidoptera, with types), Côte (Lepidoptera, Saturnidae, with types) in 200 boxes and a collection of silk butterflies (250 boxes). [1992]

MUSEUM NATIONAL D'HISTOIRE NATURELLE, 4 AVE. DU PETIT CHATEUA, 91800 BRUNOY (ESSONNE). [BFIC] [*No reply.*]

LABORATOIRE DE BIOLOGIE DES INSECTS, 118, ROUTE DE NARBONNE, F-31077, TOULOUSE. [LBIT]

Director: Prof. Dr. Jacques Bitsch. Western Palaearctic Hymenoptera, Aculeata. [1986]

(Franz Josef Land, islands of U.S.S.R.)

(French Guiana, see Guiana.)

63. FRENCH POLYNESIA, Overseas Territory of

[Includes the Society Islands, Tuamotu Islands, Marquesa Islands, Austral Islands, and Atoll Clipperton. Oceanian (Polynesia). Papeete. **Population:** 190,939. **Size:** 1,377.]

PAPEETE MUSEUM, P. O. BOX 110, RUA BREA, TAHITI. [PMFP] [*No reply.*]

(French Southern and Antarctic Lands, Overseas Territory of, includes Ile Amsterdam, Ile Saint-Paul, Iles Kerguelen, and Iles Crozet. Terre Adelie (French-claimed sector of Antarctica) is not included in this entity but rather, see ANTARCTICA. The U.S.A. does not recognize sovereignty in Antarctica; therefore Terra Adelia is excluded from this entity. *No known insect collection.*)

(Fujeira, see United Arab Emirates.)

64. GABON, Republic of

[Afrotropical. Libreville. **Population:** 1,051,937. **Size:** 103,347 sq. mi. *No known insect collection.*]

(Galapagos Islands (=Archipelago de Colon), see Ecuador.)

65. THE GAMBIA, Republic of

[Afrotropical. Banjul. **Population:** 779,488. **Size:** 4,361 sq. mi. *No known insect collection.*]

(Gambier Islands, see Mangareva Islands.)

(Gaza Strip, part of former Palestine mandate; includes city of Gaza and environs bounded by the Mediterranean, Egypt, and Israel. *No known insect collection.*)

66. GERMANY

[=East Germany, DDR, and West Germany, BRD, including all of Berlin. Palearctic. Berlin. **Population:** 77,577,077. **Size:** 137,855 sq. mi.]

Berlin

MUSEUM FÜR NATURKUNDE DER HUMBOLDT UNIVERSITAT ZU BERLIN, BEREICH ZOOLOGISCHES MUSEUM, INVALIDENSTRASSE 43, 1040 BERLIN. [ZMHB] [=ZMHU; ZMB]

Director: Not given. Professional staff: Eight curators. The Zoological Museum of the Humboldt University of Berlin contains more than 10 million specimens of insects from all over the world, especially Coleoptera, Lepidoptera, Hymenoptera (including Apidae collection of J . D. Alfken, A. Ducke [part], and H. Friese), Diptera, Orthoptera, Hemiptera, Homoptera, and other groups with many types of well known entomologists. The collection is housed in drawers. The spider collection is one of the important classical collections of the world with about 3,500 species represented with about 50,000 specimens (most of them preserved in alcohol), and with many types of L. Koch, C. L. Koch, F. Karsch, H. Zimmermann, Fr. Dahl, and the Godeffroy Museum collection. The museum publishes: "Mitteilungen aus dem Zoologischen Museum Berlin" and "Deutsche Entomologische Zeitschrift." [1986]

Bielefeld

FAKULTAT FÜR BIOLOGIE, UNIVERSITAT BIELEFELD, POSTFACH 8640, D-8400 BIELEFELD 1, NORDRHEIN-WESTFALEN. [FBUB] [*No reply.*]

Bonn

ZOOLOGISCHES FORSCHUNGSINSTITUT UND MUSEUM "ALEXANDER KOENIG," (ALEXANDER KOENIG ZOOLOGICAL RESEARCH INSTITUTE AND ZOOLOGICAL MUSEUM) ADENAUERALLE 160, 5300 BONN 1. [ZFMK]

The institution's research is focussed on the diversity of vertebrates and insects. It is sponsored by the federal state of North-Rhine-Westfalia and the government of the Federal Republic of Germany. The Department of Entomology comprises 5 sections, with the following distribution of responsibilities: Coleoptera, Strepsiptera (Dr. Michael Schmitt); Hymenoptera, Siphonaptera, Anoplura, and other parasitoid insects (Dr. K. H. Lampe); Diptera, Mecoptera, Neuroptera (Dr. H. Ulrich); Lepidoptera, Trichoptera (Dr. D. Stüning); lower arthropods, including Crustacea and Arachnida (Dr. F. Krapp). There are no sections for the hemitabolous and apterygote groups of insects, although extensive collections are held by the institution.

The arthropod collections contain about 3 million specimens, with major bulks in the Lepidoptera (about 9 million specimens) and Coleoptera (1.8 million specimens). Geographically the collection focuses on Eurasia and Africa. The majority of specimens are pinned and set in the traditional way, but alcohol collections have been set up recently in a number of taxa and for various purposes. Type material is specially important in the Lepidoptera (the famous Hoene collection of Chinese Macro- and Microlepidoptera) and Diptera (Phoridae, Microphoridae, and

Nematocera). In the Coleoptera there are important collections of myrmecophilus and termitophilous taxa (Reichensperger collection). Beside the Lepidoptera, Coleoptera, and Diptera there are important and type-holding collections in the Odonata (coll. Buchholz), the Paussidae (coll. Reichensperger), Scarabaeidae (coll. Frey), Tenebrionidae, and Meloidae (coll. Kaszab), and Curculionidae (coll. Voss).

The following serial publications are issued: "Bonner Zoologische Beiträge" (journal for systematic and organismic zoology, four issues annually) and "Bonner Zoologische Monographien" (monographs, 1-2 volumes per year issued at irregular intervals.) In addition, books and symposium volumes are published separately. [1992]

Braun-Schweig

INSECT COLLECTION, STAATLICHES NATURHISTORISCHES MUSEUM, POCKELSSTRASSE 10A, D-3300 BRAUN-SCHWEIG, NIEDER-SACHSEN. [SNMB]

Director: Prof. Dr. Otto von Frisch. Phone: 0531-331-914. Professional staff: Dr. Juergen Hevers. The collections consists of the following: Lepidoptera, about 45,000 specimens in addition to the special collection of Dr. Fritz Hartwief (1877-1962) with about 20,000 specimens; Dr. Wilhelm Wolf (1878-1963), about 3,000 specimens; Erich Wissel (1902-1979), with about 15,000 specimens. Coleoptera, about 55,000 specimens and the following special collections: R. Heinemann, about 17,000 specimens; Dr. Kirchhoff (incl. collection of Ernst Weise), about 5,000 specimens. Other insects, about 15,000 specimens, and the collection of Theodor Beling (-1898) of Tipulidae; Prof. Dr. H. Friese, 1,000 specimens of Ichneumonidae; Prof. Dr. G. Jurzitza, 1,500 specimens of Odonata. [1986]

Bremen

UBERSEE-MUSEUM, BAHNHOFSPLATZ 13, D-2800 BREMEN. [UMBB]

Director: Dr. Herbert Ganslmayr. Phone: 397-8357. Professional staff: Helmut Riemann, and 4 taxonomists support the identification of specimens. The collection consists of about 500,000 pinned specimens, 800 of which are holotypes and paratypes. The collection is stored in 3,500 insect boxes. Most of the collection is from northwestern Germany, but collections have been made in Costa Rica, Papua New Guinea, Spain, and East Africa. [1986]

Coburg

INSEKTENSAMMLUNG DES NATUR-MUSEUMS DER COBURGER LANDESSTIFTUNG, PARK 6, D-8630 COBURG, BAVARIA. [NMCL]

Director: Dr. Werner Korn. Phone: (09561) 75068 and 75069. Professional staff: Dr. Georg Aumann. The collection contains about 80,000 specimens in three separate collections: Georg Vollrath collection of macrolepidoptera from N. Bavaria; Dr. Erich Garthe collection of macro- and microlepidoptera from Bavaria; and the general insect collections, including those of the Dukes of Coburg, Czar Ferdinand of Bulgaria, and several small private collections of butterflies, beetles, and bugs. Material may be borrowed by special arrangement. [1992]

Darmstadt

HESSISCHES LANDESMUSEUM DARMSTADT, ZOOLOGISCHE ABTEILUNG, FRIEDENSPLATZ 1, D-6100 DARMSTADT, HESSEN. [HLDH]

Director: Dr. Hanns Feustel. Phone: (06151) 165703. There are five separate collections as follows: "Kaup" collection of Coleoptera in 57 boxes, with about 3,500 specimens, including a famous collection of Passalidae, with about 150 specimens, some holotypes and paratypes. "Konsul Francis Sarg" collection of Coleoptera of about 1,000 specimens from Guatemala, North America, Brazil, East Africa, India, Sumatra, and South Australia. "Wüsthoff" collection of Coleoptera with about 2,000 specimens. "Albrecht Heil" collection of Lepidoptera, 730 specimens. Several departmental collections of Diptera, Hymenoptera, Orthoptera, etc., most of them locally collected. The collection is housed in drawers and boxes. [1992]

Dortmund

MUSEUM FÜR NATURKUNDE, BALKENSTRASSE 40, D-4600 DORTMUND, NORDRHEIM-WESTFALEN. [MNNW] [*No reply.*]

Dresden

DEPARTMENT OF ENTOMOLOGY AND DEPARTMENT OF ARACHNOLOGY COLLECTIONS, STAATLICHES MUSEUM FUR TIERKUNDE, DRESDEN, FORSCHUNGSSTELLE, AUGUSTUSSTRASSE 2, 8010 DRESDEN. [SMTD]

Director: Dr. Rolf Hertel. Phone: 495-2503. Professional staff: Dipl. Biol. Regine Eck (Hymenoptera); Dr. Rainer Emmrich (Homoptera-Auchenorrhyncha); Dipl.-Biol. Uwe Kallweit (Diptera); Dr. Rüdiger Krause (Curculionidae; head of Dept.); Technical: Horst Bembenek (Lepidoptera). The estimated number of insect specimens in the collection is 5 million. Coleoptera, Lepidoptera, and Diptera are well represented. The collections of Faust, Hartmann, Penecke, Schultze, Kirsch, Ermisch, Linke, Jacobi, and Schnuse are incorporated. The Arachnida collection contains about 600 western Palearctic species. The collection is housed in drawers, boxes, and vials. Publications sponsored: "Reichenbachia, Entomologische Abhandlungen" and "Faunistische Abhandlungen." [1992]

Düsseldorf

LÖBBECKE-MUSEUM + AQUAZOO, POSTFACH 10 11 20, KAISERSWERTHER STR., 4000 DÜSSELDORF. [LMAD]

Director: Prof. Dr. M. Zahn. Phone: (0211) 899-6150. Professional staff: Dr. S. Löser (Head), Mr. D. Schulten, and Mr. H. Baumann. This collection is mostly Palaearctic Lepidoptera (400,000 specimems) and Palaearctic Coleoptera (100,000 specimens) representing several thousand species, stored in 2,000 drawers. [1992]

Eberswalde

DEUTSCHES ENTOMOLOGISCHES INSTITUT, EBERSWALDE FINOW 1, 1300. [DEIC]

Director: Dr. J. Oehlke. The collection contains over 225,000 species represented by about 2.2 million specimens. This includes the collections of: Coleoptera: Borner, Backhaus, Bennigsen, L. von Heyden, Horn, Koltze, Kraatz, Leonhard, Letzner, Liebmann, Metzler, Franklin Muller, Neresheimer, P. Pape, Rolph, S. & K. Schenkling, Schwarz, A. J. & J. C. Stern, Stierlin, Strohmeyer. Diptera: Uhmann, Hennig, Lichwardt, Oldenberg, Siebert. Heteroptera: Breddin. Hymenoptera: A. Friese, H. Haupt, Konow, von Leonhardi, Ludeke, Ratzeburg, E. Strand. Lepidoptera: Kardakoff, O. Leonhard, Pietsch, Saalmuller. Thysanoptera: von Oettingen. Insects of Taiwan: O. Sauter. There about about 23,000 types. [1992]

Erlangen

INSTITUT FUR ZOOLOGIE, UNIVERSITAT-ERLANGEN-NURNBERG, UNIVERSITATSTRASSE 19, D-8520 ERLANGEN, BAYERN. [IZUE] [*No reply.*]

Frankfurt-am-Main

FORSCHUNGSINSTITUT UND NATURMUSEUM SENC-KENBERG, ENTOMOLOGISCHE SECTION 1, SENC-KENBERGANLAGE 25, D-6000 FRANKFURT-AM-MAIN, HESSEN [SMFD].

Director: Dr. W. Ziegler. Phone: 069-754-2252. There are four entomological sections. Section I, Dr. Damir Kovac, in charge; Section II, Dr. Heinz Schröder, in charge; Section III, Dr. Jens-Peter Kopelke, in charge; Section IV, Dr. Wolgang Tobias, in charge. In addition, there is one Arachnida section, Dr. Manfred Grasshoff, in charge. Section I contains Coleoptera (*ca.* 6,500 drawers, with emphasis on world Cerambycidae and Gyrinidae, including many types; Strepsiptera (Kinzelbach, with 2,500 slides), and Thysanoptera (Dr. Richard zur Strassen) with 210,000 specimens representing 2,620 species.

[Section II, no further information.]

Section III contains Hymenoptera, with about 277,000 specimens representing *ca.* 9,800 species, especially Ichneumonidae, Apoidea, Chrysididae, and Tenthredinidae from the Palaearctic Region. Most of the insects are pinned and stored in drawers, but some additional material is preserved in alcohol vials. The collection contains types and paratypes of 690 species, especially Ichneumonidae (370 species), and Apoides (220 species). Collections from v. Heyden (1793-1866), in part), Schenck (1803-1878, *ca.* 13,750 specimens), Habermehl (1858-1940), Rebmann (1896-1970) *ca.* 10,000 specimens, and Heinrich (1903-1978) are included.

[Sections IV, and I, Arachnida, no further information.] [1992]

FORSCHUNGSINSTITUT SENCKENBERG, SECKTION FÜR LIMNISCHE ÖKOLOGIE, SENCKENBERGANIAGE 25, D-6000 FRANKFURT AM MAIN 1. [FSSF]

Director: Dr. W. Tobias. Phone: 069-7542244. Professional staff: Mrs. Irene Rademacher and Mrs. Susanne Turowski (technical assistants). Collection includes mostly Trichoptera, Plecoptera, and Diptera, including many type specimens. [1992]

Freiburg

MUSEUM FÜR NATURKUNDE, GERBERAN 32, D-7800 FREIBURG IM BREISGAU, BADEN-WURTTEMBERG. [MNFD] [*No reply.*]

Görlitz

STAATLICHES MUSEUM FÜR NATURKUNDE, PSF 425, GÖRLITZ, D-O-8900. [SMNG]

Director: Dr. W. Dunger. Phone: 24444. Professional staff: Wolfram Dunger (Collembola; Apterygota); Hans-Jürgen Schultz (Apterygota); Axel Christian (Gamasina); Bernard Seifert (Formicidae); Karlin Voigtländer (Myriapoda), and 4 laboratory technicians. Scientific research is concentrated on soil arthropods. Most of the collecting activity is restricted to Central Europe. All arthropod collections have mostly western Palearctic species. Myriapoda: approx. 150 species, 18,000 specimens. Apteryogta: approx. 450 species, 500,000 specimens; Pterygota: approx. 89,000 pinned, 13,000 processed vials of soil samples. Spiders: approx. 350 species, 25,000 specimens; Gamasina approx. 100 species, 20,000 specimens; Oribatei, approx. 400 species, 25,000 specimens; other mites approx. 10,000 specimens. Publication sponsored: "Abhandlungen und Berichte des Naturkundemuseums Gorlitz." [1992]

Gotha

MUSEUM DER NATUR GOTHA, PARKALLEE 15, PF 217, O-5800 GOTHA. [NMPG]

Director: Dr. rer. nat. Dipl. Biol. W. Zimmermann. Phone: 3167. The collection is curated by Dr. W. Zimmermann (Odonata, Ephemeroptera), R. Samietz (Chironomidae) and R. Bellstedt (Dolichopodidae). This is a general collection of Palearctic insects, including all of Central Europe, particularly Germany. About 13,000 species with 210,000 specimens are included. A few holotypes are in the collection of Ephemeroptera, Plecoptera, Trichoptera, and Lepidoptera. Specimens are available for study. No exchanges. [1992]

Greifswald

ERNST-MORITZ-ARNDT-UNIVERSITAT GREIFSWALD, ZOOLOGISCHES INSTITUT UND MUSEUM, J. S. BACH.-STR. 11/12/, 2200 GREIFSWALD. [EMAU]

Director: Prof. Dr. W. Mohrig. Phone: 2143. Scientific insect collection consists of about 2,000 boxes of all insect groups, including the collections of Von Bernuth, C. E. A. Gerstaecker, P. Heckel, O. Karl, P. Manteufel, K. Peter, J. Pfau, C. F. Pogge, K. Pogge, and O. C. F. G. Schmidt, u.a., with about 500 types. The Arachnida collection contains about 700 vials, including types of G. L. Mayr, B. Y. Sjostedt, and G. Ulmer. Publications sponsored: "Wissenschaftliche Zeitschrift der Ernst-Moritz-Arndt-Universitat Greifswald, Math.-Nat.-Reihe." [1992]

Halle

WISSENSCHAFTSBEREICH ZOOLOGIE, SEKTION BIOWISSENSCHAFTEN MARTIN-LUTHER-UNIVERSITAT HALLE, WB ZOOLOGIE, DOMPLATZ 4, 4020 HALLE (SAALE). [MLUH]

Director: Prof. Dr. sc. J. Schuh. Professional staff: Doz. Dr. sc. M. Dorn. The collection contains about 800,000 specimens of all orders from all over the world, housed in 2,000 drawers. This includes the following special collections: Burmeister, Germar, Suffrian, Roder, Taschenberg, Haupt, and Nitzsch. [1986]

Hamburg

ZOOLOGISCHES INSTITUT UND ZOOLOGISCHES MUS-EUM, ABTEILUNG ENTOMOLOGIE, UNIVERSITÄT HAMBURG, MARTIN-LUTHER-KING-PLATZ 3, D-2000 HAMBURG 13. [ZMUH]

Director: Prof. Dr. Hans Strümpel. Professional staff: Prof. Dr. Rudolf Abraham, Dr. Gisela Rack, and Dr. Hieronim Dastych. All collections of insects and Chelicerata are published in: "Mitteilungen aus dem Hamburgischen Zoologischen Museum und Institut." [*No further information supplied except for a special collection of Heteroptera from Eduard Wagner which is not included in the published catalog.*] [1992]

Hannover

NIEDERSACHSISCHES LANDESMUSEUM, AM MASCHPARK 5, D-3000 HANNOVER, NIEDERSACHSEN. [NLHD] [*No reply.*]

Jena

PHYLETISCHES MUSEUM, EBERTSTRASSE 1, 6900 JENA. [PMIG] [*No reply.*]

Karlsruhe

STAATLICHES MUSEUM FÜR NATURKUNDE KARLSRUHE, ERBPRINZENSTRASSE 13, POSTFACH 6209, D-7500 KARLSRUHE, BADEN-WURTTEMBERG. [SMNK] [=LNKD]

Director: Prof. Dr. S. Rietschel. Phone: 0049-721-175161. Professional staff: Mr. G. Ebert (Department of Entomology), Dipl. Biol. M. Verhaagh (Department of Zoology). This is mainly a collection of about 2 million Lepidoptera including many types from Europe, Near and Middle Orient, Southeast Asia, Australia, and Nearctic Region. Some Coleoptera, Hymenoptera, Hemiptera, Odonata, and Saltatoria, mostly Palearctic Region, are included. A collection of ants, mainly Neotropical and Malaysian, is currently built up, as well as a collection of Neotropical spiders. Part of the ant collection, as also, the spider collection, is stored in alcohol. The pinned collection is housed in drawers and boxes. Publication sponsored: "Carolinea - Beitrage zur naturkundlichen Forschung in Sudwestdeutschland" and "Andrias." [1992]

Leipzig

MUSEUM OF NATURAL SCIENCES, LORTZINGSTRASSE 3, LEIPZIG. [MNSL] [*No reply.*]

Mainz

ENTOMOLOGICAL COLLECTION, NATURHISTORISCHES MUSEUM MAINZ, REICHKLARASTRASSE 1+10, D-6500 MAINZ, RHEINLAND-PFALZ. [NHMM]

Director: Dr. Fr. O. Neuffer. Phone: 06131-122646. Professional staff: Dr. U. Schmidt. Phone: 06131-122647. The collection consists of about 220,000 specimens of Araneae, Opiliones, Lepidoptera, Heteroptera, Diptera, Hymenoptera, and Coleoptera (Curculionidae) housed in drawers and boxes. Publication sponsored: "Mainzer Naturwissenschaftliches Archiv"; "Mainzer Nat-urwissenschafticHes Archiv, Beiheft", and "Mitteilungen der Rheinischen Naturforschenden Gesellschaft." [1992]

München

ZOOLOGISCHE STAATSSAMMLUNG, MUNCHHAUSENSTRASSE 21, D-8000 MÜNCHEN 60, BAYERN. [ZSMC]

Director: Dr. Hubert Fechter (Deputy). Phone: 8170-0. Professional staff: Department of Entomology: Dr. E.-G. Burmeister, Dr. W. Dier; (Head), E. Diller, Dr. A. Hausmann, Dr. F. Reiss, Dr. G. Scherer. The collection consists of 13.5 million specimens in 52,560 drawers, representing *circa* 300,000 species. Some more millions are still unprepared.

Several million additional specimens in alcohol vials, and about 66,000 micro slides complete the collection. The collection contains thousands of types. Publications sponsored: "Spixiana, Zietschrift für Zoologie", "Spixiana, Supplemente," "Chironomus, Mitteilungen aus der Chironomidenkunde," and "Entomofauna, Zeitschrift für Entomologie." [1992]

Affiliated collection:

Döberl, Manfred, Seeweg 34, D-8423 Abensberg, Germany. Phone: 09443-6550. [MDGC] About 70,000 specimens of beetles, mainly Chrysomelidae, Curculionidae, Coccinellidae, and other families. [1992] (Registered with ZSMC.)

Munster

LANDESSAMMLUNGEN FUR NATURKUNDE, HIMMELREICHALLE 50, D-4400 MUNSTER, NORD RHEIN-WESTFALEN. [LNMD] [*No reply.*]

Saarbrucken

INSTITUT FÜR GEOGRAPHIE-BIOGEOGRAPHIE, UNIVERSITAT DES SAARLANDES, D-6600 SAARBRUCKEN, SAARLAND. [IGUS] [*No reply.*]

Stuttgart

STAATLICHES MUSEUM FÜR NATURKUNDE, Rosenstein 1, D-7000 STUTTGART, BADEN-WÜRTTEMBÜRG. [SMNS]

Director: Prof. Dr. B. Ziegler. Phone 0711-8936-0. Professional staff (Entomology Department): Dr. H. P. Tschorsinig (Di-ptera), Dr. T. Osten (Hymenoptera), Dr. W. Schawaller (Coleoptera), Prof. Dr. E. Möhn (Diptera, Head). The collection consists of several million specimens in about 20,000 drawers, and additional specimens in alcohol vials. The collection contains thousands of types (mainly Diptera and Coleoptera). Main resources: Diptera in general (collection of Prof. Lindner, late editor of Diptera of the Palaearctic region), Diptera, Tachinidae (Dr. H. P. Tschorsing), Saltatoria (old Krauss collection), Hymenoptera (Dr. T. Osten), Coleoptera in general (Dr. W. Schawaller). Periodicals and series sponsored: Stuttgarter Beiträge zur Naturkunde, ser. A, B, C; Mitteilungen des Entomologischen Vereins Stuttgart. [1992]

Tutzing

ENTOMOLOGISCHES INSTITUT, MUSEUM G. FREY, HOFRAT-BEISELE-STRASSE 6, D-8132 TUTZING, BEI MUNCHEN, BAYERN. [MGFT]

[This collection of Coleoptera remains intact in Tutzing, but its ultimate disposal has not taken place; both ZSMC and NHMB have interests in it. The collection is essentially closed until a final decision can be made. The holdings are especially rich in Neotropical Chrysomelidae. The museum formerly published: "Entomologische Arbiten aus dem Museum G. Frey." [1992]

Wiesbaden

NATURWISSENSCHAFTLICHE SAMMLUNG (ZOOLOGISCHE SEKTION), MUSEUM WIESBADEN, FRIEDRICH-EBERT-ALLEE 2, D-6200 WIESBADEN, HESSEN. [MWNH]

Director: Dr. Michael Geisthardt (Curator). Phone: 0611-368-2182. Professional staff: Dr. M. Geisthardt. Technical: Erhard Zenker; Gerhard Heinrich. The collection contains about 900,000 specimens of insects and spiders. The Coleoptera (about 100,000 specimens) are restricted more or less to the Palearctic Region and some parts of Africa (former German colonies), but some families are represented worldwide. Expeditions to southern Europe (especially Greece) and Cape Verde Islands enhance the collection. The Lepidoptera (about 650,000) are (except North America) represented worldwide. Of special interest are the collections of Hymenoptera, Diptera, Heteroptera, and Auchenorrhyncha (about 150,000 specimens from Germany, especially Hessen. The spider collection consists of 4,000 specimens from Germany, with small collections from Africa. There are numerous holotypes, especially in the Lepidoptera, Hymenoptera, Heteroptera, Auchenorrhyncha, and spider collections. The insects are housed in about 4,600 drawers and boxes. Specimens may be borrowed by special arrangement. [1992]

67. GHANA, Republic of

[=Gold Coast. Afrotropical. Accra. **Population:** 14,360,121. **Size:** 92,100 sq. mi.]

DEPARTMENT OF BIOLOGICAL SCIENCES, MUSEUM OF NATURAL HISTORY, UNIVERSITY OF SCIENCE AND TECHNOLOGY, KUMASI. [USTK]

Director: Dr. Emmanuel Frempong. Phone: 5351-9. Professional staff: Mr. M. A. Dawood. A small general collection including spiders. [1992]

(Gibralter, British territory. *No known insect collection.*)

(Gilbert Islands, see Kiribati.)

(Great Britain, see United Kingdom.)

(Gough Island, see United Kingdom.)

(Greater Antilles Islands, see individual islands.)

68. GREECE, Hellenic Republic of

[Palearctic. Athinai (Athens). **Population:** 10,015,041. **Size:** 50,949 sq. mi.]

GOULANDRIS NATURAL HISTORY MUSEUM, KIFISSIA. [GNHM] [*No reply.*]

(Greenland (Kalaalitt Nuneat), see Denmark.)

69. GRENADA, Commonwealth of

[Includes the southern Grenadine Islands. Neotropical. St. George's. **Population:** 84,455. **Size:** 133 sq. mi. *No known insect collection.*]

70. GUADELOUPE, Department of

[French Antilles: includes Grande-Terre, Basse-Terre, Iles des Saintes, Iles de la Petite Terre, La Desirade, Ile Saint-Barthelemy, Marie-Galante, and northern St. Martin. Neotropical. Basse-Terre. **Population:** 338,730. **Size:** 1,622 sq. mi.]

INSTITUT DE RECHERCHES ENTOMOLOGIQUE DE LA CARIBE, B.P. 119, POINTE-A-PITRE. [IREC]

Director: Fortune Chalumeau. Phone: (590) 85-71-82; FAX (590) 88-49-57. The collection currently contains about 750 drawers, with about 80,000 pinned specimens. The collection contains types. Primarily a collection of Coleoptera and Lepidoptera from the entire Neotropical and Nearctic Regions, it emphasizes the Antillean Region. Loans and exchanges are made. Books on the fauna of the Antillean Region are published. [1992]

INSTITUT NATIONAL DE LA RECHERCHE AGRONOMIQUE DE ANTILLES ET GUYANE, DOMAINE DU-CHOS, F-97170 PETIT-BOURG, GUADELOUPE. [IRAG] [*No reply.*]

(Guadalupe Island, see Mexico.)

(Guam, unincorporated territory of USA; see U.S.A., Miscellaneous Pacific Islands.]

71. GUATEMALA, Republic of

[Neotropical. Guatemala City. **Population:** 8,831,148. **Size:** 42,042 sq. mi.]

MUSEO NACIONAL DE HISTORIA NATURAL, FINCA LA AURORA, ZONE 13, GUATEMALA CITY. [MNGC] [*No reply.*]

INSECT COLLECTION, UNIVERSIDAD DEL VALLE DE GUATEMALA, GUATEMALA CITY. [UVGC]

Director: Dr. Jack C. Schuster. Phone: (502) 269-0791; FAX (502) 238-0212. The collection consists of 22,000 specimens in 30 cabinets each with 25 Cornell drawers, together with open shelving for vials, jars, and slide cabinets. At present approximately 1/3 of the space is occupied.

Emphasis is on Guatemalan species and those from surrounding areas and follows the lines of research interest of students and staff of the University. Thus, there is an emphasis on Hymenoptera, Coleoptera, and venomous arthropods; agricultural and forests pests; pollinators of Neotropical flora; medical entomology. The latter includes the Dalmat collection of Simuliidae, Clarke-Darsie collection of Culicidae (part). Also included are light trap samples from several areas of Guatemala as a continuous series begun in 1978, stored in alcohol. [1992]

(Guernsey, see Channel Islands.)

72. GUIANA, Department of

[=French Guiana. Neotropical. Conakry. **Population:** 6,909,298. **Size:** 94,926 sq. mi.]

DEPARTMENT OF ENTOMOLOGY, OFFICE DE LA RECHERCHE SCIENTIFIQUE ET TECHNIQUE D'OUTRE-MER, BOITE POSTAL 165, 97305 CAYENNE CEDEX. [ORSC] [*No reply.*]

73. GUINEA, People's Revolutionary Republic of

[Afrotropical. Conakry. **Population:** 6,909,298. **Size:** 94,926 sq. mi.]

LOCAL MUSEUM, B. P. 114, N'ZEREKORE. [LMZG] [*No reply.*]

74. GUINEA-BISSAU, Republic of

[=Portuguese Guinea. Afrotropical. Bissau. **Population:** 950,742. **Size:** 10,811 sq. mi.]

MUSEUM OF PORTUGUESE GUINEA, PRACO DO IMPERIO, BISSAU. [MPGB] [*No reply.*]

75. GUYANA, Cooperative Republic of

[=British Guiana. Neotropical. Georgetown. **Population:** 765,796. **Size:** 83,000 sq. mi.]

INSECT COLLECTION, UNIVERSITY OF GUYANA, GEORGETOWN. [UGGG] [*No reply.*]

NATIONAL AGRICULTURAL RESEARCH INSTITUTE, MON REPOS, EAST COAST DEMERARA. [NARI]

Entomologist: Dr. Leslie Munroe. Phone: (592) 020-2842, 2881, 2882, 2883. This is a collection of insects and arachnids in 160 drawers, with emphasis on economically important species. The collection has suffered deterioration for some years due to poor housing. A rehabilitation and expansion program was begun in 1989, resulting in marked, ongoing improvement. (C. Starr.) [1992]

76. HAITI, Republic of

[Neotropical. Port-au-Prince. **Population:** 6,295,570. **Size:** 10,714 sq. mi. *No known insect collection.*]

(Hawaiian Islands, see USA.)

(Heard Island and McDonald Islands, external Territory of Australia. *No known insect collection.*]

(Hispaniola, see Haiti and Dominican Republic.)

(Hokkaido, island of Japan.)

(Holland, see Netherlands.)

77. HONDURAS, Republic of

[Neotropical. Tegucigalpa. **Population:** 4,972,287. **Size:** 43,277 sq. mi.]

COLECCIÓN ENTOMOLÓGICA, ESCUELA NACIONAL DE CIENCIAS FORESTALES, APDO. 2, SIGUATEPEQUE, COMAYAGUA. [CEEF]

Director: Ing. Oscar Leverón. Phone: (504) 73-2011. The collection consists of 176,000 specimens in drawers, on slides, and in preservative. Emphasis is on forest insects, with Coleoptera predominating. (C. Starr.) [1992]

COLECCION DE INSECTOS DEL VALLE DEL GUAYAPE, OLANCHO, DIRECCION AGRICOLA REGIONAL NO. 5, MINISTERIO DE RECURSOS NATURALES, JUTICALPA, OLANCHO. [OLAN]

Director: not given. Phone: (504) 95-20-56. This is a small collection of pest and beneficial species relating to corn, sorghum, bean, and cereal crops. [1986]

ESCUELA NACIONAL DE AGRICULTURA, CATACAMAS, OLANCHO. [HENA]

Director: Not given. The collection has about 2,000 specimens, mainly in the orders Odonata, Orthoptera, Hemiptera, Coleoptera, Lepidoptera, Diptera, and Hymenoptera, collected by students, and housed in 36 drawers. These are associated with various crops. Some specimens are stored in alcohol vials. [1986]

AGROECOLOGICAL INVENTORY COLLECTION, DEPTO. DE PROTECCION VEGETAL, ESCUELA AGRÍCOLA PANAMERICANA, APARTADO 93, TEGUCIGALPA. [EAPZ] [=EAPC]

Director: Dr. Ronald D. Cave. Phone: (502) 76-61-40. Professional Staff: Ronald D. Cave, Rafael Caballero, Boris Castro, Rosa Ortega. The collection consists of 675 Cornell-type drawers containing about 60,000 specimens of all major orders; about 1500 vials with immature insects, and about 2,000 slide mounts of Thysanoptera, Homoptera, and parasitic Hymenoptera, including voucher specimens from student theses. The specimens are almost exclusively from Honduras. Specimen identification and ecological data is computerized. Material may be borrowed; usual loan rules apply. [1992]

78. HONG KONG, Colony of

[British colony, including Kowloon and the New Territories, United Kingdom dependent territory. Indomalayan. Victoria. **Population:** 5,651,193. **Size:** 413 sq. mi.]

AGRICULTURE AND FISHERIES DEPARTMENT, HONG KONG GOVERNMENT TAI LUNG FARM, LINTONGMEI, FAN KAM ROAD, SHEUNG SHUI, NEW TERRITORIES. DAFH]

Director: Unnamed. Phone: (852) 733-2100. About 1,000 species, 5,000 specimens of insects are kept in 12 cabinets, each with 12 drawers. This collection contains drawers of Lepidoptera from Hong Kong area with some donated material from Macau, and 22 drawers of other insect orders, with a minimum of slide mounted material and very little of the Araneida and Arachnida, apparently kept separate. [Information supplied in 1992 by W. J. Tennent who visited the collection.]

(Honshu, main island of Japan.)

79. HUNGARY

[Palearctic. Budapest. **Population:** 10,588,271. **Size:** 35,920.]

BAKOWYI TERMESZETTUDOMANYI MUZEUM/BAKONY MUSEUM, ZIRC, RAKOCZI TER 1, H-8420, HUNGARY. [SUEL]

Director: Dr. Sandor Toth. Professional staff: 5 scientific officers and 3 scientific assistants. This is a collection of about 200,000 specimens consisting mostly of material from the Bakony Mountains. Publication sponsored: "Folia musei historico-naturalis bakonyiensis." [1986]

ZOOLOGICAL DEPARTMENT, HUNGARIAN NATURAL HISTORY MUSEUM, BAROSS UTCA 13, H-1088 BUDAPEST. [HNHM] [=HNHB]

Director: Dr. Sandor Mahunka. Phone: 139-882. Professional staff: 26 scientific officers and 43 scientific assistants. The insect collection contains identified specimens as follows: Coleoptera, 3,000,000; Hymenoptera, 1,000,000; Lepidoptera, 1,000,000; Hemiptera and Homoptera, 500,000; Arachnoidea, 200,000. A much greater amount of material remains unidentified. The collection is housed in tens of thousands of drawers and boxes. The museum sponsors one annual and four annually appearing periodicals. [1986]

JOZSEF ATTILA TUDOMANYEGYETEN, ALLATTANITANSZEK, SZEGED, SANCSICS MJAALY U. Z. (DR. LAS-ZOL LASZLO). [JATH] [*No reply.*]

JANUS PANNONIUS MUSEUM, PECAS, P. O. BOX 158, H-7601 PECAS. [JPMP]
Director: Dr. Akos Uherkovich. Professional staff: 2 technical assistants. [No details were given about the collection.] One periodical sponsored. [1986]

80. ICELAND, Republic of

[Palearctic. Reykjavík. **Population:** 246,526. **Size:** 39,679.]

ISLANDIC MUSEUM OF NATURAL HISTORY, LAUGAVEGUR 105, P. O. BOX 5320, 125 REYKJAVIK. [NHRI]
Director: Dr. Aevar Petersen. Phone: 91-29822. Professional staff: Mr. Erling Olafsson, curator of the insect collection. This collection consists of approximately 100,000 specimens of Icelandic insects and arachnids of all orders present in the country, housed in drawers, boxes, vials, and on slides. [1992]

81. INDIA, Republic of

[Includes the Amindivi, Laccadive, Minicoy, Andaman, and Nicobar Islands, Damao, Diu, Goa, and Sikkim. Indomalayan. New Delhi. **Population:** 816,828,360. **Size:** 1,269,219 sq. mi.]

AGRICULTURAL EXPERIMENT STATION, BOGAR [AESB] [*No reply.*]

NATIONAL ZOOLOGICAL COLLECTION, ZOOLOGICAL SURVEY OF INDIA,, 34, CHITTARANJAN AVENUE, CALCUTTA 700 012. [NZSI] [=ZSCI]
Director: Dr. B. K. Tikader. Phone: 26-9248. Professional staff: Names not given. The insect collection consists of 230,847 specimens representing 23,250 named species, with 4,472 of these as types. The arachnid collection consists of 1,250 specimens, 737 named species, with 425 types. The Acarina collection contains approximately 2,000 specimens, 230 named species, and 280 types. The collection is housed in drawers, boxes, slides, and alcohol storage vials. Papers published (1972-1984): 208 research papers. [1986]

MALARIA RESEARCH CENTER, NATIONAL INSTITUTE OF COMMUNICABLE DISEASES, SHAM NATH MARG, DELHI, HARYANA. [NICD] [*No reply; probably no taxonomic collection.*]

DIVISION OF ENTOMOLOGY, NATIONAL PUSA COLLECTIONS, INDIAN AGRICULTURE RESEARCH INSTITUTE, NEW DELHI 110012, HARYANA. [INPC] [=PUSA]
Director: Dr. (Miss) Swaraj Ghai. Phone: 58-14-82. Professional staff: Dr. S. I. Farooqi, Dr. (Mrs.) Usha Ramakrishnan, Dr. R. K. Anand, Dr. S.

L. Gupta, and Mr. V. V. Ramamurthy. Indian Council of Agricultural Re search through one of its Institutes, namely, Indian Agricultural Research Institute, New Delhi, maintains insect and mite collections known as the National Pusa Collection located in the Division of Entomology. This is the largest insect and mite collection in India, and emphasizes agriculturally important groups. The collection currently contains ap proximately 200,000 labelled and identified specimens; over 10,000 processed specimens in alcohol, and approximately 11,000 slide mounts of parts and whole specimens of insects and mites. The identified collection includes about 16,000 identified species of insects and mites. In addition to this identified main collection, a large number of unidentified specimens await identification and taxonomic studies. A part of this unidentified collection has been sorted to family by specialists. Many holotypes and paratypes are also represented in the main collection. The geographic emphasis of the collection is all of India and many species are of significant economic importance. Most of these specimens are determined by specialists in the respective groups. The scientific and technical staff of the research project for the identification and taxonomic study of insects and allied arthropods of economic importance, and graduate students of entomology contribute to the continuing development of the collection. The main collection is housed in three large air conditioned rooms in 200 closed wooden cabinets, each with 20 drawers, in about 2,000 jars and vials of alcohol, and over 15 slide storing, wooden cabinets containing aluminium trays. Unidentified material is in about 2,400 wooden boxes. [1986]

DEPARTMENT OF ENTOMOLOGY, HARYANA AGRICULTURAL UNIVERSITY, HISSAR, HARYANA. [HAUH] [*No reply.*]

BIOLOGICAL CONTROL RESEARCH INSTITUTE (I.C.P.R.), C/O HORTICULTURAL RESEARCH STATION, BANGALORE, KARNATAKA. [ICPR] [*No reply; probably no taxonomic collection.*]

ENTOMOLOGY COLLECTION, DEPARTMENT OF ENTOMOLOGY, UNIVERSITY OF AGRICULTURAL SCIENCES, BANGALORE, KARNATAKA. [UASB] [*No reply.*]

ENTOMOLOGICAL COLLECTION, KERALA FOREST RESEARCH INSTITUTE, PEECHI 680 653, TRICHUR DISTRICT, KERALA. [KFRI]

Director: Dr. K. S. S. Nair. Phone: 22375 (Trichur). Professional staff: Dr. George Mathew. The collection was established in 1976. At present the collection contains over 30,000 specimens belonging to the major orders. About 800 species are identified. The collection is mainly those of Kerala State. The collection is rich in Lepidoptera and in wood boring beetles. The collection is housed in drawers and boxes. [1992]

ZOOLOGY AND BOTANY MUSEUM, MAHARAJA'S COLLEGE, ERNAKULAM, KERALA. [ZBMM] [*No reply; may not have a taxonomic collection.*]

DEPARTMENT OF ZOOLOGY, KURUKSHETA UNIVERSITY, KURUKSHETA 132119. [KUKI] [*No reply.*]

INSECT COLLECTION, BOMBAY NATURAL HISTORY SOCIETY, HORNBILL HOUSE, OPP. LION GATE, SHAGAT SINGH ROAD, BOMBAY 400023, MAHARASHTRA. [BNHS]

Director: Mr. J. C. Daniel. Phone: 243869; 225155. Professional staff: Mr. Naresh Chaturvedi. This general collection contains 80,000 insects belonging to the major orders, particularly from India. All specimens are identified and properly labelled. They are housed in drawers, boxes, and envelopes. Publication sponsored: "Journal Bombay Natural History Society." [1986]

DEPARTMENT OF ZOOLOGY, PUNJAB UNIVERSITY, CHANDIGARH, PUNJAB. [PUCP] [*No reply.*]

DEPARTMENT OF ZOOLOGY, PUNJAB AGRICULTURAL UNIVERSITY, LADHIANE, PUNJAB. [PAUP] [*No reply.*]

ENTOMOLOGY COLLECTIONS, DEPARTMENT OF ENTOMOLOGY, TAMIL NADU AGRICULTURAL UNIVERSITY, COIMBATORE, TAMIL NADU. [TNAU] [*No reply.*]

ENTOMOLOGY RESEARCH INSTITUTE, LOYOLA COLLEGE, MADRAS, TAMIL NADU. [LCMI] [*No reply.*]

ZOOLOGICAL MUSEUM, DEPARTMENT OF ZOOLOGY, MADRAS CHRISTIAN COLLEGE, TAMBA RAM, MADRAS 600 059, TAMIL NADU. [MCCM]

Director: Dr. A. Mohan Daniel. Professional staff: Mr. W. J. D. Ravi Kumar. The collection is mainly insects from the scrub-jungles of peninsular India, chiefly represented by Coleoptera, Lepidoptera, Orthoptera, Isoptera, Heteroptera, and Odonata, housed in drawers, boxes, and alcohol vials, and some spiders.

NATURAL HISTORY MUSEUM, ST. JOSEPH'S COLLEGE, TIRUCHIRAPALLI, TAMIL NADU. [MSJC] [*No reply.*]

SCHOOL OF ENTOMOLOGY COLLECTIONS, ST. JOHN'S COLLEGE, AGRA, UTTAR PRADESH. [SJCA] [*No reply.*]

DEPARTMENT OF ZOOLOGY, ALIGARH MUSLINI UNIVERSITY, ALIGARH, UTTAR PRADESH. [AMUZ] [*No reply.*]

FOREST INSECT COLLECTIONS, DIVISION OF ENTOMOLOGY, INDIAN FOREST RESEARCH INSTITUTE, NEW FOREST, DEHRA DUN, UTTAR PRADESH. [IFRI] [*No reply.*]

DEPARTMENT OF ZOOLOGY, CALCUTTA UNIVERSITY, BALLYGUNGE CIRCULAR ROAD, CALCUTTA, WEST BENGAL. [DZCU] [*No reply.*]

ENTOMOLOGY COLLECTION, SCHOOL OF TROPICAL MEDICINE, CALCUTTA, WEST BENGAL. [STMC] [*Only some specimens of mosquitoes.* Eds. 1992.]

INSECT AND SPIDER COLLECTION, NATIONAL MUSEUM OF NATURAL HISTORY, FICCI MUSEUM BLDG., BARAKHAMBA RD., NEW DELHI 110001. [NMND]

Director: Dr. S. M. Nair. Phone: 324932. Professional staff: Dr. Mammen Koshi (Curator), Mr. S. K. Saraswak (Ass't. Curator), Mr. R. Bakde (Field Collector). More than 5,000 species of insects of Delhi and adjacent areas are represented, including specimens received as donations from the Zoological Survey of India, Calcutta. These collections are maintained in drawers in cabinets, mainly for display purposes. [1986]

BENGAL NATURAL HISTORY MUSEUM, DARJEELING, WEST BENGAL. [BNHD] [*No reply.*]

82. INDONESIA, Republic of

[=Netherlands or Dutch Antilles. Includes Irian Jaya (West New Guinea) and territory formerly known as Portuguese Timor. Indomalayan. Jakarta. **Population:** 184,015,906. **Size:** 735,538 sq, mi.]

MUSEUM ZOOLOGICUM BOGORIENSE, (P. O. BOX 110), JALAN. JUANDA 3, BOGOR, JAVA [MBBJ]

Director: Dr. Soenartono Adisoemarto. Phone 24007. This is a general collection of adult insects from parts of Indonesia consisting of approx imately 30,000 specimens stored in drawers and boxes, 90% of which are pinned and pointed, and labelled. Some holotypes are present. Publication sponsored: "Treubia." [1986]

ZOOLOGICAL MUSEUM, BANDING, JAVA. [ZMBJ] [*No reply.*]

83. IRAN, Islamic Republic of

[Palearctic. Tehran. **Population:** 51,923,689. **Size:** 636,296 sq. mi.]

ZOOLOGICAL MUSEUM, FACULTE D'AGRONOMIE, UNIVERSITE DE TEHERAN, KARADJ. [UTKI] [*No reply.*]

84. IRAQ, Republic of

[=Mesopotamia. Palearctic. Bagdad. **Population:** 17,583,467. **Size:** 169,190 sq. mi.]

IRAQ NATURAL HISTORY MUSEUM, WAZIRIYA, BAGHDAD.

[INHM]

Director: Dr. Munir K. Bunni. Phone: 4165790. Professional staff: Dr. M. S. Abdul-Rasoul, Dr. H. Dawah. The collection contains 2,769 pinned specimens from various parts of Iraq, including the major orders, housed in drawers and boxes. Publications sponsored: "Bulletin of Iraq Natural History Research Center and Museum." [1986]

85. IRELAND, Irish Republic

[Excludes Northern Ireland, see United Kingdom. Palearctic. Dublin. **Population:** 3,531,502. **Size:** 26,593 sq. mi.]

INSECT COLLECTIONS, NATIONAL MUSEUM OF IRELAND, KILDANE STREET AND MERRION STREET, DUBLIN 2, COUNTY DUBLIN. [NMID] [*No reply.*]

(Irian Jaya, province of Indonesia.)

86. ISRAEL, State of

[Palearctic. Jerusalem. **Population:** 4,297,379. **Size:** 8,302 sq. mi.]

INSECT COLLECTION, ZOOLOGICAL MUSEUM, TEL AVIV UNIVERSITY, TEL AVIV 69978. [TAUI]

Director: Dr. Amnon Freidberg (Curator). Phone: 03-6408660; FAX 03-6409403. Professional staff: Prof. J. Kugler, and Mrs. T. Feler. This is a collection almost entirely restricted to Israel and adjacent areas. The Reich Collection of Lepidoptera is housed here, with 200 drawers of butterflies and Arctiidae. Holotypes are included in the collection. The collection is housed in about 2,000 drawers containing about 1,000,000 specimens, the largest collection in Israel; the Diptera section is the best represented. An effort has been made to survey the entire country. Material is available for study. [1986]

DEPARTMENT OF ENTOMOLOGY, THE VOLCANI CENTRE, BET DAGAN. [IVCB] [*No reply.*]

87. ITALY, Italian Republic

[Palearctic. Rome. **Population:** 57,455,362. **Size:** 116,324 sq. mi.]

Bologna

ISTITUTO DI ENTOMOLOGIA "GUIDO GRANDI," UNIVERSITÀ DI BOLOGNA, I-40126 BOLOGNA. [IEGG] [*No reply.*]

ISTITUTO DI ENTOMOLOGIA DELL'UNIVERSITA DEGLI STUDI, VIA FILIPPO, RE N. 6, I-40125, BOLOGNA. [IEUS]

[Contact: unknown. Holdings include Rondani collections. Eds. 1992]

Bergamo

COLLEZIONI GENERALI, MUSEO DI SCIENZE NATURALI, PIAZZA CITTADELLA 10, BERGAMO, 24100. [MBCG]

Director: Dr. Mario Guerra. Phone: 035-23-35-13. Professional staff not listed. The collection includes material from different groups of Arthropoda, best represented by Arachnida and allied groups with about 600 Mediterranean specimens, and about 500 boxes of insects. Some parts of the collection have been catalogued. [1986]

Cesena

SOCIETA PER GL STUDI NATURALISTICA DELLA ROMAGNA, C.P. 144, 47023 CESENA. [SSNR]

[Contact: Dr Giuseppe Platia. No additional information. Eds. 1992]

Firenze

DEPARTIMENTO DI BIOLOGIA, UNIVERSITA DI FIRENZE, VIA ROMANA 17, FIRENZE. [UFBI]

[Contact: Dr. Luca Bartolozzi. No additional information. Eds. 1992]

MUSEO ZOOLOGICO "LA SPECOLA," VIA ROMANA 17, 50125 FIRENZE. [MZUF]

Director: Dr. Marco Vannini. Phone: 39-55-222-451; FAX 225-325. Professional staff: Sarah Whitman Mascherini and Luca Bartolozzi (terrestrial arthropods). The collection contains approximately 800,000 (census underway 1992) specimens representing all orders, including Arachnida and allied groups, housed in boxes and in glass jars in cabinets. Special collections: Rondani collection of Diptera; Verity collection of Lepidoptera; Andreini collection of Italian and Ethiopian insects; Cavanna collection of Italian Insecta and Arachnida; Senna collection of brentid beetles; CNR collection of Somalian Insecta and Arachnida; Stefanini and Puccioni collection of Solpugida, Scorpionida, Araneida; Balzan collection of Chelonethida; Fuchs collection of European beetles; Failla collection of Italian beetles. Type catalogs exist for Coleoptera, Dermaptera, Hymenoptera, Isoptera, Heteroptera, Scorpionida, Salifuga, and Amblygygi. [1992]

Affiliated collections:

Abbazzi, Dr. Piero, Via Giovanni Duprè, 25, 50131 Florence, Italy. Collection of Italian Curculionoidea (62 boxes, 1,200 species). (Registered with MZUF.) [1992]

Bartolozzi, Dr. Luca, Entomology, Museo "La Specola", Via Romana 17, Firenze, 50125, Italy. Worldwide Lucanidae (Coleoptera), about 100 boxes (3,000 specimens of about 650 species), including many paratypes. (Registered with MZUF.) [1992]

Bordoni, Dr. Arnaldo, Via Cino da Pistoia, 10, 50133 Florence, Italy. Phone: 011 30 55 581 614. Collection of Palaearctic Staphylinidae (100 boxes, 40 HT specimens, 60 paratypes) and Zygaenidae (70 boxes). Collection of worldwide

Cicindelidae (35 boxes). (Registered with MZUF.) [1992]

Magrini, Dr. Paolo, Via Gianfilippo Braccini, 7, 50141 Florence, Italy. Phone: 011 39 55 456 381. Collection of Italian Carabidae (1,225 species in 100 boxes, including 28 holotypes). (Registered with MZUF.) [1992]

Mascagni, Dr. Alessandro, Via Giuseppe Bessi, 8, 50018 Scandiccu (FI), Italy. Phone: 011 39 55 256 821. Collection of predominantly Italian Insecta (38,000 specimens in 183 boxes), particularly Coleoptera Heteroceridae (5,000 specimens in 14 boxes). Collection of Italian Arachnida (300 specimens in alcohol). (Registered with MZUF.) [1992]

Rocchi, Saverio, Via Gran Bretagna no. 201, 50126 Firenze, Italy. Phone: 011 30 55 688147. Worldwide Dytiscidae (Coleoptera) in about 84 boxes (14,000 specimens of about 1,500 species). (Registered with MZUF.) [1992]

Terzani, Dr. Fabio, Via L. Cigoli, 12, 50142 Florence, Italy. Phone: 011 39 55 711 043. Collection of Italian Insecta (140 boxes), predominantly Coleoptera (69 boxes), Odonata (31 boxes), and Lepidoptera (25 boxes). (Registered with MZUF.) [1992]

Vanni, Dr. Stefano, Via Carlo Del Prete, 4, 50127 Florence, Italy. Phone: 011 39 55 422 3749. Collection of Italian cavernicolous Carabidae, Trechinae (Coleoptera), about 2,500 specimens in 25 boxes. (Registered with MZUF.) [1992]

Forli

MUSEO ZANGHERI DI STORIA NATURALE DELLA ROMAGNA, CORSO DIAZ N. 182, FORLI. [MZRF] [*No reply.*]

Genoa

MUSEO CIVICO DI STORIA NATURALE "GIACOMO DORIA," VIA BRIGATA LIGURIA 9, I-16121 GENOA. [MCSN]

Director: Dr. Mrs. Lilia Capocaccia. Phone: 010-56-45-67. Professional staff: Dr. Roberto Poggi (Assistant Director), Curators: Dr. Giuliano Doria, Dr. Valter Raineri. It is estimated that the collection contains over 2 million specimens, housed in about 8,500 boxes and drawers. It is the richest entomological collection in Italy, particularly for the high number of types (about 8,000). The collection of spiders consists of about 10,000 specimens in alcohol vials, with specimens from Italy, North Africa, West Africa, East Africa, Aden and South Yemen, Burma, Indomalaysia, Papua New Guinea, and North Australia, including more than 1,000 types. The largest insect collections were made, besides in Europe (particularly Italy), in South East Asia, Malaysia, Africa, and South America, by many famous collectors: O. Antinori, L. Balzan, V. Bottego, G. Bove, E. A. D'Albertis, L. M. D'Albertis, O. Beccari, G. Doria, L. Fea, R. Gestro, L. Loria, E. Modigliani, S. Patrizi, and many others. Special collections include: Bartoli (Curculionidae); G. Fiori (Byrrhidae); Binaghi (Coleoptera); Capra (Orthopteroidea, etc.); Laporte de Castelnau (Carabidae); Magretti (Chrysididae); Gribodo (Hymenoptera); Emery (Formicidae); Invrea (Mutillidae and Chrysididae); Barbera (Palearctic Lepidoptera); Toso (Palaearctic Lepidoptera); R. Doria Bombrini (Morphos); Mancini (Coleoptera), and Ferrari (Hemiptera). Publications sponsored: "Annali del Museo Civico di Storia Naturale "G. Doria" " and "Doriana." [1992]

Milan

MUSEO CIVICO DI STORIA NATURALE, CORSO VENEZIA 55, 20121 MILAN. [MSNM]

Director: Prof. Giovanni Pinna. Phone: 62085405. Professional staff: Dr. Carlo Pesarini; Dr. Carlo Leonardi. The collection consists of 1.5 million and 20,000 spiders. There are many special collections, principally those of C. Koch, A. Schatzmayr, F. Solari, G. Springer, A. Porta, A. Chiesa, E. Turati, V. Ronchetti, I. Bucciarelli, C. Nielsen, M. Bezzi, A. Fiori, C. Conci, and G. Scortecii. Also here is the L. Di Caproiacco collection of Himalayan spiders. The collection is stored in boxes and vials. [1986]

Roma

INSTITUTO NAZIONALE DI ENTOMOLOGIA, VIA CATOME 34, ROMA. [INER] [*No reply.*]

BIOLOGICAL CONTROL OF WEEDS LABORATORY-EUROPE, BEWL-E, AMERICAN EMBASSY-AGRIC, APO NY 09794 (ROMA, ITALY). [BCWL] [*No reply.*] [1992]

Sardina

MUSEO ZOOLOGICO, UNIVERSITA DI CAGLIARI, VIALE SAN BARTOLOMEO 1, COGLIARI, SARDINA. [MZUC] [*No reply.*]

Siena

ANDREUCCI COLLECTION, INSTITUTO DI ZOOLOGIA, VIA MATTIOLI 4, 53100 SIENA. [IZSI] [*No reply.*]

Torino

MUSEO, INSTITUTO DI ZOOLOGIA SISTEMATICA, UNIVERSITA DI TORINO, VIA GIOVAN NI GIDITTI 34, I-10123 TORINO. [MIZT] [*No reply.*]

SPINOLA COLLECTION, MUSEO REGIONALE SCIENZE NATURALI, VIA GIOLITTI 36, TORINO 10123. [MRSN]

Director: Dr. O. Bortesi. Phone: 011-4323001. Professional staff: Dr. P. M. Giachino (Curator), and P. L. Scarahozzino (Technician). Details about the size of the collection not supplied. Special collections: Hartig collection (part) of 300,000 Lepidoptera from Palearctic Region and Mexico, including many types; Winkler collection (from Vienna, Austria), of 400,000 Coleoptera from the Palearctic Region, except Winkler types are kept in the Vienna museum. [1992]

Venice

MUSEO CIVICO DI STORIA NATURALE, FONDACO DEI TURCHI, S. CROCE 1730, VENICE. [MCNV]

Director: Dr. Enrico Ratti. Phone: 041-5240885. Professional staff: Prof. A. Giordani Soika (Curator, Hymenoptera), Mr. S. Canzoneri (Curator, Diptera), Mr. H. Hansen (Curator, Arachnida). An estimate of the number of specimens at this time is impossible. There are about 4,000 boxes, predominately Coleoptera and Hymenoptera, with about 6,700 types and paratypes. The collection includes Araniae in fluid. Publications sponsored: "Bollettino del Museo Civico di Storia Naturale di Venezia" (Annual), and "Quaderni del Museo Civico di Storia Naturale di Venezia" (occasional). [1992]

Verona

MUSEO CIVICO DI STORIA NATURALE, LUNGADIGE PORTA VITTORIA 9, VERONA I-37129. [MSNV]

Director: Dr. Lorenzo Sorbini. Phone: 21987; 24657. Professional staff: Dr. B. G. Osella (Curator), and two technicians. The collection currently contains approximately 1.5 million pinned and labelled insects of which 1 million are identified. About 45,000 alcohol vials of specimens and 9,000 slide complete the collection. About 1,000 types are represented. The collection is housed in boxes and vials. Publications sponsored: "Bollet tino del Museo Civico di Storia Naturale," "Memorie del Museo Civico di Storia Naturale," "Quaderni didattico-divulgativi de Verona," "Memorie del Museo Civico di Storia Naturale di Verona, 2nd. ser., sect. A: Scien ze della Vita," and "Quaderni naturalistici e didattici." [1986]

MUSEO P. ZANGHERI: STORIA NATURALE DELLA ROMAGNA, C/O MUSEO CIVICO STORIA NATURALE, LUGADIJE PORTA VITTORIA 9, VERONA, I-37100. [MZRO]

Director: Dr. B. G. Osella. Phone: 045-21987. This museum is a separate part of the previous museum and is the most complete regional museum in Italy. It contains Prof. P. Zangheri's insect collection of about 10,000 species including about 90 holotypes. [1986]

88. IVORY COAST, Republic of

[Afrotropical. Abidjan. **Population:** 11,184,847. **Size:** 122,780 sq. mi. *No known insect collection.*]

89. JAMAICA

[Neotropical. Kingston. **Population:** 2,458,102. **Size:** 4,244 sq. mi.]

NATURAL HISTORY MUSEUM, INSTITUTE OF JAMAICA, 12-16 EAST STREET, KINGSTON. [IJSM]

Director: Dr. Elaine Fisher. Phone: (809) 922-0620-7. Professional staff: Dr. Thomas H. Farr. The approximately 22,000 specimens are housed in drawers, or in alcohol vials. The collection consists almost entirely of Jamaican species with a few from other West Indian islands. Specimens may be borrowed. Library facilities are limited. The C. C. Gowdey Collection of Jamaican insects is also housed. Publication: "Bulletin Institute of Jamaica, Science Series" [1992]

DEPARTMENT OF ZOOLOGY, UNIVERSITY OF THE WEST INDIES, KINGSTON, MONA, KINGSTON 7. [UWIJ]
In-charge: Dr. Eric Garraway. Phone: (809) 927-1661. The collection occupies 200 drawers and comprises mainly insects of Jamaica. It has primarily functioned in teaching, but now a taxonomic reference collection is underway. (C. Starr) [1992]

(Jan Mayen, Island Territory of Norway.)

90. JAPAN

[Includes Ryukyu, Bonin, and Volcano islands, and the islands of Minami-Tori-shima, Nishino-shima, and Okino-Tori-shima. Palearctic. Toyko. **Population:** 122,626,038. **Size:** 145,870.]

ENTOMOLOGICAL LABORATORY, FACULTY OF AGRICULTURE, KYUSHU UNIVERSITY, HAKO ZAKI, HI-GASHI-KU, FUKUOKA 812. [KUEC]
Director: Dr. K. Morimoto. Phone: 092-641-1101, ext. 6168. Professional staff: Dr. O. Tadauchi, and Miss C. Okuma. The general collection contains 2.5 million specimens of Japanese insects. The Micronesian collection includes 30,000 specimens; the Asian collection of insects occurring in paddy fields of Taiwan, Malaysia, Thailand, Pakistan, and other areas has 50,000 specimens. The Sugitani collection includes 36,000 specimens of butterflies from Korea. The spider collection consists of 50,000 specimens from Japan, southeast Asia, and the Pacific islands. The collections are housed in drawers in cabinets. Publication sponsored: "Esakia, Kyushu University Publications in Entomology." [Update 1992, Eds.]

ENTOMOLOGICAL LABORATORY, EHIME UNIVERSITY, MATSUYAMA. [EUMJ] [*No reply*]

INSECT COLLECTION, ZOOLOGICAL LABORATORY, MEIJO UNIVERSITY, TENPAKU-KU, NAGOYA 468. [ZLMU]
Professional staff: Dr. T. Okadome (Diptera); Dr. Y. Arita (Lepidoptera); Mr. K. Yamagishi (Hymenoptera). The collection contains mainly Japanese Diptera, Lepidoptera, and Hymenoptera, and some specimens of other orders from southeast Asia. The collection is housed in 1,480 drawers. Material is available for exchange and loan for study. [1986]

ENTOMOLOGICAL LABORATORY, UNIVERSITY OF OSAKA PREFECTURE, SAKAI, OSAKA 593. [UOPJ]
Director: Dr. T. Hirowatari. Phone: 0722 52 1161, ext. 2435. Professional staff: Dr. T. Yasuda, Dr. S. Moriuti, Dr. M. Ishii, and Dr. T.

Hirowatari. The collection contains 400,000 insects, including 300 holotypes and 1,000 secondary types. [1992]

ENTOMOLOGICAL INSTITUTE, FACULTY OF AGRICULTURE, HOKKAIDO UNIVERSITY, SAPPORO 060. [EIHU]

Director: Dr. Sadao Takagi. Professional staff: Dr. Masaaki Suwa, and Mr. Shin'ichi Akimoto. "Uncountable, rapidly increasing." Includes Matsumura collection, Uchida collection (Ichneumonidae), and Watanabe collection (Braconidae). Housed in boxes and on slides. Publication sponsored: "Insecta Matsumurana, new series." [1986]

LABORATORY OF INSECT SYSTEMATICS, NATIONAL INSTITUTE OF AGRO-ENVIRONMENTAL SCIENCES, KANNONDAI, TSUKUBA, IBARAKI PREF., 305. [ITLJ]

Director: Dr. T. Matsurura. Phone: 0298-38-8348; FAX 0298-38-8199. Professional staff: Dr. S. Yoshimatsu, Mr. K. Konishi. The collection contains about one million specimens of insects collected in Japan and Asia since 1901. The majority are pinned and stored in 6,000 drawers in cabinets, and others are stored in alcohol vials and mounted on microscope slides. Nearly 2,000 primary types are housed in the type material room specially designed. Type data are prepared for input into a data base of computerized system. Special collections: Habu collection (type series of Carabidae and Chalcididae), Ishii (Chalcidoidea), Katsuya (Hymenoptera), Kurosawa (Thysanoptera), Kuwana (Coccoidea), Kuwayama (Neuroptera, Trichoptera, and other insects of Kurile Islands), Nobuchi and Niijima (Scolytidae and Platypodidae), Shiraki (Syrphidae, Tephritidae, and Tabanidae), and Tsuneki (Formicidae). [1992]

ENTOMOLOGICAL COLLECTIONS, NATIONAL SCIENCE MUSEUM (NATURAL HISTORY), HYA KUNIN-CHO 3-23-1, SHINJUKU-KU, TOKYO 160. [NSMT] [*No reply.*]

DEPARTMENT OF ENTOMOLOGY, NATIONAL INSTITUTE OF HEALTH OF JAPAN, 10-35 KAMIOSAKI, 2-CHOME, SHINAGAWA-KU, TOKYO 141. [ENIH] [*No reply.*]

ENTOMOLOGY LABORATORY, FACULTY OF AGRICULTURE, KAGOSHIMA UNIVERSITY, UERA TA-CHO, KA-GOSHIMA 890. [KUIC]

[Contact: Dr. Jun-ichi Yukawa. Professional staff includes Dr. Akira Nagatomi. The collection is strong in Diptera. Eds. 1992.]

OSAKA MUSEUM OF NATURAL HISTORY, NAGAI PARK, HIGASHI-SUMIYOSHI-KU, OSAKA 546. [OMNH] [*No reply.*]

LABORATORY OF ENTOMOLOGY, TOKYO UNIVERSITY OF AGRICULTURE, SAKURAGAOKA, SETAGAYA-KU, TOKYO 156. [TULE] [*No reply.*]

DEPARTMENT OF MEDICAL ZOOLOGY, TOKYO MEDI-CAL

AND DENTAL UNIVERSITY, YUSHI MA, BUNKYO-KU, TOKYO. [TMDU] [*No reply.*]

(Java (Jawa), Island of Indonesia.)

(Jersey, see Channel Islands.)

(Johnston Atoll, see U.S.A., Miscellaneous Pacific Islands.)

91. JORDAN, Hashemite Kingdom of

[Palearctic. Amman. **Population:** 2,850,482. **Size:** 37,738 sq. mi. *No known insect collection.*]

(Juan Fernandez Islands, see Chile.)

(Kalaalitt Nunaat (=Greenland), see Denmark.)

(Kalimantan, Indonesian province of Borneo.)

(Kampuchea, see Cambodia.)

92. KENYA, Republic of

[Afrotropical. Nairobi. **Population:** 23,341,638. **Size:** 219,788 sq. mi.]

UNIVERSITY OF MOI, EL DORET [UMED]
[Contract: Dr. Marc DeMeyer. Eds. 1992.]

WILDLIFE ADVISORY AND RESEARCH SERVICE, LTD., P. O. BOX 929, NAIROBI. [WARS] [*No reply.*]

SECTION OF ENTOMOLOGY, NATIONAL MUSEUM OF KENYA, P.O. BOX 40658, NAIROBI. [NMKE]
Director: Richard E. Leakey. Phone: 742-161. Professional staff: Dr. J. Mark Ritchie. The collection contains approximately 1.5 million specimens especially from Kenya, Uganda, and Tanzania. It includes the largest butterfly collection in Africa, and good representation of Coleoptera. Special collections: Van Someren collection of East African insects; Tephritidae determined by H. K. Munro; Khamala collection of East African Culicoides (on slides); Darlington collection of East African termites. The collection is housed in 4,000 drawers in cabinets, and some in alcohol vials. Special interests include Acridoidea, Agromyzidae, and Isoptera. Publication sponsored: "Journal of the East Africa Natural History Society and National Museum." [1986]

K.A.R.I. FOREST INSECTS REFERENCE COLLECTION, KENYA AGRICULUTRAL RESEARCH INSTITUTE (KARI), P. O. BOX 57811, NAIROBI. [KARI]
Director: Dr. B. N. Majisu. Phone: 0154-32883. The collection con-

sists of about 42,000 identified pinned insects plus specimens preserved in alcohol. The whole collection contains about 6,000 species housed in drawers. [1986]

NATIONAL AGRICULTURAL RESEARCH LABORATORIES (NARL), P. O. BOX 14733, NAIROBI. [NARL]

Professional staff: Dr. G. N. Kibata, J. M. Ongaro, C. N. Kironji. Phone/FAX: 0254 2 444144. A collection of about 50,000 pinned, identified specimens of insects, 500 speciesin vials, and 70 species on slides, representing about 4,500 species, housed in 240 collection boxes. [1992]

(Kerguelen Island, see France.)

(Kermadec Islands, see New Zealand.)

(Khmer Republic, see Cambodia.)

93. KIRIBATI, Republic of

[=Gilbert Islands, former British colony, which includes the Gilbert Islands; Fanning Atoll and Washington Island in the Line Islands; Ocean Island; and those islands claimed by the U.S.: Caroline, Christmas, Flint, Malden, Starbuck, and Vostok in the Line Islands; and Birnie, Gardner, Hull, McKean, Phoenix (incl. Enderbury and Kanton [Canton]) Sydney in the Phoenix Islands. Oceanian (Polynesia). **Population:** 67,638. **Size:** 378 sq. mi. *No known insect collection.*]

94. KOREA, Democratic People's Republic of

[=North Korea. Palearctic. Pyongyang. **Population:** 21,983,795. **Size:** 46,540 sq. mi. *No known insect collection.*]

95. KOREA, Republic of

[=South Korea. Palearctic. Seoul. **Population:** 42,772,956. **Size:** 38,291 sq. mi.]

DEPARTMENT OF PLANT PROTECTION, COLLEGE OF AGRICULTURE, KANGWEON NATIONAL UNIVERSITY, KANGWEON PROV. [KNUC] [*No reply.*]

NATIONAL SCIENCE MUSEUM, 2 WARYONG-DONG, CHON-GNO-KU, SEOUL. [NSMK] [*No reply.*]

DEPARTMENT OF BIOLOGY, KON-KUK UNIVERSITY, 93-1, MOJIN-DONG, SUNGDONG-KU, SEOUL. [KKUK] [*No reply.*]

KOREAN ENTOMOLOGICAL INSTITUTE, KOREA UNIVERSITY, SEOUL, KOREA. [KEIU] [*No reply.*]

(Kuril Islands, islands of U.S.S.R.)

96. KUWAIT, State of

[Palearctic. Kuwait City. **Population:** 1,938,075. **Size:** 6,880 sq. mi.]

THE EDUCATIONAL SCIENCE MUSEUM, MINISTRY OF EDUCATION, P. O. BOX 7, SAFAT 13001. [KNHM] [*Reply; no insect collection.*]

(Kyushu, island of Japan.)

(Laccadive Islands, see Lakshadweep.)

(Lakshadweep, see India.)

97. LAOS, Lao People's Democratic Republic

[Indomalayan. Vientaine. **Population:** 3,849,752. **Size:** 91,400 sq. mi. *No known insect collection.*]

98. LATVIA

[Palearctic. Riga. **Population:** 2,681,000. **Size:** 24,710 sq. mi. *No known insect collection.*]

99. LEBANON, Republic of

[Palearctic. Beirut. **Population:** 2,674,385. **Size:** 4,036 sq. mi.]

PEYRON COLLECTION, MUSEUM OF NATURAL HISTORY, BIOLOGY DEPARTMENT, AMERICAN UNIVERSITY OF BEIRUT, BEIRUT. [AUBL]

Director: Mrs. Nada Itani Raad. Phone: 34074-, ext. 2431. Approximately 3,000 genera represented. No further details. [1986]

(Leeward Islands, see Antigua, St. Christopher, Nevis, Anguilla, and Montserrat.)

100. LESOTHO, Kingdom of

[=Basutoland. Afrotropical. Maseru. **Population:** 1,666,012. **Size:** 11,720 sq. mi. *No known insect collection.*]

(Lesser Antilles, various islands, see Leeward and Windward islands.)

101. LIBERIA, Republic of

[Afrotropical. Monrovia. **Population:** 2,463,190. **Size:** 37,743 sq. mi. *No known insect collection.*]

102. LIBYA, Socialist People's Libyan Arab Jamahireyia

[Palearctic. Hun (no longer Tripoli). **Population:** 3,956,211. **Size:** 685,524 sq. mi.]

NATURAL HISTORY MUSEUM, THE CASTLE, TRIPOLI. [NHML]
[*Mail returned by post office without comment.*]

103. LIECHTENSTEIN, Principality of

[Palearctic. Vaduz. **Population:** 27,825. **Size:** 61.8 sq. mi.]

PRINCIPALITY OF LIECHTENSTEIN, FL-LANDESFORSTAMT (MICHAEL FASEL), FL-9490 VADUZ. [PLFV]

Director: Vacant. Phone: 075 66111. Collection of night-active Lepidoptera, collected 1980 by light trap. About 20 show-case type drawers. [1992]

(Line Islands, see Kiribati/U.S.A.)

(Lord Howe Island, see Australia.)

(Loyalty Islands, see New Caledonia.)

104. LITHUANIA

[Palearctic. Vilna. **Population:** 3,690,000. **Size:** 25,100 sq. mi. *No known insect collection.*]

105. LUXEMBOURG, Grand Duchy of

[Palearctic. Luxenbourg-Ville. **Population:** 366,232. **Size:** 999 sq. mi.]

MUSEUM D'HISTOIRE NATURALLE DU GRAND-DUCHY DE LUXEMBOURG, 24, RUE MUNSTER, L-2460 LUXEMBOURG. [MGDL]

Director: Mr. Marc Meyer. Phone: +3521-146-2233-200. Professional staff: Mr. Alphonse Pelles. This is a local collection of Lepidoptera, Coleoptera, Odonata, Saltatoria, Diptera (part), Hymenoptera (part), and Arachnida, stored in 1,000 drawers and about 10,000 alcohol vials. The international collection contains most insect orders, mainly from South America and Africa, housed in 2,000 drawers. Series sponsored: "Travaux Scientifiques du Musée national d'histoire naturelle de Luxembourg." [1992]

106. MACAU, Province of

[=Macao; Portuguese overseas province. Indomalayan. Macau. **Population:** 432,232. **Size:** 6.5 sq. mi.]

CAMARA MUNICIPAL DAS ILAS (ISLAND GOVERNMENT OF MACAU), DEPT. DE SERVIÇOS AGRÁRIOS, PARQUE DA RESERVA DE SIAC PAI VAN COLOANE ISLAND. (via Hong Kong). [MACA]

Director: Mr. Pun Wing Wah. Phone: 328277. Approximately six museum drawers containing Lepidoptera, mainly butterflies, of Macau constitutes this collection, but additional museum drawers have been ordered in order to house materials that are being collected for a faunal inventory of the zoological fauna of Macau. [1992]

107. MACEDONIA

[=Makedonija, formerly a part of Yugoslovia. Palearctic. Skopje. *The changing situation in this newly formed country prevents further information at this time.*]

(Macquarie Island, see Australia.)

108. MADAGASCAR, Democratic Republic of

[=Malagasy Republic. Afrotropical. Antananarivo. **Population:** 11,073,361. **Size:** 224,532 sq. mi.]

MUSEUM OF FOLKLORE, ARCHAEOLOGY, AND PALAEONTOLOGY, ORSTOM, PARC DE TSIMBAZAZA, TANANARIVE. [MFAP] [*No reply.*]

LABORATOIRE DE ZOOLOGIE, E.E.S.S. ANKATSO, UNIVERSITY OF ANTANANARIVO. [ZUAC] [*No reply.*]

PARC BOTANIQUE ET ZOOLOGIQUE DE TSIMBAZAZA, B.P. 561, ANTANANARIVO. [PBZT] [*No reply.*]

(Madeira Islands, Province of Portugal.)

(Malagasy Republic, see Madagascar.)

109. MALAWI, Republic of

[=Nyasaland. Afrotropical. Lilongwe. **Population:** 7,679,368. **Size:** 45,747 sq. mi.]

DEPARTMENT OF AGRICULTURAL RESOURCES COL-LECTION, MINISTRY OF AGRICULTURAL AND NATURAL RESOURCES, BVUMBWE EXPERIMENT STATION, P. O. BOX 5748, LIMBE. [BESM]

Director: Mr. A. Gadabu. The pinned collection totals about 14,000 specimens, representing about 3,000 genera or species, housed in drawers and cabinets, and alcohol vials of about 6,000 unidentified specimens, and about 750 slides. [1986]

MUSEUM OF MALAWI, P. O. BOX 30360, CHICHIRI, BLANTYRE. [MMCM]

Director: Mr. M. Kumwenda. The collection's total number of specimens, all pinned, is nearly 3,000, of which at least 500 are identified to species. [1986]

DEPARTMENT OF BIOLOGY COLLECTION, CHANCELLOR COLLEGE, UNIVERSITY OF MALA WI, P. O. BOX 280, ZOMBA. [CCCZ]

Director: Dr. C. O. Dudley. Approximately 18,000 specimens in the collection representing about 4,000 species, with about 1,000 identified, including 14 types. [1986]

DEPARTMENT OF FORESTRY COLLECTION, MINISTRY OF AGRICULTURAL AND NATURAL RESOURCES, FOREST RESEARCH INSTITUTE, P. O. BOX 270, ZOMBA. [DFCZ]

Director: Mr. A. Majawa. The pinned collection totals about 9,000 specimens representing 2,000 identified genera or species, housed in boxes and 200 alcohol vials. [1986]

110. MALAYSIA

[Indomalayan. Kuala Lumpur. **Population:** 16,398,306. **Size:** 127,320 sq. mi.]

SABAH MUSEUM, JALAN MUZIUM, KOTA KINABALU 88000. [SMJM]

[Contact: Dr. Jelius Gantor. Phone: 088-535-51. Eds. 1992.]

NATURAL HISTORY DIVISION, SARAWAK MUSEUM, KUCHING, 93566 KUCHING, SARAWAK. [SMSM]

Director: Mr. Peter M. Kedit. Phone: 082-244232. Professional staff: Charles Leh, Curator of Natural History. The collection consists of 136 families of Sarawak insects, with over 5,000 species and about 30,000 specimens. Most specimens are over 70 years old. The spider collection is poor. Latest collection of insects from Mulu, Sarawak, by the Royal Geographic Society (1978-1981), is still in the British Museum (Natural History). The collection is housed in 336 drawers and in 12 boxes. Types are segregated. Publication sponsored: "Sarawak Museum Journal, Annual." [1992]

NATIONAL MUSEUM, DAMANSARA ROAD, KUALA LUMPUR. [NMKL]

Director: Dr. Dato' Shahrum bin Yub. Professional staff: Mr. Roy Loudres. The arachnid collection consists of 70 species of spiders housed in drawers and boxes. No further information was submitted. [1986]

SELANGOR MUSEUM, KUALA LUMPUR. [SMKM] [*No reply.*]

FOREST RESEARCH CENTRE, SANDAKAN. [FRCS]

[Contact: Dr. V. K. Chen. Eds. 1992.]

FOREST RESEARCH INSTITUTE MALAYSIA, KEPONG, KUALA LUMPUR 52109. [FRIM] [*No reply.*]

111. MALDIVES, Republic of

[Indomalayan. Male. **Population:** 203,187. **Size:** 115 sq. mi. *No known insect collection.*]

112. MALI, Republic of

[=French Sudan. Afrotropical/Palearctic. Bamako. **Population:** 8,665,548. **Size:** 471,042 sq. mi.]

SCIENCE EDUCATION CENTRE, RAHDHEBAI MAGU, GALDHU, MALÉ 20-04. [SECM]
Phone: 321330. [*No information received about the insect collection.*]

(Malpelo Island, see Colombia.)

113. MALTA, Republic of

[Palearctic. Valletta. **Population:** 369,240. **Size:** 58 sq. mi. *No known insect collection.*]

(Man, Isle of, British Crown Dependency.)

(Mangareva Islands (=Gambier Islands), see French Polynesia.)

(Marcus Island (=Minami Tori Shima), see Japan.)

(Mariana Islands, see U.S.A.)

(Marion Island, part of the Prince Edward Islands, South Africa.)

(Marquesas Islands, see French Polynesia.)

114. MARSHALL ISLANDS

[Saipan; UN trusteeship administered by U.S.A. Oceanian (Micronesia). Majuro. **Population:** 40,609. **Size:** 70 sq. mi.]

MID-PACIFIC RESEARCH LABORATORY, ENEWETAK ATOLL. [MPRL]

The insect collection assembled by L. D. Tuthill and others during the 1950's was moved to Bishop Museum [BPBM] for inclusion in studies reported in "Insects of Micronesia."

115. MARTINIQUE, Department of

[French overseas province. Neotropical. Fort-de-France. **Population:** 351,105. **Size:** 180 sq. mi. *No known insect collection.*]

(Mascarene Islands, see Mauritius/France.)

116. MAURITANIA, Islamic Republic of

[Afrotropical/Palearctic. Nouahibou. **Population:** 1,919,106. **Size:** 397,840 sq. mi.*No known insect collection.*]

117. MAURITIUS

[Includes Rodrigues island, Agalega islands, and Cargados Carajos Shoals. Indomalayan. Port Louis. **Population:** 1,099,983. **Size:** 788 sq. mi.]

NATURAL HISTORY MUSEUM, MAURITIUS INSTITUTE, P. O. BOX 56, PORT LOUIS. [MIMM]

Director: Mr. R. Gajeelee. Insects on display: approximately 500 specimens. There is also a reserve collection of local insects kept in 200 insect drawers. Publication sponsored: "Mauritius Institute Bulletin." [1992]

MAURITIUS SUGAR INDUSTRY RESEARCH INSTITUTE, REDUITT. [MSIR]

Director: Dr. John R. Williams. Phone: (230) 454-1061. Professional staff: Mr. M. A. Rajabalee. Insects of the Mascarene Islands, mainly Lepidoptera, Hemiptera. and Coleoptera; approximately 500 identified specis, 50,000 specimens. [1992]

(Mayotte, Administrated by France.)

(Mbini, part of Equatorial Guinea.)

(Melilla, see Spain.)

(McDonald Island, see Australia.)

118. MEXICO, United Mexican States

[Nearctic/Neotropical. Ciudad de México. **Population:** 83,527,567. **Size:** 756,066 sq. mi.]

COLECCION ENTOMOLOGIA, CENTRO DE ENTOMOLOGICA Y ACAROLOGIA, COLEGIO DE POSTGRADUADOS, 56230 CHAPINGO. [CEAM]

Director: Dr. Alejandro Gonzales. Phone: 5-85-45-55, ext. 5106. Professional staff: Mr. Armando Equihua, Mr. Nestor Bautista, Ms.

Socorro Anaya. The collection consists of 50,000 pinned specimens, of agriculturally important groups, with emphasis on Coleoptera, housed in drawers. [1986]

COLECCION CENTRAL DE INSECTOS, INSTITUTO NAC-IONAL DE INVESTIGACIONES FORESTALES Y AGRO-PECUARIAS, APDO. POSTAL 112, CELAYA, GTO. [INIA]

Director: Antonio Marin J. Phone: 5-85-45-55, ext. 5140. Professional staff: Ing. Angelica Paez Lamadrid. This collection contains about 85,000 specimens as a synoptic collection of pests of the basic food crops in Mexico, especially corn, wheat, and beans, but has been expanded to include species which may become pests in the future. It contains the Charles C. Plummer collection of Mexican Membracidae. [1992]

COLECCION NACIONAL DE INSECTOS, MUSEO DE HISTORIA NATURAL DE LA CIUDAD DE MEXICO, APDO. POSTAL 18845, DELEGACION MIGUEL HIDALGO, MEXICO 11800, D.F. [MCMC]

Director: Dr. Pedro Reyes-Castillo. Phone: 2-71-18-71, ext. 133. Professional staff: M. En C. Miguel Angel Moron R., Biol. M.A. Eugenia Diaz B., Biol. Gemma Quintero G. The collection contains approximately 80,000 specimens representing about 10,000 species. The collection is housed in drawers and boxes, with some in alcohol vials. Special collections: Halffter, Scarabaeidae; Muller, Lepidoptera; Young, aquatic Coleoptera; and Alfredo Barrera, ectoparasites. [1992]

INSTITUTO DE ECOLOGIA, APDO. 18845, MEXICO 11800, D.F. [IEMM]

Director: M. en C. Pedro Reyes-Castillo. Phone: 2-71-18-71. Professional staff: Dr. Gonzalo Halffter; M. S. Miguel Angel Moron; Biol. Gemma Quintero; Biol. Camelia Castillo; Mr. Javier Villalobos; Mr. C. Deloya. Only Coleoptera Lamellicornia are contained in this collection which includes about 120,000 specimens from all of the world. About 250 types of species described from Mexico are included. This collection is housed in 14 drawers and over 1,000 boxes. Publications sponsored: "Acta Zoologica Mexicana (nueva serie)" and "Publicationes del Instituto de Ecologia." The latter is a series of handbooks. [1992]

ESTACIÓN DEL BIOLOGÍA "CHAMELA," UNIV. NAC. AUTÓNOMA DE MEXICO, APTO. 21, 48980 SAN PATRICIO, JALISCO. [EBCC]

Curator: Ricardo Ayala. Phone/FAX: (333) 7-02-00. Professional staff: Enrique Ramirez (Diptera), and Felipe A. Moguera (Coleoptera). About 70,000 pinned specimens in 7 cabinets. It contains the largest bee collection in Mexico, 12,000 specimens, 500 species; Cerambycidae, 8,000 specimens, 400 species; wasps, Vespidae and flies, Syrphidae. Some secondary types. [1992]

COLECCION ENTOMOLOGICA, INSTITUTO DE BIOLOGIA, UNIVERSIDAD NACIONAL AUTONOMA DE MEXICO, APDO. POSTAL 70133, 04510 MEXICO, D.F. [UNAM]

Director: Dr. Harry Brailovsky Alperowitz (Curator). Phone: 550-5884. Professional staff: Dr. Harry Brailovsky A. (Hemiptera-Heteroptera), Dr. Joaquin Bueno (Trichoptera), Dr. Roberto Johansen (Thysanoptera), Dr. Carlos Beutelspacher (Lepidoptera), Dr. Alfonso Garcia Aldrete (Psocoptera), Dra. Silbia Santiago (Aquatic Coleoptera), M. en C. Santiago Zaragoza (Coleoptera), M. en C. Enrique Gonzalez (Odonata), Dra. Leonila Vazquez (Lepidoptera), M. en C. Hector Perez (Lepidoptera), Dra. Julieta Ramos Elorduy (Insect-like food proteins), and 12 technicians. The collection is the largest in Mexico with about 1.5 million pinned specimens and approximately 60,000 slide mounts. There are over 1 million specimens in alcohol or envelopes to be mounted. These are sorted to family as studies are made. Many holotypes and paratypes are included, particularly from Mexico, but also from Central and South America and the Pacific islands. The collection is housed in 120 cabinets with 4800 drawers. More than 600 primary types and several paratypes are in the collection. Material is available for exchange and loan. Publication spon sored: "Anales del Instituto de Biologia UNAM. Serie Zoologia." [1992]

COLECCION DE INSECTOS, INSTITUTO TECNOLOGICO DE MONTERREY, ESCUELA DE AGRICULTURA, MONTERREY, N.L. [ITMM]

Director: Dr. Luis Orlando Tejada. Phone: 48-20-00, ext. 436. Graduate students help maintain the collection. It is a small collection used primarily for teaching, housed in 50 drawers and in vials. Some mites are on slides. Most of the material is identified to family, few to species. Some genera of Scarabaeidae and Meloidae are well represented. Specimens are available for exchange and loan. [1992]

COLECCION DE INSECTOS, CENTRO DE INVESTIGACIONES AGRICOLAS NOROESTE, APDO. POSTAL 515, CIUDAD OBREGON, SONORA. [CIAN]

Director: Dr. Francisco Pacheco. Phone 4-01-01. The collection consists of nearly 3,000 pinned specimens and about 6,600 alcohol specimens. Included in the collection are 157 paratypes of 6 species. A catalog of the collection is frequently published. [1986]

ESTACION EXPERIMENTAL DE AGRICOLAS DE LA CAMPARA, CHIHUAHUA, CHIHUAHUA. [SEAC] [*No reply.*]

COLECCION DE TERMITAS MEXICANAS, INIREB, LACITEMA-INIREB, APDO. 63, XALA PA, VERACRUZ. [INIR]

Director: Mr. Victor Perez-Morales, M.Sc. Phone: 7-53-35. Professional staff: Mr. Rene Ramirez Ramirez (Biologist), Miss Socorro del Angel Blanco (Technician). This collection of Isoptera is mainly from Mexico, but other American localities are represented. Specimens are housed in alcohol vials. [1986]

119. MICRONESIA, Federated States of

[Oceanian (Micronesia). A new country which includes 607 Pacific islands comprising the island groups of Ponape, Yap, Truk, and Kosrae, until recently a trust territory of the U.S.A. Kolonia. **Population:** 86,094. **Size:** 271 sq. mi. *No known insect collection.*]

(Midway Islands, see U.S.A. Miscellaneous Pacific Islands.)

(Moluccas, Island Group of Indonesia.)

120. MONACO, Principality of

[Palearctic. Monaco. **Population:** 28,917. **Size:** 1.21 sq. mi.]

MUSÉE OCEANOGRAPHIQUE DE MONACO, AVENUE SAINT-MARTIN, 98000 MONACO. [MONA]

Director: Dr. C. Carpine, Conservateur des Collections. Phone: 93 15 36 25. The collection consists of Insecta and Arachnida, mostly marine, totaling 250 species, including 7 types. Series: "Résultats des Campagnes scientifiques du Prince Albert Ier de Monaco." [1992]

121. MONGOLIA, Mongolian People's Republic

[=Outer Mongolia. Palearctic. Ulan Bator. **Population:** 2,067,624. **Size:** 604,250 sq. mi. *A collection is known at the Academy of Sciences in Ulan Bator, but we do not have a contact or address. The collection contains voucher specimens from the* "Insects of Mongolia" *series.*]

122. MONTSERRAT, Colony of

[British colony. Neotropical. Plymouth. **Population:** 12,078. **Size:** 102 sq. mi. *Reply: no insect collection.*]

123. MOROCCO, Kingdom of

[Includes Ifni. Palearctic. Rabat. **Population:** 24,976,168. **Size:** 274,461 sq. mi.]

INSTITUT SCIENTIFIQUE CHERIPEN, AVENUE BIAR-NAY, RABAT. [ISCM] [*No reply.*]

124. MOZAMBIQUE, People's Republic of

[Afrotropical. Maputo. **Population:** 14,947,554. **Size:** 302,739 sq. mi.]

DR. ALVARO DE CASTRO PROVINCIAL MUSEUM, LARGO DO PODRAO DAS DESCOBERTAS, LOURENCO MARQUES. [CPMM] [*No reply.*]

(Myammar, see Burma.)

125. NAMIBIA

[=South West Africa. Afrotropical. Windhoek. **Population:** 1,301,598. **Size:** 317,873 sq. mi.]

STATE MUSEUM OF NAMIBIA, P. O. BOX 1203, WINDHOEK 9000. [SMWN]

Director: Dr. J. M. Mendelson, Curator (spiders). Phone: (061) 293376. Professional staff: Mrs. R. E. Griffin (insects); Mr. E. Marais; Ms. C. Roberts. The insect collection contains about 500,000 specimens, with emphasis on Coleoptera from the African southwestern arid biogeographical region which includes Namibia, southern Angola, and the northwestern Cape Province of South Africa. The collection is mostly South West Africa, with some from southern Angola and northwestern Cape Province of South Africa. Special collections include part of the Gaerdes collection, especially Tenebrionidae and Thysanura. More than 1500 primary and secondary types are housed in the collection. The spider collection contains about 10,000 specimens of Arachnida, Myriapoda, and Chilopoda. Crustacea are kept in a separate collection curated by Ms. B. Curtis. The collection is housed in drawers and steel cabinets. Publication sponsored: "Cimbebasia" (Journal of the State Museum). [1992]

DESERT ECOLOGICAL RESEARCH UNIT OF NAMIBIA (AT GOBABEB) P. O. BOX 1592, SWAKOPMUND. [GRSW]

Director: Dr. M. K. Seely. This is a reference collection of various arthropods related to ecological research carried out in the desert, and eventually will be turned over to the State Museum at Windhoek. It is stored in drawers and alcohol vials. [1992]

126. NAURU, Republic of

[=Pleasant Island. Oceanian (Polynesia). Yaren district. **Population:** 8,902. **Size:** 8.2 sq. mi. *No known insect collection.*]

(Navassa Island, Territory of U.S.A., uninhabitated.)

127. NEPAL, Kingdom of

[Indomalayan/Palearctic. Kathmandu. **Population:** 18,252,001. **Size:** 56,827 sq. mi.]

NATIONAL MUSEUM OF NEPAL, P.O. BOX 180, CHHAUNI, KATHMANDU. [NMNK] [*No reply.*]

128. NETHERLANDS, Kingdom of the

[=Holland. Palearctic. Amsterdam. **Population:** 14,716,100. **Size:**

INSTITUUT VOOR TAXONOMISCHE ZOOLOGIE, ZOOLOGISCH MUSEUM, UNIVERSITEIT VAN AMSTERDAM, PLANTAGE MIDDENLAAN 64, 1018 DH AMSTERDAM. [ZMAN] [=ZMUA]

Director: Dr. J. P. Duffels. Phone: 020-5223240. Professional staff: Dr. W. N. Ellis, Dr. Th. van Leeuwen, Dr. P. Oosterbroek, Dr. L. Botosaneanu, and 8 technical assistants. The collection contains about 8 million labelled specimens, 85% of which are identified. The majority are pinned specimens stored in about 32,000 drawers. The slide collection, mainly Collembola and Acari, contains about 85,000 slides. Alcohol material is preserved in about 8,000 jars. Geographical regions best represented are in general the Palearctic and Indonesia (former Dutch East Indies) with types in almost all orders. The collections of these regions contain many type specimens. Research is concentrated in these two areas. The oldest part of the collection is the former "Natura Artis Magistra" collection, with material, especially Coleoptera, of 1830's. Special collections include: Myriapoda (Jeekel, worldwide); Acari (van Eyndhoven, Palearctic); Collembola (Ellis, Palearctic); Orthoptera (Kruseman, Jeekel, one of the largest Palearctic collections); Hemiptera (Duffels, Indo-Australian Cicadidae); Lepidoptera (large collection from the Netherlands, and Palearctic, also Caron (Zygaenidae, with type material from several authors); van Oorschot (Lepidoptera, Rhopalocera from Turkey), and Lepidoptera from Indonesia; *Toxopeus* (Lycaenidae); van Groenendael (general collections from Java, Flores, Sulawesi, Papua New Guinea, and other islands); Trichoptera: (Botosaneanu, Palearctic and Central America, many types); Diptera (de Meijere, the Netherlands and Indonesia, many types; Van der Wulp, Central America, including types; Nijveldt, Palearctic Cecidomyiidae; van de Goot, Brachycera, mainly Syrphidae, including types; van Leeuwen, Oosterbroek, largest Palearctic Tipulidae collection, including types); Hymenoptera: (Oudemans, Palearctic Symphyta; Gijswijt, Palearctic Chalcididae; Vogt, Bombidae; Cameron, Indonesian Parasitica and Aculeata, including types; Leclercq, Indonesian Sphecidae, including types; extensive recent Parasitica and Aculeata collections from Africa collected by Schulten); Coleoptera (large Palearctic, Nearctic, and Indonesian collections, especially: Corporaal, Cleridae, worldwide, including types; Van Nidek, Cicindelidae, worldwide, including types; Vogt, Roeschke, Palearctic Carabidae, including types; Roeschke, Carabidae, including types; Bradshaw, former Natura Artis Magistra collection from Central Africa, including types. Material may be borrowed for study. Publications sponsored: "Bijdragen tot de Dierkunde," "Beaufortia," "Bulletin Zoologisch Museum Universiteit van Amsterdam," and "Verslagen en Technische Gegevens Instituut voor Taxonomische Zoologie." Library facilities are among the largest and most complete in Europe. [1986]

MUSEON, DEPARTMENT OF BIOLOGY, DEPARTMENT OF BIOLOGY, STADHOUDERSLAAN 41, 2517 HV THE HAGUE. [MUDH]

Head of Department: Drs. Arno L. van Berge Henegouwen. Phone: 31 (0)70-3381405; FAX 31 (0)703541820. Professional staff: Drs. Rob T. A. Schonten. The collection contains about 80,000 specimens representing all orders, but mainly Coleoptera (P. P. Everts), also strong in Heteroptera, from Europe, Indonesia, Suriname, and Kenya, housed in 1,000 drawers. The collection contains four holotypes and 11 paratypes. [1992]

Affiliated Collections:

Drs. R. T. A. Schouten, Simon Vestdyklaan 15, 2343 KW Oegeest, The Netherlands. [RTAS] Phone: 31 (0)71-173015; FAX 31 (0)70-3541820. This is a collection of about 8,000 specimens of Lepidoptera (90% Heterocera) from Europe, Ivory Coast, Kenya, Mali, and Suriname, particularly Crambinae (Pyralidae) and Pyraloidea. The collection contains 9 holotype and 38 paratypes. (Registered with MUDH.) [1992]

Drs. Arno L. van Berge Henegouwen, Scheveningsebos 45, 2716 HV Zoetermeer, The Netherlands. [ALBH] Phone: 31 (0)79-211952; FAX 31 (0)70-3541820. This collection contains 7,000 Coleoptera (Hydrophilidae) from Europe, Africa (Kenya, Ivory Coast, Uganda), Pakistan, Indonesia (Java, Bali, Sumbawa, Flores), and South America (Colombia). The collection contains 1 holotype and 39 paratypes. (Registered with MUDH) [1992]

NATIONAAL NATUURHISTORISCHE MUSEUM, RAAMSTEEG 2, LEIDEN 2311 PL. [RMNH]

Director: Dr. P. J. van Helsdingen. Phone: 31 (0)71-143844; FAX 31 (0)71-133344. Professional staff: 6 entomologists. The collections were recently (1991) estimated as housing 270,000 species, represented by 5,200,00 specimens. The collections are stored in large sized drawers of which at present there are *ca.* 22,500. Registration of types is in progress. Important collections from the former Dutch East Indies (now Indonesia) and Surinam (formerly Dutch Guyana)." Publications sponsored: "Zoologische Verhandelingen," "Zoologische Mededelingen," "Zoologische Bijdragen," and "Zoologische Monografieen." [Largest collection in the Netherlands. Eds.] [1992]

Affiliated Collection:

van der Zanden, G., Jongkindstraat 2, 5645 JV Eindhoven. [GZHC] Phone: 040-111359. This is a collection of a bout 15,000 Palaearctic Hymenoptera (Aculeata), housed in 60 glass-topped boxes (30 x 40 cm), all named by European specialists, particularly Megachilidae, with a number of primary and secondary types. (Registered with RMNH.) [1986]

NATUURHISTORISCH MUSEUM, M. H. TROMPLAAN 19, 7511 ENSCHEDE. [MVEN]

Director: Mr. P. Venema. Phone: 053-323409. Professional staff: Mr. F. van Stuivenberg and Mrs. Drs. R. E. van Maarle. The collection contains about 350 boxes of insects. Publication sponsored: "Natuur en Museum." [1986]

INSECT COLLECTION, FRIES NATUURHISTORISCH MUSEUM, HEERESTRAAT 13, LEEWARDEN. [FNML] [*No reply.*]

INSECT COLLECTION, NOORDBRABANTS NATUURMUSEUM, SPOORLAAN 434, 5038CH, TILBURG. [NNKN]

Director: Dr. F. Ellenbroek. Professional staff: Ir. E. Bouvy. The collection contains about 175 boxes of specimens, including a collection of Lepidoptera donated by Mr. Verhaak. Publication sponsored: "De Oude Ley." [1992]

NATURAL HISTORY MUSEUM, DE BOSQUETPLEIN 6-7, POSTBUS 882, 6200 AW, MAASTRICHT. [NHME]

Director: None given. Phone: 043-293064. Professional staff: Drs. F. N. Dingemans-Bakels. In this collection mainly local fauna is represented. It is housed in about 1,000 drawers. Groups best represented are: Lepidoptera, Coleoptera, Hymenoptera, and Diptera. There are also some special collections: the Wasmann collection of ants and ant guests from all over the world, housed in 175 drawers and 120 jars of alcohol; the Willemse collection with over 400 drawers of Orthoptera of the world; the Van Boven worldwide collection of ants with emphasis on those of the Netherlands and tropical Africa housed in 75 drawers. Material is available for loans to other institutions. There is a moderate library. [1992]

INSECT COLLECTION, NATUURHISTORISCH MUSEUM, VAN SERSSENLAAN 4, ROTTERDAM, [NHMR]. [*No reply.*]

129. NETHERLANDS ANTILLES

[Includes Curaçao, Bonnaire, St. Eustatius, Saba, and St. Maarten. Neotropical. Willemstad (Curaçao). **Population:** 182,676. **Size:** 309 sq. mi. *No insect collections.*]

(New Britain, island of Papua New Guinea.)

(Nevis, see St. Christopher (Kitts)-Nevis.)

(New Amsterdam Island, see France.)

130. NEW CALEDONIA, Territory of, and Dependencies

[French territory; includes Ile des Pins, Iles Loyaute, Ile Huon, Iles Belep, Isles Chesterfield, and Ile Walpole. Oceanian (Melanesia). Nouméa. **Population:** 150,981. **Size:** 7,376 sq. mi.]

LABORATOIRE DE ZOOLOGIE APPLIQUEE, ORSTON, BOITE POSTALE A 5, NOUMEA. [ONNC]

Senior Scientist in Charge: J. Chazeau. Phone: 26-10-00. Professional staff: Dr. J. Chazeau, and Dr. L. O. Brun. The collection contains over 200 boxes of insects, as a small reference collection of New Caledonia and other islands. Types are deposited in the Paris Museum. [1992]

(New Guinea, island comprising Irian Jaya (West Irian), and part of Papua New Guinea.)

(New Hebrides, see Vanuatu.)

131. NEW ZEALAND

[Includes Stewart Island, Chatham Island, and Kermadec Islands (Polynesia). Antarctic. Wellington. **Population:** 3,343,339. **Size:** 103,883 sq. mi.]

Auckland

AUCKLAND INSTITUTE AND MUSEUM, PRIVATE BAG 92018, AUCKLAND 1. [AMNZ]

Director: Mr. John W. Early. Phone: 64-9-3090443. Professional staff: Ms. Rosemary F. Gilbert. About 270,000 specimens of New Zealand insects (78,000 pinned, 192,000 in alcohol) and 70,000 non-New Zealand insects (42,000 pinned, 28,000 in alcohol) comprises this collection which includes 251 primary types, mostly from New Zealand and offshore islands, including the subantarctic. Strengths are New Zealand Lepidoptera and Coleoptera. Important historical collections now incorporated are those of C. E. Clark, T. Brown, W. Hemmingway, C. Geissler (Apoidea), and recently the collection of Eric D. Pritchard. Non-New Zealand insects contains a substantial number of Australian Coleoptera and worldwide Papilionoidea and Antarctic and subantarctic Collembola. Publications: "Records of the Auckland Institute and Museum." [1992]

NEW ZEALAND ARTHROPOD COLLECTION, KO TE AITANGA PEPEKE O AOTEAROA, LANDCARE RESEARCH NEW ZEALAND LTD, PRIVATE BAG 92170, AUCKLAND. [NZAC]

Director: Mr. J. F. Longworth. Phone: +64-9-849-6330; FAX: +64-9-849-7093; e-mail: 1ratkc@marc.cri.nz. Professional staff: Ms. J. A. Berry, Dr. R. C. Craw, Dr. T. K. Crosby, Mr. J. S. Dugdale, Mr. C. T. Duval, Dr. P. A. Maddison, Dr. M.-C. Lariviere, Mr. D. W. Helmore, Mrs. G. Hall, Mrs. R. C. Henderson, Mrs. M. Belsten, and Mrs. L. Cluniue. Approximately 1 million pinned and 5.5 million specimens in ethanol; all labelled and sorted at least to family level comprises this collection. About 60,000 slide mounts complete the collection. Nearly all material is from New Zealand, with holdings from nearby Pacific islands. It is the largest collection of New Zealand insects. About 2,500 primary types are separately housed in drawers. All specimens are available to specialists under normal loan conditions. Special collections include the Maskell collection of Homoptera; UNDP/FAOD-SPEC collection of insects of the South Pacific, from the Survey of Pests and Diseases of Agriculture. This collection is housed in boxes on shelves as in a library; some specimens in drawers. Ethanol-stored specimens are in vials in about 8,500 jars. Slides are stored in 120 cabinets. Sponsored publications: "Fauna of New Zealand." [1992]

Affiliated Collections:

Andrew, I. C., Entomology Division, Private Bag, Palmerston North, New Zealand. [IANZ] Phone: 69099, ext. 7404; 74355. About 2,000 specimens, mostly collected in Palmerston North, comprise this collection. It represents most of the common species of all orders, but especially Diptera and Lepidoptera, all housed in boxes. (Registered with NZAC.) [1986]

Barratt, Barbara I. P., 47 Franklin St., Dunedin, New Zealand. [BIPB] Phone: DU 738-391; MSI 3809. This is a collection of about 2,000 specimens of Coleoptera, mainly Carabidae, Scarabaeidae, Curculionidae, and with many other families represented. Most specimens from Otago and Southland, housed in 20 drawers. (Registered with NZAC.) [1986]

Chambers, Francis Dudley, Namu Rd., Opunake, R.O. 31, Taranaki, New Zealand. [FCNZ] Phone: 8515. This collection consists of 31 plastic boxes of New Zealand Hymenoptera, including 8 boxes of Proctotrupidae. (Registered with NZAC.) [1986]

Davies, Thomas H., 84 Beach Road, Haumoana, Hawkes Bay, New Zealand. [THDC] Phone: Hastings 750681. The collection consists of 2,000 specimens of Papilionoidea from Papua New Guinea, New Zealand Lepidoptera, especially of Hawkes Bay (2,000), and about 4,000 specimens representing all orders. There is a special collection of *Delias* (Pieridae), 650 specimens. The collection is housed in 55 drawers and 50 boxes. (Registered with NZAC.) [1986]

Fox, Dr. Kenneth J. Deceased. [KFNZ] Deposited in the National Museum of New Zealand collection (NMNZ). [1992]

McLellan, I. D., P. O. Box 95, Westport, New Zealand. [IDMC] Phone: WP8541. This is a collection of about 9,000 specimens of most of the species of New Zealand Plecoptera, and over 200 specimens of Thaumaleidae (Diptera) representing most of the New Zealand species. Representative collections of Australian and South American Plecoptera, totaling about 200 species. Most of the specimens are stored in alcohol vials and housed in drawers in cabinets. (Registered with NZAC.) [1992]

PLANT PROTECTION CENTRE COLLECTION, LYNFIELD AGRICULTURAL, P. O. BOX 41, AUCKLAND 1. [PPCC]

Director: Mr. R. G. Sunde. Phone: 676026. Professional staff: Mrs. O. R. Green, Mrs. D. H. Cooper. The insect collection, consisting of 50 drawers, 5 display boxes, alcohol vials, and slide collections of aphids, mites, *etc.*, is basically a reference collection of species which are regularly submitted for identification of specimens relating to problems in agriculture, horticulture, or intercepted at ports.

Affiliated Collection:

Green, C. J., 40 Preston Avenue, Henderson, Auckland 8, New Zealand. [CGNZ] Phone: 836-5888. The collection consists of about 10,000 adult insects of most orders in 150 boxes in cabinets. Special collections include Tortricidae and Formicidae of New Zealand. (Registered with PPCC.) [1986]

Christchurch

CANTERBURY MUSEUM, ROLLESTON AVE., CHRISTCHURCH 1. [CMNZ]

Director: M. M. Trotter. Phone: 68377. Professional staff: R. Anthony Savill (Curator), John B. Ward (Research Associate, Trichoptera), and Terry Hitchings (Research Associate, Ephemeroptera). The collection contains over one million specimens representing the New Zealand and subarctic fauna and includes a large number of types. Emphasis is on Dictyoptera, Orthoptera, Lepidoptera, Carabidae (Coleoptera), Dermaptera, Diptera, Ephemeroptera, and Trichoptera. It also includes a good representation of World insects. Current research is on Trichoptera and Cicindelidae. The collection is housed in drawers, in alcohol vials, and on slides. [1992]

DEPARTMENT OF ZOOLOGY, UNIVERSITY OF CANTERBURY, PRIVATE BAG, CHRISTCHURCH. [UCNZ]

Director: Head, Zoology Department. There are about 10,000 specimens, stored in boxes, particularly Carabidae and Tipulidae. No type specimens. [1986]

Dunedin

OTAGO MUSEUM, GREAT KING ST., DUNEDIN. [OMNZ]

Director: Dr. R. Cassels. Phone: 4772-372. Professional staff: Director Emeritus Dr. R. R. Foster, and A. C. Harris. The collection consists of about 163,000 Arachnida in the main collection, and over a million in the alcohol collection; 100,000 pinned insects and about 1 million in alcohol for processing. The spider collection contains many primary types. Special collections: Fulton collection of Coleoptera; Howes collection of all orders; Harris collection of Coleoptera and Hymenoptera, with primary types. [1992]

Affiliated Collections:

Patrick, Brian H., 38 St. Albans St., Dunedin, New Zealand. [BHPC] Phone: (0 3) 4770677. This is a collection of 42,000 specimens representing 1,800 species, mainly from southern New Zealand, stored in 5 cabinets with about 220 boxes. The collection includes Acrididae and Cicadidae, with a small collection of N.Z. alpine Coleoptera. (Registered with OMNZ.) [1992]

Lyford, Mr. Brian, 67 Wynyard Cres., Queenstown 9692. [BLPC] Phone: (03) 4426189. The collection contains about 5,100 specimens, representing 450 species of moths, and a general collection of other orders: 2,000 specimens. (Registered with OMNZ.) [1992]

Lincoln

ENTOMOLOGICAL MUSEUM, DEPARTMENT OF ENTOMOLOGY, LINCOLN UNIVERSITY, P. O. Box 84, CANTERBURY. [LCNZ]

Director: Dr. Rowan Emberson. Phone: (64) (3) 325-2811, ext. 8384. Professional staff: Mr. John Marris. The collection consists of over 100,000 pinned specimens, mostly identified to genus, stored in 1,300 boxes. Alcohol collection of larvae and non-insect material, and litter sample residues is maintained. Material is from throughout New Zealand, with emphasis on the South Island mountain areas; some exotic material, particularly Carabidae. No primary types are held, but secondary type material of various New Zealand insects is included. [1992]

LINCOLN PLANT HEALTH STATION, MINISTRY OF AGRICULTURE AND FISHERIES, P.O. BOX 24, LINCOLN. [PCNZ]

Director: Dr. Kenneth G. Somerfield. Phone: 252-811. Professional staff: Lindsay M. Emms, Barney P. Stephenson, Maurice O'Donnell. The collections are for diagnostic and reference purposes and consist of 13,000 pinned specimens representing 752 species, 588 genera, and 155 families. These are housed in 60 drawers and 192 boxes. Most species represented occur in New Zealand, with limited material from elsewhere. A collection of slides of scale insects, and herbarium of insect and mite damaged plants, and a collection of spiders is maintained. [1986]

Mosgiel

NEW ZEALAND PASTORAL AGRICULTURE RESEARCH INSTITUTE, INVERMAY AGRICULTURAL CENTRE, PRIVATE BAG, MOSGIEL. [IARC]

Director: B. I. P. Barratt (Curator). Phone: (03) 489-3809. Professional staff: Dr. B. I. P. Barratt and Mr. C. M. Ferguson. The collection consists of 60 boxes, 6,000 specimens of mainly agriculture pests, especially Lepidoptera and Coleoptera. [1992]

Palmerston North

DEPARTMENT OF ECOLOGY, MASSEY UNIVERSITY, PRIVATE BAG, PALMERSTON NORTH. [MUNZ]

Head of Department: Dr. Robin Fordham. Phone: (06) 356-9099. Professional staff: Elizabeth A. Grant, Dr. Ian Stringer, Dr. Murray Potter, and Dr. Robin Fordham. The collection consists of about 30,000 specimens housed in boxes and also display trays; some in alcohol vials, representing general reference collection, display specimens, and African butterflies. [1992]

Rotorua

FOREST RESEARCH INSTITUTE, PRIVATE BAG, ROTORUA. [FRNZ]

Director: Mr. M. J. Nuttall. Phone: +64 7 347 5899. Professional staff: J. Bain, D. J. Cross, G. P. Hosking, M. K. Kay, R. H. Milligan, G. R. Sandlaut, and R. Zondag. This collection contains approximately 100,000 specimens of forest insects and insects affecting timber in use. Parasitoids are well represented. An important part of the collection features insects discovered during quarantine inspections of imported timber. Major holdings are in Coleoptera, Lepidoptera, Hymenoptera, with host records for reared material. No material is available for exchange, but specimens are loaned for taxonomic study. The collection is stored in 118 drawers and 1,000 boxes. Publication sponsored: "New Zealand Journal of Forestry Science." [1992]

Wellington

ENTOMOLOGICAL COLLECTION, MUSEUM OF NEW ZEALAND, PRIVATE BAG, WELLINGTON. [MONZ] [=NMNZ]

Director: A. N. Baker. Phone: 385-9609. Professional staff: R. G. Ordish, and R. L. Palma. Principally a collection from the New Zealand subregion including New Zealand subantarctic and incorporates the private collection of T. Cockroft (Coleoptera), C. Fenwick (Lepidoptera, Diptera), G. Howes (Lepidoptera), G. V. Hudson (Lepidoptera, Coleoptera), J. H. Lewis (Lepidoptera, Coleoptera), A. C. O'Connor (Coleoptera), J. T. Salmon (Collembola), M. N. Watt (leaf mining insects), R. L. C. Pilgrim (Mallophaga). It has over 1,000 primary types and is not an exchange collection. It has major holdings in Mallophaga, Lepidoptera, Orthoptera, and Cicadidae, most groups obtainable through Berlese funnel extraction, and some families of Coleoptera. The collection is housed in unit trays in drawers in 21 wooden cabinets, plus shelving for alcohol vials and microslides. [1992]

Affiliated Collections:

Hornabrook, Dr. R. W., 27 Orchard St., Wadestown, Wellington, New Zealand. [RWHI] Phone: 735-320. This collection of Coleoptera specializes in the species from the South Pacific and Papua New Guinea, which includes 15,000 specimens of weevils, including various paratypes, a large collection of Staphylinidae, Carabidae, Cleridae, and most other families. There is a large collection of New Zealand Coleoptera of all families and a smaller collection of Lepidoptera from Papua New Guinea. Much of this material is identified. There is a small representative collection of Palearctic insects. The collection is stored in 110 drawers in cabinets, and over 40 boxes. The butterfly collection is in 75 drawers, and other Lepidoptera in boxes. (Registered with MONZ.) [1992]

SCHOOL OF BIOLOGICAL SCIENCES, VICTORIA UNIVERSITY OF WELLINGTON, P. O. Box 600, WELLINGTON. [VUWE]

Director: G. W. Gibbs. Phone: 64 4 472-1000. This is a small teaching and research collection of 36 drawers plus slide mounts and alcohol collections, contains all New Zealand insects including a research collection of Lepidoptera, especially butterflies, Micropterigidae and Mnesarchaeidae. [1992]

Whangarei

NORTHLAND REGIONAL MUSEUM, P. O. BOX 1359, WHANGAREI. [NRNZ]

Director: Bruce Young. Phone: (09) 43-89630. The collection consists of approximately 10,000 specimens of Coleoptera collected by Mr. E. R. Fairburn. It is housed in 112 drawers. [1992]

132. NICARAGUA, Republic of

[Neotropical. Managua. **Population:** 3,407,183. **Size:** 46,430 sq. mi.]

CENTRO EXPERIMENTAL DE NUEVA GUINEA, NUEVA GUINEA, ZELAYA. [CENG]

About 2,000 specimens representing about 400 species of the Nueva Guinea zone of Nicaragua. (C. Starr.) [1992]

CENTRO NACIONAL DE PROTECCIÓN VEGETAL [CENAPROVE], MUSEO DE ENTOMOLOGÍA, CENTRO NATIONAL DE PROTECCIÓN VEGETAL, MINISTERIO DE AGRICULTURA Y GANADERÍA, MANAGUA. [CENA]

Curators: Ms. Aminta Romero and Ms. María Pedrina Córdoba. Phone: (505) 025-8536. About 50,000 specimens, representing about 2,000 species of Nicaragua insects, with emphasis on economically important species. The collection is in two parts, one arranged taxonomically, and the other according to commodity.

SERVICIO ENTOMOLÓGICO AUTÓNOMO, MUSEO ENTOMOLÓGICO, SEA, A.P. 527, LEÓN. [SEAN]

Director: Juana Tellez. Phone: (505) 0311-6586; FAX (505) 0311-5700 (specify A.P. 527). Curator: Jean-Michel Maes. Roughly 200,000 specimens, about one-half identified, these representing about 6,600 species. Greatest strength is in Coleoptera, which accounts for 65% of tabulated specimens and 50% of the species. The collection has two main thrusts, toward a synoptic representation of insects and arachnids of Nicaragua and toward maximal global representation in areas of taxonomic research by staff.

This is effectively then national collection of Nicaragua, with a leadership role among the various other collections. It is located outside the US-Contra war zones and so was not directly affected by the war. However, the earlier economic blockade and the fact that the blocade has not been actively revered have taken their toll. SEA is a private organization, ineligible for many kinds of funding available to government organizations. There is minimal computerized record-keeping.

The SEA collection has no networking agreement with collections outside of Nicaragua. However, it makes numerous loans to individual specialists and is relatively well integrated into the international biosystematic community. (C. Starr.) [1992]

MUSEO ENTOMOLÓGICO, UNIVERSIDAD NACIONAL AGRARIA, MANAGUA. [UNAC]

Curator: Ms. Alba de la Llana. Phone: (505) 02-3-619, or 02-31968. Taxonomically arranged collection of about 3,000 specimens, representing about 650 species, and a collection of about 1,000 economically important specimens arranged according to commodity. (C. Starr.) [1992]

MUSEO ENTOMOLÓGICO, UNIVERSIDAD NACIONAL AUTÓNOMA DE NICARAGUA, LEÓN. [UNAN]

Curator: Ms. Miriam Corrales. Phone: (505) 0311-2612. A research collection and a teaching collection (possibly not rigorously separated) with a combined total of 10,000-15,000 specimens, representing a little over 1,000 species. (C. Starr.) [1992]

(Nicobar Islands, see India.)

133. NIGER, Republic of

[Afrotropical/Palearctic. Niamey. **Population:** 7,213,045. **Size:** 489,076 sq. mi. *No known insect collection.*]

134. NIGERIA, Federal Republic of

[Afrotropical. Lagos. **Population:** 111,903,502. **Size:** 351,649 sq. mi. *No known insect collection.*]

135. NIUE

[=Savage Island. Self-governing country in free association with New Zealand. Oceanian (Polynesia). Alofi. **Population:** 17,995. **Size:** 91.5 sq. mi. *Replied: no insect collections*. 1992.]

(Norfolk Island, Territory of Australia. *Reply: no insect collection*.]

(North Cyprus, Turkish Federated State of, see Cyprus)

(North Yemen, see Yemen.)

136. NORTHERN IRELAND, Commonwealth of

[U. K. Commonwealth. Palearctic. Belfast. **Population:** 1,490,288. **Size:** 5,459 sq. mi. *No known insect collection.*]

137. NORTHERN MARIANA ISLANDS, Commonwealth of

[Oceania (Micronesia). Saipan. **Population:** 20,591. **Size:** 293 sq. mi. *No known insect collection.*]

138. NORWAY, Kingdom of

[Excludes Svalbard and Jan Mayen. Palearctic. Oslo. **Population:** 4,190,758. **Size:** 125,050 sq. mi.]

ENTOMOLOGICAL COLLECTION, ZOOLOGICAL MUSEUM, UNIVERSITY OF BERGEN, MUSÉPLASS, 5007 BERGEN. [ZMUB]

Director: changes every year. Phone: (05) 21 30 50. Professional staff: Lita Greve Jensen (Curator). The collection contains 360,000 pinned specimens, over 30,000 alcohol vials of insects and spiders, and 13,200 slides, mostly of Chironomidae. There are 150 holotypes. The collection also contains the late Dr. Andreas Strand collection of Coleoptera, Prof. O. A. Saether's and Dr. E. Willassen's collection of Chironomidae, Dr. A. Løken's collection of bumblebees. The collection is stored in cabinets and alcohol vials. Sponspored publication: "Fauna of Hardangervidda." [1992]

ZOOLOGICAL MUSEUM, UNIVERSITY OF OSLO, SARS GT. 1, OSLO 0562. [ZMUN]

Curators: Dr. A. Lillehammer (Hemiptera, Coleoptera, and spiders), Dr. J. E. Raastad (Holometabola, excl. Coleoptera). The collection contains 143,500 pinned specimens mostly from Norway, and also Europe, Indo-Australia, and other places. Types of 453 species are held in the collection. The alcohol collection is separate. [1992]

(Novaya Zemlya, island of Russia.)

(Ocean Island, see Banaba.)

(Ogasawara Archipelago, see Bonin Islands.)

(Okinawa, island prefecture of Japan.)

139. OMAN, Sultanate of

[=Muscat. Palearctic. Muscat. **Population:** 1,265,382. **Size:** 120,000 sq. mi.]

NATIONAL INSECT COLLECTION, NATURAL HISTORY MUSEUM, P. O. BOX 668, MUSCAT. [SOIC]

Director: Said Ali Said Al-Farsi; Phone: 605 400; FAX 602 735. Curator: Mr. M. D. Gallagher. Eight cabinet each of 12 Cornell-type glass-topped drawers store representatives collections of all species of insects known in Oman, most pinned, spread, identified, and labelled, esp. Orthoptera, Dictyoptera; actively expanding collections of Hemiptera, Lepidoptera, Diptera, and less than complete Odonata, Hymenoptera, and Coleoptera. The liquid collection includes Isoptera, ants, ticks, and Thysanura. Siphonaptera are mounted on slides. Publication sponsored: "Journal of Oman Studies" ceased at vol. 10. [1993]

140. PAKISTAN, Islamic Republic of

[=West Pakistan. Indomalayan/Palearctic. Islamabad. **Population:** 107,467,457. **Size:** 310,403.]

DEPARTMENT OF ZOOLOGY, UNIVERSITY OF PUNJAB, LAHORE. [ZMLP]

Director: Dr. Muzaffer Ahmad, Professor Emeritus. Phone: 854096; 415931. It is not certain whether this is a public or private collection. It consists of termites which includes 550 species, mostly of the Oriental region, including 80 holotypes. The collection is preserved in alcohol. [1986]

141. PALAU, Republic of

[Palau Islands. (Oceanian [Micronesia]). UN trusteeship administered by USA. Koror. **Population:** 15,000. **Size:** 196 sq. mi. *No reply. No known insect collection.*]

142. PANAMA, Republic of

[Includes the Canal Zone. Neotropical. Ciudad de Panamá. **Population:** 2,323,622. **Size:** 29,762 sq. mi.]

MUSEO DE INVERTEBRADOS GRAHAM B. FAIRCHILD, UNIVERSIDAD DE PANAMA, ESTAFETA UNIVERSITARIA, PANAMA. [MIUP] [=GBFM]

Director: Dr. Diomedes Quintero Arias. Phone: 64-0582. Professional staff: Lic. Roberto Cambra (Hymenoptera, Aculeata), Lic. James Coronado (Pompilidae), Professors Yolanda Aguila (Trichoptera), Evidelio Adames (Diptera), hematophaga), Ivan Luna (Lonchaeidae), Viodela L. Chong (Agromyzidae). The collection consists of over 103,000 pinned insects and about 45,000 arachnids (paratypes of H. W. Levi), housed in drawers and vials. Most of the collection is from Panama, but some Amblypygi (paratypes of D. Quintero, worldwide), Mutillidae and Gomphidae (all the Americas, paratypes of Jean Belle), Pompilidae and Ichneumonidae (Peru). [1992]

MUSEO DE ENTOMOLOGIA FACULTAD DE AGRONOMIA, UNIVERSIDAD DE PANAMA, OCTAVIO MENDEZ PEREIRA, PANAMA. [MEUP]

Director: Ing. Baltazar Gray. Phone: 239652. Professional staff: Dr. Cheslavo Korytkowski, Marta Best. This is a small collection of insects that was originally used in the School of Agriculture in the course of introductory entomology as a part of the curriculum for a degree in agriculture sciences. Now the University of Panama has created a graduate program in entomology, hence the collection is being used for taxonomical support in the program and for thesis aid. It is now located in the School of Agriculture in Panama City. It consists of 60 drawers, 64 boxes, 700 vials, and 300 slides, approximately 15,000 specimens. [1986]

SMITHSONIAN TROPICAL RESEARCH INSTITUTE, P. O. Box 2072, BALBOA, PANAMA, or UNIT 0948, APO AA 34002-0948. [STRI]

Collection Manager: Dr. Annette Aiello. Phone: (507) 27-6022. This collection was begun recently and still is quite small, occupying only two cabinets (Cornell drawers). It is anticipated that the collection will serve mainly as a synoptic collection for some time to come, and that it will include voucher material donated by the many people who visit STRI to do research projects. The collection is housed at STRI, Tupper Center, room 219. [1992]

Affiliated collections:

Aiello, Dr. Annette, Research Associate, Smithsonian Tropical Research Institute, Unit 0948, APO AA 34002-0948. Phone: (507) 27-6022. The collection mainly consists of reared Panamanian insects, with overwhelming emphasis on Lepidoptera. The material reared from 1977 through 1980 is housed at the NMNH, and material reared from 1981 through the present is at STRI, Tupper Center, room 219. The latter is in 30 CAS drawers. (Registered with STRI.) [1992]

Barios, Dr. Hector, Smithsonian Tropical Research Institute, Unit 0948, APO 34002-0948. Phone: (507) 33-5259. This is a collection of weevils from light trap samples collected over many years at BCI by Henk Wolda (retired), some of which are being mounted in short series. The remainder will be maintained in alcohol storage. The number of species represented is estimated at 1090. (Registered with STRI.) [1992]

Gillogly, LTC. Alan R., Research Associate, PSC 02, Box 3067, APO AA 34002. [AAAG] Phone: (507) 87-6360. About 25 Cornell drawers (4,000 specimens,

1500 Panamanian) of Passalidae, with 200 vials of larvae and eggs The collection is housed at STRI, Tupper Center, room 219. (Registered with STRI.) [1992]

Roubik, Dr. David, Staff Scientist, Smithsonian Tropical Research Institute, Unit 0948, APO AA 34002-0948. Phone: (507) 27-6022. About 40,000 specimens of Neotropical and Paleotropical bees, primarily Meliponinae; also all major bee families, with approximately 500 species from Panama and Costa Rica, 400 species from South America, and 80 species from Malaysia and Thailand. The collection contains approximately 50 type species of Meliponinae and Halictidae, with a total of approximately 1,000 type specimens. The collection is maintained at STRI, Tupper Center, room 530, where it serves as a reference collection and repository for genera being revised, and is stored in 55 CAS drawers. (Registered with STRI.) [1992]

Stockwell, Dr. H. P., Research Associate, PSC 03, Box 1527, APO AA 34003. Phone: (507) 27-6022. [HPSC]. This is a collection of Coleoptera in about 100 Cornell drawers, and Curculionidae in about 80 drawers, including 100 paratypes of 15 species. The collection is housed at STRI, Tupper Center, room 219. (Registered with STRI.) [1992]

Windsor, Dr. Donald M., Staff Scientist, Smithsonian Tropical Research Institute, UNIT 0948, APO AA 34002-0948. Phone: (507) 27-6022. This is a collection of Panamanian Cassidinae (Coleoptera, Chrysomelidae), phasmids, and Panamanian wasps and their nests. It is housed in a Cornell cabinet and stored at STRI, Tupper Center, in rooms 219 and 229. (Registered with STRI.) [1992]

Wolda, Dr. Henk, 1625 106th Street SE, Bellevue, WA 98004 (retirement address). [HWIC] Phone: (206) 455-1443. Curated by Sr. Miguel Estribi and Sr. Saturnino Martinez. This collection of Panama insects was assembled from light traps set at various station in Panama, especially BCI, and has been the basis for the numerous publications on insect seasonality by Henk Wolda and collaborators. It is estimated that the collection includes about 2,500 specimens of Homoptera (Auchenorrhyncha only), 1,000 Cerambycidae, 300 Blattodes, and unspecified, but smaller, numbers of Scarabaeidae, Elateridae, Chrysomelidae, and various other coleopteran families. The collection is housed at STRI, Tupper Center, room 227, in a mixture of Cornell and CAS drawers. (Registered with STRI.) [1992]

143. PAPUA NEW GUINEA

[Includes Bismarck Archipelago (Admiralty Islands, New Britain, New Ireland, *etc.*), Louisiade Archipelago, D'Entrecasteaux Islands, northern Solomon Islands (Bougainville, Buka, *etc.*). Trobriand Islands, and Woodlark Island. Oceanian (Melanesia). Port Moresby. **Population:** 3,649,503. **Size:** 178,704 sq. mi.]

CENTRAL REFERENCE INSECT COLLECTION, DEPARTMENT OF PRIMARY INDUSTRY, P. O. BOX 2141, BOROKO, N.C.D. [TPNG]

Director: Dr. R. Kumar. Phone: 217899. This collection contains over 100,000 specimens of Papua New Guinea insects. It is a reference collection aimed at the identification of crop insects. However, there is included a fine butterfly collection as well as other non-pest insects. It is the largest collection in the country, but it does not hold long series. Specimens are loaned to specialists for systematic work. [Updates, Eds. 1992]

NATIONAL FOREST INSECT COLLECTION, FOREST RESEARCH CENTRE, P.O. BOX 314, LAE, MOROBE PROV. [FICB]

Director: (Vacant) Phone: 675-42-4188. Professional staff: Mr. P. P. Daur, Mr. John Dobunaba. Included in the collection are about 40,000 named pinned insects and 5,000 in alcohol vials. The collection is mainly of the native insect fauna of Papua New Guinea forests, as well as all forest insect pests. The collection is housed in drawers in metal cabinets in an air conditioned room. [Updates, Eds. 1992]

ENTOMOLOGY COLLECTION, BIOLOGY DEPARTMENT, UNIVERSITY OF PAPUA NEW GUINEA, P. O. BOX 320, UNIVERSITY, N.C.D. [UPNG]

Director: Chairman, Department of Biology. Phone: 24-5210; FAX 24-5187. Professional staff: Dr. E. J. Brough, Mr. T. Mala. Primarily a teaching collection, it places emphasis on crop pests. There are approximately 10,000 specimens; no types. It is housed in drawers in cabinets. It includes the Stan Christian mosquito collection. Publication sponsored: "Science in New Guinea." [1986]

WAU ECOLOGY INSTITUTE, P. O. BOX 77, WAU. [WEIC]

Director: Mr. Harry Sakulas. Phone: 44-6341. Professional staff: Dr. Mr. Max Kuduk, Mr. Joseph Somp. Bishop Museum (BPBM) liasion: Dr. Allen Allison. The collection contains over 15,000 specimens, including many phytophagous insect and their plant hosts, stored in metal cabinets in an air conditioned room. Publications sponsored: the "Wau Ecology Institute Handbook" series, which includes numbers on beetles, butterflies, and various vertebrate groups; also miscellaneous publications treating conservation, *etc.* [1992]

(Paracel Islands, disputed islands in the South China Sea. *No known insect collection.*]

144. PARAGUAY, Republic of

[Neotropical. Asunción. **Population:** 4,386,024. **Size:** 157,048 sq. mi.]

INVENTORIO BIOLOGICO NACIONAL, MINISTERIO DE AGRICULTURA Y GANADERIA, SUCURSAL 19 CAMPUS, CUIDAD UNIVERSSITARIA, CENTRAL XI. [INBP]

Director: John A. Kochalka. Phone: (595-21) 505-075. Professional staff: Blanca Barrio, Carlos Aguilar, Belia Torres, and Patricia France. Insects and spiders of Paraguay, about 30,000 insect specimens, pinned, 13,000 spiders (including those of John Kochalka from Sierra Nevada de Santa Marta, Colombia, *see below*) in alcohol vials, and 50,000 unsorted specimens. There are Paraguayan Anobiidae holotypes of species described by Toskima, and Paraguayan and Colombian spider holotypes and paratypes from Levi. Periodical sponsored: "Boletín del Museo Nacional de Historia Natural del Paraguay."

Affiliated collection:

Kochalka, John A.. Approximately 8,000 spiders in alcohol mostly from the Sierra Nevada de Santa Marta, Colombia make up this collection, but about one-half of these are temorarily stored with Dr. Jon Reiskind, Department of Biology,

University of Florida, Gainesville, FL 32611. Some species published by Levi are at the MCZC. Some species described by Platnick are at the AMNH. (Registered with INBP.) [1992]

145. PERU, Republic of

[Neotropical. Lima. **Population:** 21,269,074. **Size:** 496,255 sq. mi.]

MUSEO DE ENTOMOLOGIA, ESTACION EXPERIMENTAL AGRICOLA DE LA MOLINA, MINISTERIO DE AGRICULTURA Y ALIMENTACION, LIMA. [EELM] [*No reply.*]

COLECCION DEL DEPARTMENTO DE ENTOMOLOGIA, MUSEO DE HISTORIA NATURAL, UNIVERSIDAD NACIONAL MAYOR DE SAN MARCOS, AV. ARENALES 1267, APARTADO 14-0434, LIMA-14. [MUSM]

Director of Museum and Head of the Department of Entomology: Dr. Gerardo Lamas. Professional staff: Mr. Pedro Lozada (Homoptera), Ms. Diana Silva (Arachnida), Mr. Ruben Tejada (Ephemeroptera), Mr. Enrique Perez (Diptera), Ms. Helen Blancas (aquatic insects), and Mr. Joachim Hoffman (Odonata). The collection consists of about 250,000 mounted and 150,000 unmounted specimens, plus over 50,000 arachnids in alcohol. It is strong in butterflies, Hymenoptera, Coleoptera, Homoptera, Odonata, Diptera, and some groups of aquatic insects. The collection includes that of Antonio Raimondi (19th century), and the Renan Garcia collection, both merged with the general collection. Besides this, a very extensive collection (over 1.5 million specimens) of tropical canopy insects is currently on extended loan to the Department of Entomology, National Museum of Natural History, Washington, DC (USNM). The collections in Lima are housed in wooden drawers in metal cabinets. Publications sponsored: "Publicaciones del Museo de Historia Natural (Lima), UNMSM, Serie A (Zoologia)," and "Memorias del Museo de Historia Natural, UNMSM." [1992]

MUSEO DE ENTOMOLOGIA, UNIVERSIDAD NACIONAL AGRARIA, APARTADO 456, LA MOLINA, LIMA. [UNAD] [*No reply.*]

DEPARTAMENTO DE FITOTECNICA, MUSEO DE ENTOMOLOGIA, UNIVERSIDAD NACIONAL "PEDRO RUIZ GALLO," APARTADO 3, LAMBAYEQUE. [UPRG] [*No reply.*]

(Peter Island, see Norway.)

146. PHILIPPINES, Republic of the

[Indomalayan. Manila. **Population:** 63,199,307. **Size:** 115,831 sq. mi.]

ST. LOUIS UNIVERSITY MUSEUM, BAGUIO CITY. [SLUB] [*No reply.*]

ENTOMOLOGY COLLECTION, UNIVERSITY OF SAN CARLOS, CEBU CITY 6401. [USCP]

Director: Ms. Milagros M. Tumilap. This is a general collection of Philippine insects containing about 10,000 pinned identified specimens and about 10,000 unidentified pinned specimens. More than 30,000 specimens are stored in pill boxes and envelopes to be sorted and studied. The collection is housed in 7,500 drawers in cabinets. Specimens are loaned for study. [1986]

DEPARTMENT OF ENTOMOLOGY, UNIVERSITY OF THE PHILIPPINES, LAGUNA [UPPC]

[Contact: Dr. Clare Baltazar, Professor. Eds. 1992.]

BIOLOGICAL MUSEUM, VISAYAS STATE COLLEGE OF AGRICULTURE, BAYBAY, LEYTE. [VSCA]

Director: Dr. Leonila A. Corpuz-Raros. Professional staff: Ms. Rosemarie T. Rosario, Ms. Ma. Juliet P. Canete. The collection consists of about 30,000 pinned specimens, 2,000 slide mounts, and 3,000 vials of alcohol specimens, including insects, spiders, mites, and other arthropods. The collection of mites is emphasized. The collection is housed in drawers and boxes, slide boxes, and vials. [1986]

REFERENCE COLLECTION OF ARTHROPODS, DEPARTMENT OF ENTOMOLOGY, INTERNATIONAL RICE RESEARCH INSTITUTE, P. O. BOX 933, MANILA. [IRRI]

Director: Dr. M. S. Swaminathan. Phone: 9-211. Professional staff: Dr. James A. Litsinger, Mr. Alberto T. Barrion. This is a collection of Philippine arthropods from rice and rice-based crops, consisting of about 88,800 specimens in cabinets and boxes or on slides. It includes rice insects from Africa, Bangladesh, and Thailand. Publications sponsored: "Philippine Entomologist," "IRRI Research Paper Series," and "International Rice Research Newsletter." [1986]

NATIONAL MUSEUM OF THE PHILIPPINES, EXECUTIVE HOUSE, P. BURGOS STREET, MANILA. [MPMP] [*No reply.*]

(Phoenix Islands, see Kiribati.)

147. PITCAIRN

[Oceanian (Polynesia). British dependency island group, includes Henderson Island, Ducie, and Oeno atolls. Adamstown. **Population:** 55. **Size:** 1.75 sq. mi. *No known insect collection.*]

148. POLAND, Polish People's Republic

[Palearctic. Warszawa. **Population:** 37,958,420. **Size:** 120,727 sq. mi.]

MUSEUM OF THE INSTITUTE OF ZOOLOGY, POLISH ACADEMY OF SCIENCE, WILCZA 64, 00-679 WARSZAWA. [ZMPA]

Director: Mr. Eugeniusz Kierych. Phone: 29-32-21. Professional staff: Mr. Adam Kedziorek, Mr. Stanislaw Adam Slipinski, and Ms. Anna Slojewska. The insect collection contains over 3 million specimens, approximately 100,000 species, including about 7,500 types. It is the largest collection in Poland, containing Palearctic species, and representatives from the Neotropical, Ethiopian, and Oriental Regions. Some parts of the collection of the former Museum of Natural History of Szczecin have been incorporated. The specimens are stored in boxes on shelves and in cabinets, slide mounts, and in alcohol vials. Special collections include types of G. Enderlein and G. Heinrich in Hymenoptera. The following special collections of Coleoptera are included: R. Kleine, M. Nunberg, M. Liebke, R. Bielawski, M. Mroczkowski, Fr. Kessel, Sz. Tenenbaum, J. Kinel, G. and E. Mazur, K. Galewski, P. Franck, M. Wegrzecki, and various Polish coleopterists. The Arachnida collection contains 500,000 identified specimens, including about 800 types. It includes the collection of Wladyslaw Kulczynski, Wladyslaw Taczanowski, and E. Keyserling. Specimens may be borrowed for study. [1986]

DEPARTMENT OF ENTOMOLOGY, UNIVERSITY OF WROCLAW, CYBULSKIEGO 32, WROCLAW 21. [UWCP] [*No reply.*]

INSTITUTE OF SYSTEMATIC ZOOLOGY, EXPERIMENTAL, POLISH ACADEMY OF SCIENCES, SLAWKOWSKA 17, 31-016 KRAKOW. [ISZP] [*No reply.*]

INSTITUTE OF BIOLOGY, UNIWERSYTET MIKOLAJA KOPERNIKI, GAGARINA 9, 87-100 TORUN. [IBUN] [*No reply.*]

149. PORTUGAL, Republic of

[Palearctic. Lisboa. **Population:** 10,388,421. **Size:** 33,549 sq. mi.]

INSTITUTO DE ZOOLOGIE E ESTACAO DE ZOOLOGIA MARITIMA "DR. AUGUSTO NOBRE," UNIVERSIDADE DO PORTO, PORTO. [IZPC] [*No reply.*]

MUSEU E LABORATORIO ZOOLOGICO DA FACULDADE DE CIENCIAS, UNIVERSIDADE DE COIMBRA. COIMBRA. [MZCP] [*No reply.*]

LABORATORIO DE ZOOLOGIA, FACULDADE DE CIENCIAS, RUA DE ESCOLS POLETECNICO, LISBOA 2. [LZLP] [*No reply.*]

CENTRO DE ZOOLOGIA DO I.I.C.T., DIV. ARAENOENTOMOLOGIA, R. DE JUNQUEIRA 14, 1300 LISBOA. [IICT]

Curator: Prof. Luis F. Mendes. Phone: (01) 3637055; FAX (01) 3631460. This is a very large collections of arthropods. [1992]

(Portuguese Guinea, see Guinea-Bissau.)

(Prince Edward Islands, see South Africa.)

150. PUERTO RICO, Commonwealth of

[Associated with U.S.A. Neotropical. San Juan. **Population:** 3,358,879. **Size:** 3,459 sq. mi.]

ENTOMOLOGICAL PIONEERING RESEARCH LABORATORY, UNIVERSITY OF PUERTO RICO, MAYAGUEZ, PR 00708. [EPRL] [*No reply.*]

151. QATAR, State of

[Palearctic. Doha. **Population:** 328,044. **Size:** 4,416 sq. mi. *No known insect collection.*]

(Ras al Khaimah, see United Arab Emirates)]

(Rapa Nui Island, see Easter Island.)

(Redonda, see Antigua.)

152. REUNION, Department of

[French overseas department: includes Ile Europa, Bassas da India, Ile Juan de Nova, Iles Glorieuses, and Ile Tromelin. Indomalayan. Saint-Denis. **Population:** 557,441. **Size:** 970 sq. mi.]

MUSEUM OF NATURAL HISTORY, JARDIN DE L'EST. SAINT-DENIS. [NHST] [*No reply.*]

(Revillagigedo Island, see Mexico.)

(Rhodesia, see Zimbabwe.)

(Rio Muni, see Mbini.)

(Rodriguez, dependency of Mauritius.)

153. ROMANIA

[Palearctic. Bucuresti. **Population:** 23,040,883. **Size:** 91,699 sq. mi.]

MUZEUL DE ISTORIA NATURALA "GRIGORE ANTIPA," L. CHAUSSEE KISSELEF 1, BUCHAREST. [MGAB] [*No reply.*]

MUZEUL DE ISTORIA NATURALA, BDUL INDEPENDENTEI 16,7600, IASI. [MINI] [*No reply.*]

MUZEUL ORASULUI SF. GHEORGHE, STRADA 16 FEBRUAR-UUL 10, SF. GHEORGHE. [MOSG] [*No reply.*]

MUZEUL BRUKENTHAL, PIATA REPUBLICII 4, SIBIU. [MBSR] [*No reply.*]

COLECTIA DE INSECTE A MUZEULUI JUDETEAN COVASNA, STRADA 16 FEBRUARIE NR. 10, 400 SFINTU GHEORGHE, JUDETUL COVASNA. [CIJC]

Director: Dr. Szekely Zoltan. Professional staff: Mrs. Kocs Iren. The collection is limited to Lepidoptera and Curculionidae, with special representation of species west of Transylvania and some parts of eastern Europe and Asia. The Lepidoptera collection consists of the Dioszeghy Laszlo Collection of 24,000 specimens, 2,085 species in 100 drawers. The Curculionidae collection consists of 886 specimens, 305 species, in 7 drawers. Publication sponsored: "Aluta." [1986]

(Ross Dependency, Antarctic Territory of New Zealand, see New Zealand.)

154. RUSSIA, Commonwealth of Independent States

[Excludes Estonia, Latvia, and Lithuania; but apparently includes southern Kuril Islands, Habomai Islands, southern Sakhalin, and northern East Prussia. The following republics belong to the Commonwealth: Russian Federation, Byelorussia, Ukraine, Kazakhstan, Armenia, Azerbaijan, Georgia, Moldavia, Kirgizia, Turkmenistan, Tadzhikistan, and Uzbekistan. It is very difficult to get mail into Russian unless the address is written in the Russia alphabet. Palearctic. Moscow. **Population:** 286,434,844. **Size:** 8,599,228 sq. mi.]

INSTITUTE OF ZOOLOGY, UKRAINIAN ACADEMY OF SCIENCE, LENIN STREET, 15,252650 KIEV 30, UKRAINE. [UASK] [=ZIKU] [*No reply.*]

ZOOLOGICAL MUSEUM, VLADIMIRSKAYA SS, KIEV, UKRAINE. [ZMKU] [*No reply.*]

ZOOLOGICAL MUSEUM, ACADEMY OF SCIENCE, UNIVERSITETSKAYA, NABERZHNAYAL, B-164, St. PETERSBURG. [ZMAS] [=ZIL, ZIAS] [*No reply.*]

MOSCOW STATE SPIDER COLLECTION, DEPARTMENT OF INVERTEBRATE ZOOLOGY, ZOOLOGICAL MUSEUM, UNIVERSITY OF MOSCOW, HERZEN STR. 6, MOSCOW 103009. [ZMUM]

Director: Dr. K. Mikhajlov. The collection, established in the middle of the nineteenth century, contains more than 20,000 specimens including 150 types (nearly 40 species). This is the largest spider collection in the Soviet Union. [1986]

INSTITUTE FOR EVOLUTION, MORPHOLOGY, AND ECOLOGY OF ANIMALS, LENINSKY PR. 33, MOSCOW 117071. [IEME] [*No reply.*]

INSTITUTE OF ZOOLOGY, ACADEMY OF SCIENCE OF ARMENIAN S.S.R., 7 PARUYR SEVEH STREET, 375044 YEREVAN, ARMENIA SSR. [ASAY] [*No reply.*]

INSTITUTE OF ZOOLOGY, ACADEMY OF SCIENCE, ACADEMICHESKAYA STR., 27, MINSK 220072, BYELORUSSIA. [ISAR] [*No reply about insect collection, but curator reports below about personal collection.*]

Associated collection:

Karasjov, Dr. V. P., Academicheskaya Str. 23-24, 220012 Minsk, Byelorussia. Phone: 39-49-73. Over 1,500 species of Curculionidae from former USSR, Western Europe, North America, and some tropical regions. There is type material of *Tychius, Smicronyx, Othiorhynchus*, and some others. There are also about 600-700 species of Cerambycidae, *Carabus, Cicindela*, Cetoniinae, Buprestidae, and some other families mainly from Russia. [1992]

PLANT PROTECTION INSTITUTE, TBILISI, GEORGIA. [GPPT] [*No reply.*]

155. RWANDA, Republic of

[Afrotropical. Kigali. **Population:** 7,058,350. **Size:** 10,169 sq. mi. *No known insect collection.*]

(Ryukyu Islands, islands of Japan.)

(Sabah, see Malaysia.)

156. ST. CHRISTOPHER-NEVIS-ANGUILLA

[United Kingdom associated state. Anguilla is technically still part of St. Christopher-Nevis-Anguilla even though the British govern Anguillan affairs. (St. Christopher=St. Kitts.) Neotropical. Basseterre. **Population:** 36,738. **Size:** 101 sq. mi. *No known insect collection.*]

157. ST. HELENA

[United Kingdom Dependent Territory, includes Ascension, Gough, Inaccessible, Nightingale, and Tristan islands. Afrotropical. Jamestown. **Population:** 8,624. **Size:** 47 sq. mi. *No known insect collection.*]

(St. Kitts Island, see St. Christopher-Nevis-Angilla.)

158. ST. LUCIA

[Commonwealth Nation cf United Kingdom. Neotropical. Castries. **Population:** 136,564. **Size:** 238 sq. mi.]

RESEARCH & DEVELOPMENT STATION, WINDWARD ISLANDS BANANA GROWER'S ASSOCIATION, P. O. BOX 115, CASTRIES. [WIBG]

Director: Mr. Guy Mathurin. The main part of this small collection is mainly contained in 34 drawers. Its emphasis is on Lepidoptera of St. Lucia. The material is in variable condition, as the collection room is not air-conditioned. After about a decade of inactivity, regular collecting was resumed recently, with emphasis on light-trapping. (C. Starr) [1992]

(St. Paul Island, see France.)

159. SAINT-PIERRE AND MIQUELON, Department of

[Overseas Territory of France. Nearctic. St. Pierre. **Population:** 6,274. **Size:** 93 sq. mi. *No reply; no known insect collection.*]

160. SAINT VINCENT and THE GRENEDINES

[British associated state; includes northern Grenadine islands. Neotropical. Kingstown. **Population:** 107,425. **Size:** 150 sq. mi. *No known insect collection.*]

(Sakhalin, island of U.S.S.R.)

(Sala y Gomez Island, see Chile.)

(Samoa, see American or Western.)

(San Ambrosio Island, see Chile.)

(San Felix Island, see Chile.)

161. SAN MARINO, Republic of

[Palearctic. San Marino. **Population:** 22,986. **Size:** 23.4 sq. mi. *No known insect collection.*]

(Santa Cruz Island, see Solomon Islands.)

162. SÃO TOMÉ AND PRINCIPE, Democratic Republic of

[Afrotropical. São Tomé. **Population:** 117,430. **Size:** 372 sq. mi. *No known insect collection.*]

(Sarawak, see Malaysia.)

(Sardinia, island region of Italy.)

163. SAUDI ARABIA, Kingdom of

[=Arabia. Palearctic. Riyadh. **Population:** 15,452,123. **Size:** 864,869 sq. mi. *No known insect collection.*]

(Savage Island, see Niue Island.)

(Scotland, see United Kingdom.)

164. SENEGAL, Republic of

[Afrotropical. Dakar. **Population:** 7,281,022. **Size:** 74,206 sq. mi. *No known insect collection.*]

(Serbia [=Srbija}, see Yugoslavia.)

165. SEYCHELLES, Republic of

[Includes Aldabra Islands, Alphonse, Bijoutier, and St. Francois islands, Amirante Isles, Cosmoledo Group, Farquhar Group, Ile Desroches, and St. Pierre Islet. Indomalayan. Victoria. **Population:** 68,615. **Size:** 175 sq. mi.]

NATIONAL ARCHIVES AND MUSEUM DIVISION, P. O. BOX 720, LA BASTILLE, UNION VALE, MAHÉ. [SMVM]

Director: H. J. McGaw. Phone: 405. The collection consists of 10 boxes, over 600 specimens of pinned insects, sorted to family. [1986]

(Sharjah, see United Arab Emirates.)

(Shetland Islands, see United Kingdom.)

(Shikoku, island prefecture of Japan.)

(Sicily, island region of Italy.)

166. SIERRA LEONE, Republic of

[Afrotropical. Koidu. **Population:** 3,963,289. **Size:** 27,653 sq. mi. *No known insect collection.*]

167. SINGAPORE, Republic of

[Indomalayan. Singapore. **Population:** 2,645,443. **Size:** 240 sq. mi.]

ZOOLOGICAL REFERENCE COLLECTION, DEPARTMENT OF ZOOLOGY, NATIONAL UNIVERSITY OF SINGAPORE, KENT RIDGE, S-0511 SINGAPORE. [NMSC]

Director: Mrs. Chang Man Yang. Phone: 7722875. FAX: 7792486.

Professional staff: Mrs. C. M. Yang (aquatic and semiaquatic bugs). The collection consists of 30,000 pinned specimens in 500 drawers, and 8,000 alcohol vials, including 20 holotypes and 17 paratypes. These were collected mainly from Peninsula Malaysia since the 1900's. It contains material formerly at the Raffles Museum, the Singapore National National Museum, and the University of Singapore. Some specimens from H. M. Pendlebury, H. C. Abraham, C. A. Gibson-Hill (from Christmas Island) are also deposited in this collection. Periodical sponsored: "Raffles Bulletin of Zoology." [1992]

GLOBAL COLOSSEUM, P. O. BOX 11, TANJONY PAGOR, SPORE 2, SINGAPORE. [GCTP] [*No reply, probably a dealer.*]

168. SLOVAKIA, Socialist Republic of

[*Due to unsettled condition in this area, the status of this country is not certain, but it is presumed to be an independent republic as a part of the former Czechoslovakia.*]

NATURAL HISTORY COLLECTIONS, SLOVENSKÉ NÁRODNÉ MUZEUM, PRÓRODOVEDNÉ MÜZEUM, ZOOLOGICKÉ ODDELENIE, VAJANSKÉHO NÁBREZIE 2, 814 36 BRATISLAVA [SNMC].

Director: RNDr. Branislav Matousek, C. Sc. Phone: 332-985. Professional staff: Milan Rybecky, CSc., Chairman, Department of Zoology, and RNDr. Vladimir Jansky. This large collection of insects, with over 1,260,000 specimens (Coleoptera, 539,271 specimens, Lepidoptera 136,840 specimens, Homoptera 133,818 specimens, Diptera 121,644 specimens, Hymenoptera 94,628 specimens) and a spider collection of 30,900 specimens. The collection is stored in boxes and vials. The museum sponsors: "Zborník Slovenského národného múzea - Prírodnée vedy" and "Annotations Zoologicae et botanicae." [1992]

(Society Islands, see French Polynesia.)

(Socotra Island, see South Yemen.)

169. SLOVENIA

[Palearctic. Ljbljana. [*Due to unsettled conditions, the status of this country is not certain; formerly part of Yugoslovia.*]

SLOVENIAN NATURAL HISTORY MUSEUM, PRIRODOSLOVNI MUZEJ SLOVENIJE, PRESER NOVA 20, 61000 LJUBLJANA. [PMSL]

Director: Mr. Ignac Sivec. Phone: 003-861-218-846. The collection contains about 900 cardboard boxes containing pinned, labelled, and identified specimens of beetles from the Palearctic Region; 70 boxes of pinned Hymenoptera, Yugoslavian; 80 boxes of pinned Orthoptera, Yugoslavian; 550 drawers of Lepidoptera, Palearctic; about 50,000 undetermined specimens of Diptera in alcohol, Palearctic; 2,500 vials of Plecoptera, Palearctic, Oriental; about 22,000 slides of identified Mallophaga and Siphonaptera of the world. Publication sponsored: "Scopolia," journal of the museum. [1986]

170. SOLOMON ISLANDS

[=British Solomon Islands. Solomon Islands Protectorate of the United Kingdom includes southern Solomon Islands, primarily Guadalcanal, Malaita, San Cristobal, Santa Isabel, Choiseul. The Northern Solomon Islands constitute part of Papua New Guinea. Oceanian (Melanesia). Honiara. **Population:** 312,196. **Size:** 10,639 sq. mi.]

MINISTRY OF NATURAL RESOURCES, HONIARA. [HSIC] [*No reply.*]

171. SOMALIA, Somali Democratic Republic

[Afrotropical. Mogadishu. **Population:** 7,990,085. **Size:** 246,201 sq. mi. *No known insect collection.*]

172. SOUTH AFRICA, Republic of

[Includes Walvis Bay, and Marion, and Prince Edward islands. Afrotropical. Cape Town. **Population:** 35,093,971. **Size:** 471,445 sq. mi.]

INSECT COLLECTION, NATIONAL MUSEUM BLOEMFONTEIN, P. O. BOX 266, BLOEMFONTEIN 9300. [BMSA]

Director: None given. "Emphasis on Coleoptera, currently numbering about 82,000 specimens, with other orders about 7,000 specimens." The collection is stored in drawers in steel cabinets. Publications sponsored: "Navorsinge Nasionale Museum, Bloem fontein," and "Memoirs National Museum, Bloemfontein." [1986]

ENTOMOLOGY DEPARTMENT, SOUTH AFRICAN MUSEUM, P. O. BOX 61, QUEEN VICTORIA STREET, CAPE TOWN 8000. [SAMC]

Head, Department of Entomology: Dr. H. G. Robertson. Phone: 22-3330; FAX (021) 24-6716. Professional staff: Dr. S. Van Noort, Dr. V. B. Whitehead, and Ms. M. Cochrane (Collections Manager). The collection is strongest in Southern African Coleoptera and Hymenoptera. It contains 458,000 pinned insects, 24,600 alcohol vials (mainly spiders and scorpions), and 2,560 microscope slides. Specimens date back to 1863. There are over 7,200 primary type series mainly the product of work by R. Trimey (butterflies) W. F. Purcell (arachnids), A. J. Hesse (bombyliids), R. F. Lawrence (arachnids and myriapods), and G. Arnold (aculeate Hymenoptera, particularly ants). Publications sponsored: "Annals of the South African Museum." [1992]

DURBAN MUSEUM, P. O. BOX 4085, CITY HALL, SMITH STREET, DURBAN, NATAL. [DMSA]

Director: Dr. Q. B. Hendey. Professional staff: C. D. Quickelberge. The general collection consists of about 250,000 specimens. Special collections: British moths and butterflies from Mr. P. J. Rogers; Major C. Harford collection from India, Malta, and the Rhone Valley. The collection is housed in drawers primarily, some boxes. Publications sponsored: "Durban Museum Novitates."

ALBANY MUSEUM, SOMERSET STREET, GRAHAMSTOWN, CAPE PROVINCE. [AMGS] [*No reply.*]

ALEXANDER MCGREGOR MEMORIAL MUSEUM, P.O. BOX 316, KIMBERLEY, 8300. [MMKZ]

Director: Mrs. E. A. Voigt. Phone: 0531-32645. Professional staff: Mr. P. C. Anderson (Senior Museum Natural Scientist), Ms. B. Y. Wilson (Collection Manager). The collection consists of about 120 bottles containing approximately 300 specimens of spiders. The collection at present is unaccessioned. [1992]

DEPARTMENT OF MEDICAL ENTOMOLOGY, SOUTH AFRICAN INSTITUTE FOR MEDICAL RESEARCH, P. O. BOX 1038, JOHANNESBURG, 2000. [SAIM]

Director: Dr. Richard Hunt. Phone: 27-11-725-0511. Professional staff: Dr. Maureen Coetzee, Mr. Anthony Cornel, and Mr. Anton van Rensburg. This is a collection of 30,000 slides of medically important insects, and 123 drawers of pinned insects, including about 386 types of Culicidae, Ceratopogonidae, Simuliidae, Psychodidae, Pthiraptera, and Siphonaptera. [1992]

DEPARTMENT OF ZOOLOGY AND ENTOMOLOGY, UNIVERSITY OF NATAL, P.O. BOX 375, PIETERMARITZBURG 3200 NATAL. [UNSA]

Director: Prof. Denis J. Brothers. Phone: 27-331-63320. Professional staff: Dr. R. M. Miller. This is a teaching collection only. [1986]

Affiliated Collection:

Brothers, Prof. Denis J., Department of Zoology and Entomology, University of Natal, P. O. Box 375, Pietermaritzburg, 3200 R. South Africa. [DJBC] This is a collection of Aculeate Hymenoptera (about 15,000 Mutillidae, including a few types, mainly Afrotropical, but also Neotropical, Nearctic, and Australian); other uncurated Aculeata, housed in drawers in cabinets. (Registered with UNSA.) [1986]

NATAL MUSEUM, PRIVATE BAG 9070, PIETERMARITZBURG 3201, NATAL. [NMSA]

Director: Dr. B. R. Stuckenberg. Phone: (0331) 451404. Professional staff: Dr. J. G. H. Londt, Mr. D. Barraclough, Mr. A. Whittington. The collection is heavily Diptera orientated, with over 100,000 specimens, and contains mostly Afrotropical species. Specimens of Diptera are mostly pinned, but a good collection of larvae are alcohol preserved. There are almost 6,000 type specimens. There is also the largest Arachnida collection in Africa (over 90,000 specimens including acarines, myriapoda, scorpions, and spiders) preserved in alcohol. The collection is available for study. The museum publishes the "Annals of the Natal Museum." [1992]

TRANSVAAL MUSEUM, P.O. BOX 413, PRETORIA, 0001 [TMSA]

Director: Dr. I. L. Rautenbach. Phone: (012) 322-7632; FAX (021) 322-7939. Professional staff: Dr. S. Endrody-Younga, Mr. R. Toms, Mr. M. Kruger, and Dr. C. K. Brain. The collection consists of the following: Approximately 700,000 specimens of Coleoptera including the C. Koch collection of Tenebrionidae; 5,000 drawers of Southern Africa Lepidoptera with leaf miners and Rhopalocera particularly well represented, including the D. A. Swanepoel and K. Pennington collections; general entomology, the B. I. Balinsky Odonata collection and a collection of insect sound tapes; 15,000 specimens of Arachnida and Myriopoda, with Mygalomorpha and scorpions well represented. Publications sponsored: "Annals of the Transvaal Museum," and "Monographs of the Transvaal Museum." [1992]

NATIONAL COLLECTION OF ARACHNIDA AND INSECTS, PLANT PROTECTION RESEARCH INSTITUTE, AGRICULTURE BLDG., BEATRICE ST., PTE. BAG X134, PRETORIA 0001. [PPRI]

Director: Dr. A. S. Dippenaar. Phone: (012) 21311, ext. 679. Professional staff: Mrs. A. van der Berg. The collection includes 54 spider families collected in the southern part of Africa, the family Thomisidae is especially well represented. The collection currently contains 9,000 labelled and identified specimens and about 6,000 unidentified specimens. The specimens are stored in alcohol vials housed in seven specially designed steel cabinets with drawers. Publication sponsored: "Phytophylactica." [1986]

DEPARTMENT OF ENTOMOLOGY, UNIVERSITY OF PRETORIA, PRETORIA 0002. [UPSA]

Director: Prof. E. Holm. Phone: 012-4203232. Professional staff: Prof. E. Holm, and Prof. C. H. Scholtz. The collection contains about 100,000 specimens, half of which are Coleoptera, with 30 holotypes. About 25% of the total collection has been identified by specialists. Special collections include world Trogidae, world Buprestidae, and Namib Coleoptera. The collection is housed in drawers. Publications sponsored: "Journal of the Entomological Society of Southern Africa" and "Cimbebasia." [*This collection is apparently now a part of the Transvaal Museum.*] [1992]

QUEENSTOWN MUSEUM, P.O. BOX 296, QUEENSTOWN, CAPE PROVINCE 5820. [QFMQ]

Director: Mrs. J. Preston, B.A. (Curatrix). Phone: 0451-5860. This is a small collection of insects and spiders stored in alcohol vials, most all are identified. [1992]

SOUTH AFRICAN NATIONAL COLLECTION OF INSECTS, PRIVATE BAG X134, PRETORIA 0001. [SANC]

Director: Dr. G. L. Prinsloo. Phone: 012-28-5140. Professional staff:

Dr. M. W. Mansell, Mr. C. D. Eardley, Mr. R. G. Oberprieler, Mr. I. M. Millar, Miss V. M. Swain, Mrs. O. Neser, and Mrs. N. C. Pienaar. This collection contains over 560,000 insect specimens distributed over 22 orders and 416 families. It consists largely of Afrotropical material with the emphasis on the fauna of southern Africa. The most important groups in the collection, *i.e.*, those which have received attention from specialists, are Isoptera (33,400 colony samples); Orthoptera, mainly Acrididae (31,120 specimens); Homoptera (100,943 specimens and slides), Thysanoptera (50,000 slides); Neuroptera (6,590 specimens); Coleoptera (152,730 specimens; Diptera, mainly Trephritidae (48,390 specimens), and Hymenoptera, mainly certain Parasitica (128,670 specimens). In addition, it contains a reference collection of Lepidoptera (12,000 specimens), and small collections of most other insect orders. The type collection contains 1,232 holotypes and 478 cotypes. The collection is stored in drawers, slide boxes, and alcohol vials. The type collection is stored in unit trays in steel cabinets. [1986]

(South Arabia, see South Yemen.)

(South Georgia Island, see Falkland Islands.)

(South Orkney Islands, see British Antarctic Territory.)

(South Sandwich Islands, see Falkland Islands.)

(South Shetland Islands, see British Antarctic Territory.)

(South West Africa, see Namibia)

(South Yemen, see Yemen.)

(Southern and Antarctic Territories, see France.)

(Southern Rhodesia, see Zimbabwe.)

(Soviet Union, see Russia)

173. SPAIN, Spanish State of

[Includes the Balearic and Canary Islands, Ceuta, Melilla, Islas Chafarinas, Penon de Alhucemas, and Penon de Velez della Gomera, and see Spanish North Africa. Palearctic. Madrid. **Population:** 39,209,765. **Size:** 194,897 sq. mi.]

MUSEO ZOOLOGIA, APARTADO DE CORREOS 593, BARCELONA. [MZBS] [*No reply.*]

MUSEO NACIONAL DE CIENCIAS NATURALES, PASEO DE LA CASTELLANA 84, MADRID. [MNMS]
Director: Dr. Emiliano Aguirre. Phone: 91-261-8600. Professional

staff: Uncertain. It is our understanding that the Instituto Espanol de Entomologia merged with the Museo Nacional to form a Department of Entomology. The figures given here combine the two collections, with about 2 million specimens housed in about 19,000 boxes. Publications sponsored: "Trabajos del Museo Nacional de Ciencias Naturales," "EOS, Revista espanola de Entomologia," and "GRAELLSIA, Revista de Entomologios Ibericos." [1986]

ESTACION BIOLOGICA DE DONANA, C/PARAGUAY 1, SEVILLA. [EBDS] [*No reply.*]

CATEDRA DE ENTOMOLOGIA, E.T.S. INGENIEROS DE MONTES, UNIVERSIDAD POLITECNICA, CIUDAD UNIVERSITARIA, MADRID (3). [MINC] [*No reply.*]

ZOOLOGY DEPARTMENT, UNIVERSIDAD DE LA LAGUNA, TENERIFE, CANARY ISLANDS. [ULCI]
[Contact: Dr. Marcos Baez. Eds. 1992]

(Spanish North Africa, several islands, possession of Spain. [*No known insect collection.*]

(Spanish Sahara, see Western Sahara.)

(Spitsbergen, Main Island of Svalbard, see Svalbard and Jan Mayen.)

(Spratly Islands, disputed islands in the South China Sea. *No known insect collection.*]

174. SRI LANKA, Democratic Socialist Republic of

[=Ceylon. Indomalayan. Colombo. **Population:** 16,639,695. **Size:** 24,886 sq. mi.]

NATIONAL MUSEUM, SIR MARCUS FERNANDO MAWATHA, COLOMBO 7. [CNMS] [*No reply.*]

175. SUDAN, Democratic Republic of the

[Afrotropical/Palearctic. Khartoum. **Population:** 24,014,495. **Size:** 967,500 sq. mi.]

SUDAN NATURAL HISTORY MUSEUM, UNIVERSITY OF KHARTOUM, GAMA AVE., P.O. BOX 321, KHARTOUM. [UKMS]
Director: Dr. Dawi M. Hamed. Phone: 81873. The insect collection contains about 1,200 specimens housed in boxes. [1986]

(Sumatra (Sumatera), island of Indonesia.)

176. SURINAME

[=Dutch Guiana. Neotropical. Paramaribo. **Population:** 395,000. **Size:** 63,037 sq. mi.]

CENTER FOR AGRICULTURAL RESEARCH, UNIVERSITY OF SURINAM, PARAMARIBO. [CARS] [*No reply.*]

BOTANICAL AND ZOOLOGICAL COLLECTION, SURINAME STATE MUSEUM, P.O. BOX 2306, MEDEDELINGEN, PARAMARIBO. [SSMS]

Director: Drs. Gloria C. Leurs. Phone: 99744. The insect collection is housed in 90 drawers. Only a few taxa are identified. There is a small collection of spiders. [1986]

UNIVERSITY OF SURINAME, TECHNOLOGISCHE FACULTEIT, LEYSWEG PARAMARIBO. [USTF] [*No reply.*]

NATIONAL ZOOLOGICAL COLLECTION OF SURINAME, DEPARTMENT OF ZOOLOGY, UNIVERSITY OF SURINAME, LEYSWEG, P. O. BOX 9212, PARAMARIBO. [NZCS]

Head curator: Dr. P. E. Ouboter. Invertebrates: Dr. B. De Dijn. Phone: (597) 465558, ext. 320; FAX: (597) 462291. The land arthropods comprise about 20,000 specimens from Suriname, with strength in Odonata, butterflies, and stingless bees. The collection has several functions in research, reference, and public education. It suffers acutely from shortage in equipment and supplies and is in need of new space and upgrading. Computerized record-keeping has begun, but the collection has, as yet, no formal networking with others. Improved connections with other collections are very much desired. (C. Starr.) [1992]

(Svalbard and Jan Mayen, island Territory of Norway, includes Spitsbergen and Bear Island. *No known insect collection.*]

177. SWAZILAND, Kingdom of

[Afrotropical. Mbabane. **Population:** 735,302. **Size:** 6,704 sq. mi. *No known insect collection.*]

178. SWEDEN, Kingdom of

[Palearctic. Stockholm. **Population:** 8,393,071. **Size:** 170,250 sq. mi.]

DEPARTMENT OF ENTOMOLOGY, NATURHISTORISKA MUSEET, BOX 7283, S-402, 35 GÖTEBORG. [GNME]

Director: Dr. Ted von Proschwitz. Phone: 46-31-145609. Professional staff: Vacant. This is primarily a collection of Swedish insects, myriapods, and spiders. There are about 800,000 specimens. Most groups have some world-wide representation. The collection is housed in boxes and vials. [1992]

MUSEUM OF ZOOLOGY, LUND UNIVERSITY, HELGONAV. 3, S-223, 62 LUND. [MZLU] [=ZILS]

Director: Prof. S. A. Bengtsson. FAX +46/46 10 45 41. Professional staff: Dr. Lennart Cederholm and Dr. Roy Danielsson. The collection contains about 2 million specimens pinned, labelled, and identified, mostly Scandinavian material. In addition there is a large amount of unidentified material. Special collections: Fallen (Diptera, Hemiptera), Dahlbom (Hymenoptera), Thomson (Hymenoptera, Coleoptera), Lindroth (Carabidae), Princis (Blattodea), Danielsson (Aphidoidea). There are 200 cabinets with about 10,000 drawers. It is estimated that there are 10,000 types. [1992]

NATURHISTORISKA RIKSMUSEET, SEKTIONEN FUR ENTOMOLOGI, S-10405 STOCKHOLM. [NHRS] [=NRS, =NREA]

Director: Prof. Edvard Sylven. Phone: 08-15-0240. Professional staff: Dr. Per Lindskog, Dr. Torbjorn Kronestedt, Dr. Per Inge Persson. The collection, including Insecta, Arachnida, and Myriapoda, contains 6 to 7 million specimens of which more than 90% are insects. The main part is pinned and systematically arranged in about 16,000 drawers in cabinets. The collection is composed of two parts, one covering the Swedish fauna and the other built up on a worldwide basis. The Swedish collection covers fairly well the Swedish fauna in most groups, especially Lepidoptera, Coleoptera, Hemiptera, and Araneae. The world collection is the greatest part, especially Coleoptera, Lepidoptera, Hemiptera, and Orthoptera. There are about 20,000 type specimens. Special collections: (Expedition material) Wahlberg, J. A., South Africa, 1838-1854; "Eugenie's" voyage around the world, 1851-1853; Nordenskjold, A. E., North Polar region and Siberia, 1878-1883; Sjostedt, Y., West Africa, Camer-oun, 1890-1892, and East Africa, Kilimandjaro-Meru, 1905-1913; Mjoberg, E., north west and east Australia, 1910-1913, and Sumatra, 1923-1924; Roman, A., South America, Amazon, 1914-1915, and 1923-1924; Malaise, R., Kamtschatka, 1920-1922, and Burma, 1934; Hedin, S. and Hummel, D., northwest China, 1927-1930. Special collection and important type holdings: Chevrolat, A. (1799-1884) collection of Curculionidae; De Geer, C. (1720-1778); Paykull, G. (1759-1826); Fallen, C. F. (1764-1830), Diptera; Schonherr, C. J. (1772-1848); Dalman, W. (1787-1828); Boheman, C. H. (1796-1868); Holmgren, A. E. (1829-1888), esp. Ichneumonidae; Thorell, T. Th. (1830- 1901), Arachnida; Stal, C. (1833-1878), Hemiptera and Orthoptera; Aurivillius, C. (1853-1928), especially Lepidoptera and Coleoptera; Sjostedt, Y. (1866-1948), African Isoptera and Orthoptera; Tullgren, A. (1874-1958), Swedish Araneae; Tragardh, I. (1878-1951), Swedish Acari; Lundblad, O. (1890-1970), Hydracaridae; Malaise, R. (1895-1978), Tenthredinidae; Forsslund, K. -H. (1900-1973), Swedish Acarida, Trichoptera, and Formicidae; Brundin, L. (1907-), Chironomidae; Sylven, E. (1920-), Cecidiomyiidae; Lundblad, O., aquatic Hemiptera.

UPPSALA UNIVERSITY, ZOOLOGICAL MUSEUM, P. O. Box 561, S-75122 UPPSALA. [UZIU]

Director: Dr. Lars Wallin. Phone: 018-182668. Professional staff: Dr.

Sten Jonsson. This is a general collection of about 500,000 specimens. It includes a Linnaean collection of 420 specimens; C. P. Thunberg's collection of 30,000 (including 1,500 types), L. Gyllenhaal's collection of 10,000 (incl. unchecked number of types), Å. Holm's spider collection, 20,000 tubes (about 100,000 specimens, including 500 types). [1992]

179. SWITZERLAND, Swiss Confederation

[Palearctic. Berne. **Population:** 6,592,558. **Size:** 15,943 sq. mi.]

ENTOMOLOGY DEPARTMENT, NATURHISTORISCHES MUSEUM, AUGUSTINERGASSE 2, 4001 BASEL. [NHMB]

Director: Dr. Michel Brancucci. Phone: 061-25-82-82. Professional staff: Dr. Richard Heinertz. The insect collection contains 2.5 million specimens; (The spider collection 2,600 is housed in the Zoological Department; it contains 30,000 specimens in alcohol representing 2,600 species. *Contact: Dr. Ambros Hänngi in the Zoology Department.*) The insect collection is housed in drawers. Special collections: Coleoptera collections from: Lautner, 40,000 Central Europe; Mandl, 30,000 Carabinae; Klapperich, 40,000 Dominican Republic; Gehrig, 40,000 Europe; Allenspach, 35,000 Switzerland; Gottwald, 35,000 Central Europe; Wittmer, 150,000 Cantharidae, Malachiidae, Phengodidae, Drilidae, Dasytidae, Karumiidae, and Cleridae of the world; Hymenoptera: Santschi, Formicidae of the world; Diptera: Keiser, world; Lepidoptera: Courvoisier, 10,000 Lycaenidae world; Beuret, 21,000 Lycaenidae, Holarctic; Paravicini, Heterocera, Holarctic; Gutzwiller, butterflies, world. Publications sponsored: "Fauna of Saudi Arabia," and "Entomologica Basiliensia." [1992]

NATURISTORISCHE MUSEUMS, BERNASTRASSE 15, CH-3005 BERN. [NMBS]

Director: Prof. Dr. Marcel Güntert. Phone: (41-31) 48-22-22. Professional staff: Dr. Charles Huber. The collection contains about 1.25 million specimens pinned and 120,000 alcohol vials (60,000 insects and 60,000 spiders). Special collections include: Emil Goeldi, Carl Vorbrodt, Hans Pochon, Anton Schmidlin, August Raetzer, Hans Bangerter, Otto P. Wenger, Werner Moser, and Emil Frey-Gessner. Publications sponsored: "Jahrbuch, Naturhistorisches Museum Bern." [1992]

MUSÉE d'HISTOIRE NATURELLE, CH-2300 LA CHAUX-DE-FONDS. [MHNC]

Director: Marcel S. Jacquat. Phone/FAX: 41 39 23 39 76. The collection consists of about 50,000 specimens of insects and spiders, mostly from Europe, Asia, and Oceaniana.

MUSÉUM, d'HISTOIRE NATURELLE, CASE POSTALE 434, CH-1211 GENEVA 6. [MHNG]

Director: Dr. Volker Manhnert. Phone: 022-735-91-30. Professional staff: Dr. Daniel Burckhardt, Dr. Ivan Löbl, Dr. Bernd Hauser, Dr. Charles Lienhard. The collection is housed in 20,000 boxes, with a total of over 2.5 million specimens of insects. Among the special collections of insects are in Coleoptera those of G. Benick, L. C. Genest, H. John, A. Melly, J. Ochs, R. Petrovitz, L. Reiche; in Lepidoptera of G. Audéoud, C. Blachier, B. Laporte, J. Plante, J.-L. Reverdin, J. Romieux; in Hymenoptera of A. Forel, E. Frey-Gessner, H. de Saussure, H. Tournier; in Orthoptera of K. Harz, A. Nadig; in Siphonaptera of F. Peuss; and in Arachnida of R. de Lessert. Publications sponsored: "Revue Suisse de Zoologie," and "Archives des Sciences." [1992]

ENTOMOLOGICAL COLLECTION, MUSÉE ZOOLOGIE PLACE DE LA RIPONNE 6, LAUSANNE 17. ZERN. [MZLS]

Director: Dr. Pierre Goeldlin. Phone: 21-312-8336;FAX 21-23-6840. Professional staff: Dr. Daniel Cherix, and Michel Sartori. The collection is housed in about 3,500 boxes. Among the spider collections are those of A. Aeschlimann (Ixodoidea), and C. E. Ketterer (Arachnida). Special collections include: de la Harpe (Microlepidoptera), V. Nabokov (Lepidoptera), P. Basilewsky and P. Letellier (Coleoptera), N. Cerutti (Heteroptera), J. Aubert (Plecoptera), Ch. Degrange and M. Sartori (Ephemeroptera), P. Goeldlin (Diptera) J. de Beaumont, J. -F. Aubert, and H. Kutter (Hymenoptera). The main part of the collection is constituted by material from Switzerland and Europe. Publication sponsored: "Bulletin Romand d'Entomologie." [1992]

INSECT COLLECTION OF NATUR-MUSEUM LUZERN, KASERNENPLATZ 6, CH-6003, LUZERN. [NMLS]

Director: Dr. Peter Herger. Phone: 041-24-54-11. The collection contains about 250,000 pinned and labelled specimens housed in 2,000 drawers. About 80,000 of these are identified, and there are over 1,000 vials of specimens and 20,000 envelopes of processed specimens; approximately 500 slides and half a million specimens awaiting preparation. Some holotypes and paratypes are housed in the collection. The main part of the collection is material from central Switzerland, especially Lepidoptera and Coleoptera. Publication sponsored: "Entomologische Berichte Luzern." [1986]

ENTOMOLOGISCHES INSTITUT, EIDGENOSSISCHE TECHNISCHE HOCHSCHULE-ZENTRUM, UNIVERSITATSSTRASSE 2, CH-8006, ZURICH. [ETHZ]

Director: Prof. Dr. W. Sauter. Phone: 01-256-39-22. This collection of insects is mainly from Switzerland and the western Palearctic Region, but some is from other parts of the world. Special collections by A. von Schulthess Rechberg, E. Escher-Zollikofer, Max Banninger, G. Huguenin, G. Schoch, and R. Biedermann are represented. [1986]

180. SYRIA, Syrian Arab Republic

[=United Arab Republic. Palearctic. **Population:** 11,569,659. **Size:** 71,043 sq. mi. *No known insect collection.*]

(Tahiti, see French Polynesia.)

181. TAIWAN, Republic of China

[Present administration includes island of Taiwan (=Formosa), P'eng-Hu (Pescadores) as well as small islands along the mainland coast. Note: Even though we have made direct contact with some of these institutions, we have been unable to get any information about most of these collections from the directors. Indomalayan. Taipei. **Population:** 20,004,391. **Size:** 13,900 sq. mi. Most of the following information was kindly supplied by Christopher K. Starr.]

NATIONAL MUSEUM OF NATURAL SCIENCE, 1 KUAN CHIEN RD., TAICHUNG 400. [NMNS]

Curator: Dr. Cheng-Shing Lin. Phone: (04) 322-6940. Professional staff: Dr. Chung-Chu Chiang, Mr. Kuen-woei Huang, and Mr. E. C. Haio. This is the most recent collection in Taiwan, started in 1985, but it is also the most ambitious. The pinned, sorted collection contains about 36,000 specimens, with probably another 10% in the process of rapid curation. It is strongest in Homoptera, moths, and aculeate Hymenoptera. The facilities are modern, including steel cabinets mounted on a compactor track in a very large, climatically controlled room. The collection is actively being increased. Publications: "Bulletin of the National Museum of Natural Science" and "Special Publications of the National Museum of Natural Sciences." [1992]

TAIWAN FORESTRY RESEARCH INSTITUTE, 53 NAN-HAI RD., TAIPEI. [TFRI]

Curator: Mrs. Yu-cheng Chang. In this collection of about 15,000 specimens of forest insects, about 10,000 are Lepidoptera. Another major component of the collection is Homoptera, especially aphids and scale insects. A majority of the species of scales in Taiwan are believed to be represented here. Among the research priorities is the establishment of a data bank on forest insects, making collecting in different forest habitats a continuing effort. [1992]

DEPARTMENT OF ZOOLOGY-ENTOMOLOGY, NATIONAL TAIWAN UNIVERSITY, TAIPEI. [NTUC]

Curator: Dr. Tung-ching Hsu. The pinned collection of about 238,000 specimens occupies 2,300 drawers in a single very large room. The collection is used both for teaching and research; probably two-thirds of the total is for research. The type collection contains about 2,000 specimens. Most the collection was made by Japanese collectors or students since the occupation. The bulk of the collection of Coleoptera, Lepidoptera, and Hymenoptera has been transferred to TARI. This collection and that of TARI include the only significant holding of historically important specimens in Taiwan. [1992]

INSECT COLLECTION, TAIWAN AGRICULTURAL RESEARCH INSTITUTE, 189 CHUNGCHENG ROAD, WUFENG, TAICHUNG 41301. [TARI]

Curator: Mr. Liang-yih Chou. Phone (04) 3302301. Professional staff includes Mr. Shan-jen Fang. Collection size approximately 1.8 million specimens, with 1.2 million of these collected since 1980. About 70% of the recent collections are Hymenoptera, mostly parasitic. Homoptera is next in size, over 120,000 specimens, followed by Coleoptera at over 58,000 and Hemiptera at over 50,000. The main collection occupies two very large rooms, with the type collection separated, in a smaller adjoining room. This is the largest insect collection in Taiwan. [1986, 1987, updated by Eds. 1992.]

DEPARTMENT OF ENTOMOLOGY, NATIONAL CHUNG HSING UNIVERSITY, 250 KUO-KUANG RD., TAICHUNG. [NCHU]

Curator: Dr. Chung-tu Yang. The collection is in three distinct, separately maintained parts: A general collection under Dr. Yang's supervision, a research collection of Homoptera of about 5,000 specimens under Dr. Yang, and a currently inactive research collection of longhorn beetles (information may be obtain from the Department Chair. The general collection of about 76,000 taxonomically useful specimens obtained from the teaching collection, and tends to be mainly lowland specimens. [1992]

BIOLOGY INSTITUTE, TUNGHAI UNIVERSITY, BOX 851, TAICHUNG. [TUTC]

Director: Dr. Chin-Seng Chen. Phone: 886-4-359-0249. Professional staff: Mr. Y. T. Hwang, Mr. T. Y. Chen, Mr. L. P. Hsu, Ms. H. J. Lee. The collection consists of the following: Protura, 20 species (including 4 holotypes); Collembola, 50 species; Trichoptera, 50 species (including 6 holotypes), and Diptera, Ceratopogonidae, 60 species of *Culicoides*, Psychodidae, 50 species (including 3 holotypes), Culicidae, 60 species. [1992]

TAIWAN PROVINCIAL MUSEUM, NO. 2 SIANG-YANG ROAD, TAIPEI. [TMTC]

Curator: Mr. Hsiau-yue Wang. An active collection of unknown size, but contains material from "Orchid Island" (=Botel Tobago), an extremely interesting oceanic island east of the southern tip of Taiwan. [1992]

DIVISION OF PLANT PROTECTION, TAIWAN PROVINCIAL GOVERNMENT, DEPARTMENT OF AGRICULTURE, NAN-TOU. [DPPC] [*No reply.*]

182. TANZANIA, United Republic of

[Includes Zanzibar and Pemba. Afrotropical. Dar es Salaam. **Population:** 24,295,250. **Size:** 342,102 sq. mi.]

UNIVERSITY OF DAS ES SALAAM, P. O. BOX 35064, DAR ES SALAAM. [UDSM]

Director: Head, Department of Zoology and Marine Biology. Phone: 49058. The collection contains a few Lepidoptera and Acrididae housed in drawers and boxes.

183. THAILAND, Kingdom of

[=Siam. Indomalayan. Bangkok. **Population:** 54,588,731. **Size:** 198,115 sq. mi.]

ENTOMOLOGY AND ZOOLOGY DIVISION, DEPARTMENT OF AGRICULTURE, PATHOYOTHIN ROAD, BANGKHEN, BANGKOK 10900. [EMBT]

Director: Dr. Angoon Lewvanich. Phone: (02) 579-4128. Professional staff: Mrs. Waree Hongsaprug, Mrs. Boopa Laosinchai, Mrs. Sommai Chunram, and Mrs. Wipada Vungsilabutr. The collection consists of approximately 200,000 insect specimens representing 7,000 identified species, and 10,000 unidentified species. The spider collection consists of 10,000 specimens, 100 identified species, and 500 unidentified. [1992]

(Tibet, see People's Republic of China.)

(Tierra del Fuego, see Argentina.)

(Timor, island of Indonesia.)

(Tobago Island, see Trinidad and Tobago.)

184. TOGO, Republic of

[Afrotropical. Lomé. **Population:** 3,336,433. **Size:** 21,925 sq. mi. *No known insect collection.*]

(Tokelau Islands, New Zealand territory, includes Atafu, Fakaofu, and Nukunonu atolls. *No known insect collection.*]

185. TONGA, Kingdom of

[Oceanian (Polynesia). Nuku'alofa. **Population:** 99,620. **Size:** 289 sq. mi. *No known insect collection.*]

(Torres Islands, see Vanuatu.)

(Transkei, see South Africa.)

186. TRINIDAD AND TOBAGO

[Neotropical. Port of Spain. **Population:** 1,279,920. **Size:** 1,980 sq. mi.]

CARIBBEAN EPIDEMIOLOGY CENTRE, 16 JAMAICA BLVD. PORT OF SPAIN. [CARE]

Entomologist: Samuel Rawlins. Phone: (809) 622-4261, 622-4262, 662-4745. This is a reference collection of haematophagous and other medically important insects and arachnids, occupying a standard cabinet, vials, and slides. (C. Starr.) [1992]

INTERNATIONAL INSTITUTE OF BIOLOGICAL CONTROL, GORDON STREET, CUREPE. [CIBC]

Director: Dr. Peter Baker. Phone: 662-4173. Professional staff: Mr. Michael Morais. The collection contains about 30,000 specimens as a general collection of Trinidad species with emphasis on agricultural pests and their natural enemies plus stronger representation of Lepidoptera, Chalcidoidea, bees, and Coccinellidae assembled by staff members with special interests. The collection is housed in unit trays in drawers in cabinets. [1992]

NATIONAL MUSEUM AND ART GALLERY, 117 FREDERICK ST., PORT-OF-SPAIN. [NMTT] [*No reply.*]

INSECT COLLECTION, DEPARTMENT OF BIOLOGICAL SCIENCES, UNIVERSITY OF THE WEST INDIES, ST. AUGUSTINE. [UWIC]

Curator: Dr. Christopher K. Starr. Phone: (809 663-1334, ext 2046 or 2047. The collection consists of 118 drawers of mainly Caribbean insects, with emphasis on Trinidad, in addition to the Bacart butterfly collection. The collection was until recently largely inactive for many years, although good preservation was maintained. The material includes a number of paratypes, mainly Michel's species of Mutillidae. However, the collection does not maintain primary types but deposits elsewhere any primary types received or arising out of existing holdings. [1992]

INSECT COLLECTION, CENTRAL EXPERIMENT STATION, MINISTRY OF AGRICULTURE, LANDS AND FOOD PRODUCTION, CENTENO, VIA ARIMA PO. [CEST] [*No reply.*]

(Tristan da Cunha, Dependency of St. Helena.)

(Trust Territory of the Pacific Islands, see Belau, Federated States of Micronesia, Marshall Islands, and Northern Mariana Islands.)

(Tuamotu-Gambier Archipelago, see French Polynesia.)

(Tubuai Islands, see Austral Islands, French Polynesia.)

187. TUNISIA, Republic of

[Palearctic. Tunis. **Population:** 7,738,026. **Size:** 63,170 sq. mi.]

COLLECTION DR. NORMAND, DEPARTMENT OF AGRICULTURE, TUNIS. [NDAT] [*No reply.*]

188. TURKEY, Republic of

[Palearctic. Ankara. **Population:** 54,167,857. **Size:** 300,948 sq. mi.]

ATATURK UNIVERSITESI, ERZURUM. [AUTC] [*No reply.*]

189. TURKS AND CAICOS ISLANDS

[British colony. Neotropical. Jamestown. **Population:** 9,295. **Size:** 166 sq. mi. *No known insect collection.*]

190. TUVALU

[British dependency, includes the islands of Nanumanga, Nanumea, Nui, Niutao, and Vaitupu; and those islands claimed by the U.S.A: Funafuti, Nukufetau, Nukulailai, and Nurakita. Formerly known as Ellice Islands. (Oceanian (Polynesia). Funafuti. **Population:** 8,475. **Size:** 10 sq. mi. *No known insect collection.*]

191. UGANDA, Republic of

[Afrotropical. Kampala. **Population:** 16,446,906. **Size:** 76,084 sq. mi.]

NATURAL HISTORY, UGANDA MUSEUM, 5-7 KIRA ROAD (P. O. BOX 365), KAMPALA. [UMKU]

Director: Augustine Wanzama. Phone: 041-32707. Professional staff: L. G. Nkata, E. R. Kateeba, S. Chalo, S. Musango, J. Ssebadduka, J. Nzabonimpa. The collection consists of about 1,000 specimens housed in drawers and boxes. These are mostly for display purposes. [1986]

HERBARIUM AND INSECT COLLECTION, KAWANDA AGRICULTURAL RESEARCH STATION, P. O. BOX 7065, KAMPALA. [KARS] [*No reply.*]

TSETSE CONTROL DEPARTMENT MUSEUM COLLECTION, MINISTRY OF ANIMAL INDUSTRY AND FISHERIES, P. O. BOX 7033, KAMPALA. [TCDU]

Director: Dr. T. N. Kangwagye. Professional staff: S. C. U. Bikingi-Wataaka, Joseph Serunjoji, and L. D. Semakula. The collection contains families of biting flies. The bulk of the collection of about 200,000 specimens are Glossinidae and Tabanidae which are housed in drawers in cabinets and in boxes. [1986]

(Ukraine, now separate, but included here under Russia.)

(Umm al Qaiwain, see United Arab Emirates (Palearctic).)

(Union of Soviet Socialists Republics, see Russia.)

192. UNITED ARAB EMIRATES

[Includes Abu Dhabi, Dubai, Sharjah, Ajman, Umm al Qaiwan, Ras al Khaimah, and Fujeira. Palearctic. Abu Dhabi. **Population:** 1,980,354. **Size:** 30,000 sq. mi. *No known insect collection.*]

193. UNITED KINGDOM of Great Britian and Northern Ireland

[Includes Scotland, Wales, Orkney, Shetland, and Channel islands and the Isle of Man. Palearctic. London. **Population:** 56,935,845. **Size:** 94,249 sq. mi. *The museum collections in Great Britian are extensive, and are being described in several volumes. Those listed here are a sample, but we believe we have included all of the major collections.*]

Bolton

BOLTON MUSEUM, LE MANS CRESCENT, BOLTON, LANCASHIRE, BLI ISE. [BMUK]

Director: Dr. J. R. A. Gray. Phone: (0204) 22311, ext. 379/361. Professional staff: Mr. E. G. Hancock. The collection contains about 200,000 specimens, the large majority of them British, though not particularly local to the Bolton area. Special collections include those of Dr. P. B. Mason, Francis Walker, Benjamin Cooke, Alfred Beaumont, F. W. Edwards, George E. Hyde, C. H. Schill, F. C. Adams, A. L. Montandon, E. A. Brunetti, G. H. Verrall, A. Ford, J. Gray, R. Hamlyn-Harris, F. Smith, O. Schmeiderknechte, E. Saunders, Dr. Capron, Rev. F. D. Morice, P. Cameron, Dr. R. Brauer, Edwin Brown, G. R. Crotch, C. C. Dupre, Prof. A. Foerster, G. Lewis, A. Matthews, W. G. Pelerin, E. C. Rye, and D. Sharp. The collection is stored in cabinets and drawers in unit trays, and some in boxes, or alcohol vials. [1992]

Brambler

NATIONAL BUTTERFLY MUSEUM (SARUMAN MUSEUM), BRAMBLER. [NBME] [*No reply; may no longer exist.*]

Brighton

BRIGHTON BOROUGH COUNCIL, BOOTH MUSEUM OF NATURAL HISTORY, 194 DYKE ROAD, BRIGHTON, EAST SUSSEX BN1 5AA. [BMBN]

Phone: 0273-552586. Keeper of Biology, Dr. Gerald Legg. Professional staff: Mr. J. Cooper, Mr. J. M. Adams. The collection contains over 500,000 specimens of insects and Arachnida housed in cabinets and drawers, including about 650 types of foreign macrolepidoptera. The collection is dominated by butterflies and contains specimens of the various families from most regions of the world. There is strong emphasis on the Nymphalidae, particularly those specimens in the collection of the late Arthur Hall made between 1900 and 1939. [1992]

Bristol

DEPARTMENT OF NATURAL HISTORY, THE CITY MUSEUM AND ART GALLERY, QUEEN'S ROAD, BRISTOL BS8 1RL. [CMBK]

Curator of Natural History: Mr. Ray Barnett; Conservator of Natural History Objects, Mr. Derek Foxwell. Phone: 0272-223571. Approximately 1600 cabinet drawers and 450 boxes comprise this insect collection, 60% of which is British Lepidoptera. The major collections are those of C. Bartlett, Rev. J. W. Metcalfe, G. B. Consy, and I. R. P. Heslop, much of which is of local significance to the Bristol region. Other important collection are: Diptera of H. L. F. Audcent, again mainly from around Bristol and tropical Coleoptera of Stephen Barton. The latter containing Buprestidae types (12 species). [1992]

Cambridge

UNIVERSITY MUSEUM OF ZOOLOGY INSECT COLLECTION, DOWNING STREET, CAMBRIDGE, CAMBRIDGESHIRE CB3 2EJ. [CUMZ]

Director: Dr. K. A. Joysey. Phone 358717. Professional staff: Dr. R. H. L. Disney (Phoridae), Dr. W. A. Foster. The collection includes 660,000 specimens, including the Crotch collection of Coccinellidae and Erotylidae, the Percy Sladen collection from an expedition to the Indian Ocean (1909), and the Haviland collection of termites. They are housed in drawers, boxes, and vials. [1986]

Cardiff

SUBDEPARTMENT OF ENTOMOLOGY, DEPARTMENT OF ZOOLOGY, NATIONAL MUSEUM OF WALES, CATHAYS PARK, CARDIFF, S. GLAMORGAN, WALES CF1 3NP. [NMWC]

Director: Dr. Douglas A. Bassett. Phone: (0222) 397951. Professional staff: Adrian F. Amsden, head of Entomology, and John C. Deeming. The collection of over 575,000 specimens is divided into two series, British and non-British, including a small collection of insects imported through the docks and elsewhere maintained separately. The collection is housed in cabinets and drawers. The library includes the Willoughby Gardner bequest of early printed books mainly on entomology from 1481-1800. Insect specimens may be borrowed for study. [1986]

Dorchester

DORSET COUNTY MUSEUM, NATURAL HISTORY RE-SEARCH COLLECTION, HIGH WEST STREET, DORCHESTER, DORSET DT1 1XA. [DORC]

Director: Mr. Richard De Peyer, MA, FMA. Phone: 0305-262735. Professional staff: Ms. Kate Hebditch, B. S. The collection contains over 100,000 specimens, mostly from Dorset, housed in drawers in cabinets or in boxes. Publication sponsored: "Proceedings of the Dorset Natural History and Archaeological Society." [1992]

Edinburgh

INSECT COLLECTION, DEPARTMENT OF NATURAL HISTORY, NATIONAL MUSEUMS OF SCOTLAND, CHAMBERS ST.,

EDINBURGH EH1 1JF, SCOTLAND. [RSME]

Director: Dr. Mark R. Shaw (Keeper of Natural History), Dr. Graham E. Rothway (Curator of Entomology). Phone: 031-225-7534. Professional staff: Miss Suzanne I. Baldwin. The collection contains about 1.5 million curated specimens about half of which are British. Most of the non-British material is from the Old World and particular strengths are Odonata, Neuroptera, and Lepidoptera; other groups include a high proportion of the unidentified and sometimes unsorted material. The British collections are comprehensive and rich in Scottish specimens. There are about 500 types. The spider collection comprises about 2,000 specimens, representing about 60% of the British species, with only about 300 non-British specimens and no types. There is also a collection of about 2,000 British mites, 250 British pseudoscorpions, and 400 British opiliones. Special collections: K. J. Morton, W. A. F. Balfour-Browne, J. R. Malloch, E. C. Pelham-Clinton, C. G. M. de Worms, A. Richardson, H. J. Adams, and L. Dufresne. Unusual strengths include reared parasitic wasps (especially Ichneumonidae and Braconidae) and the larval stages of hover flies. The collection is mostly housed in cabinets and boxes, with slides, and alcohol vials, as appropriate. [1992]

Glasgow

HUNTERIAN MUSEUM: ZOOLOGY SECTION, DEPARTMENT OF ZOOLOGY, UNIVERSITY OF GLASGOW, GLASGOW G12 8QQ, SCOTLAND. [HMUG]

Director: Mr. Malcolm McLeod. Phone: 041-339-8855, ext. 4772 (assistant curator's number). Professional staff: Dr. Ron Dobson (retired), Honorary Curator, Entomology; Dr. Roy Crowson (retired), past Honorary Curator, Entomology; Ms. Margaret Reilly (assistant curator of all collections). The collection probably exceeds 1.5 million specimens of all orders of insects and arachnids. It is strong in British insects, European and tropical Coleoptera, Lepidoptera, and Diptera There is an alcohol collection (*ca.*500 jars), including spiders from the Crowson collection. Special collections: William Hunter collection which contains many Fabrician types (part catalog of the Coleoptera types published by R. A. Staig); J. J. F. X. King collection of British insects; F. V. Theobald mosquitoes; and the T. G. Bishop collection of Coleoptera (contains types). The collections are housed in drawers, boxes, and alcohol vials. [1992]

Haslemere

HASLEMERE EDUCATIONAL MUSEUM, HASLEMERE, SURREY. [HEMS] [*No reply.*]

Ipswich

IPSWICH MUSEUM, HIGH STREET, IPSWICH IP1 3QH. [IPSM]

Curator: Mrs. S. Muldoon. Phone (0473) 213761. Professional staff: Mr. H. Mendel. The collection houses about 250,000 specimens, including many types and voucher specimens, mostly of British insects, but about

2,000 specimens of exotic Lepidoptera are included. Special collection: C. Morley collection of British Insects. The collection is housed in drawers and boxes. [1986]

London

DEPARTMENT OF ENTOMOLOGY, AND DEPARTMENT OF ZOOLOGY [FOR SPIDERS], THE NATURAL HISTORY MUSEUM, LONDON SW7 5BD. [BMNH]

Head, Collections Management Division: Dr. Mike Fitton Phone: 071-938-9446. Professional staff: About 40 curators. The collection, in one 6 story building comprises 30 million specimens, representing one-half million species, of which 250,000 are represented by primary types. [This is the third largest collection in the world. Eds.] The largest orders represented are Lepidoptera (10 million) and Coleoptera (6 million), but the Homoptera, Diptera, and Hymenoptera in particular have been expanded considerably in recent years (1.8 million each). [There are, of course, many special collections represented here, but a list was not supplied. Obviously, it would be very long, and this information is readily available elsewhere. Eds.] Publications sponsored: "Bulletin of the British Museum (Natural History), Entomology Series." [1992]

Affiliated Collection:

Gowing-Scopes, Eric, Rosewood, Stonehouse Road, Halstead, Kent. TN14 7HN, England. [EGSC] This collection is primarily Curculionidae and Lucanidae of the World, with 2,000 species of weevils of the world, including 50 paratypes, 1,000 species of Australian weevils, and 100 species of Lucanidae of the world. The collection is housed in 100 drawers in 10 cabinets. (Registered with BMNH.)

BRITISH ENTOMOLOGICAL AND NATURAL HISTORY SOCIETY, ALPINE CLUB, 74, SOUTH AUDLEY STREET, LONDON W1. [BENH] [*No reply.*]

LINNEAN SOCIETY, BURLINGTON HOUSE, PICCADILLY, LONDON WIV OLQ. [LSUK]

Director: No staff particularly associated with the collections. The current zoology curator is Dr. M. G. Fitton of the British Museum of Natural History. Phone: 071-434-4479. The insect collection of Carolus Linnaeus and associated material added in the late 18th and early 19th centuries by J. E. Smith and others are held here. As well as insects, the collection includes arachnids, etc., the Aptera of Linneaus. The collection is housed in glass-topped boxes. Details about the use of the collection may be obtained from the Society. [1992]

Manchester

DEPARTMENT OF ENTOMOLOGY, MANCHESTER MUSEUM, THE UNIVERSITY, MANCHESTER M13 9PL. [MMUE]

Director: Colin Johnson (Keeper). This is the third largest collection (about 3 million specimens) in the British Isles. The collection of British insects covers all orders and is extremely comprehensive (about 1.25 million specimens), with extensive material from northern England. Alcohol material, which also includes immature stages, is also represented. The foreign collections (about 1.75 million specimens) are world-wide in scope, with Coleoptera (about 950,000 specimens) and Lepidoptera (with about 750,000 specimens) best represented. The collection is housed in about 4,250 drawers, boxes, and cartons, plus plastic boxes containing tubes of alcohol material. There are about 10,500 types representing over 2,350 species, half of which are Spaeth Cassidinae. [1992]

Norwich

NORWICH CASTLE MUSEUM, NORWICH, NORFOLK. [NCMK] [*No reply.*]

Nottingham

NOTTINGHAM NATURAL HISTORY MUSEUM, WALLOTON HALL, WALLOTON PARK, NOTTINGHAM, NOTTINGHAM NG8 2AE. [NHMN]

Professional staff: Dr. Sheila Wright. Phone: 0602-281333 and 0602-281130. The collection consists of approximately 360,000 specimens, mostly British Coleoptera, Lepidoptera, Diptera, and Hymenoptera, but also contains worldwide systematic collection, including spiders. [1992]

Oxford

HOPE ENTOMOLOGICAL COLLECTIONS, UNIVERSITY MUSEUM, PARKS ROAD, OXFORD OXI 3PW. [OXUM] [=UMO]

Director: Dr. George McGevin. Phone: 0865-272950. Within Britain, the collections are second only in size and importance to the collection in the Natural History Museum. The British collections are arranged synoptically. The foreign collections contain 20,000 types, especially of Lepidoptera (F. Walker). Many other special collections are housed. [1986]

Perth

PERTH MUSEUM AND ART GALLERY, GEORGE ST., PERTH, TAYSIDE, PHI 5LB. [PMAG]

Director: Mr. J. A. Blair. Phone: 0738-32488. Professional staff: M. A. Taylor. The collection consists of about 100,000 specimens, including a few types, mainly Lepidoptera and Heteroptera, mainly from Scotland, housed in drawers. [1986]

Priestgate

PETERBOROUGH MUSEUM AND ART GALLERY, PRIESTGATE, PETERBOROUGH, PEI ILF. [PMAU]

Director: Martin Howe. Phone: 0773-347329. Professional staff: Dr.

Gordon Chancellor. Number of specimens approximately 60,000, housed in 30 columns of drawers. The bulk of the collection is local Lepidoptera, with a smaller collection of beetles. There are also some bugs, Ephemeroptera, etc. and some spiders. [1992]

South Croydon

CROYDON NATURAL HISTORY AND SCIENTIFIC SOCIETY, 96A BRIGHTON ROAD, SOUTH CROYDON, SURREY CR2 6AD. [CNHS] [*No reply.*]

York

THE YORKSHIRE MUSEUM, MUSEUM ST., YORK, YORKSHIRE. [YMUK] [*No reply.*]

194. UNITED STATES OF AMERICA

[Includes fifty states and the District of Columbia. See under name of possession or territory for details about other collections. Nearctic. Washington, D.C. **Population:** 247,818,000. **Size:** 3,618,770 sq. mi.]

Several of the large insect collections have been listed in order of rank for number of specimens. It is doubtful if actual counts of any of these have been made. However, those collections with one million or more specimens (as reported in the description of collection; some of these did not send in updated collection descriptions, hence they date back to 1986) are listed here in order:

30,000,000 National Museum of Natural Science
16,000,000 American Museum of Natural History
13,000,000 Bishop Museum
9,000,000 Field Museum of Natural History
7,000,000 Harvard University
7,000,000 California Academy of Sciences
6,500,000 Florida State Collection of Arthropods
6,500,000 Carneige Museum of Natural History
6,000,000 Illinois Natural History Survey
6,000,000 University of California, Davis
5,000,000 Los Angeles County Museum
4,500,000 University of Michigan
4,500,000 University of California, Berkeley
4,000,000 Cornell University
4,000,000 Ohio State University
3,000,000 University of Kansas
3,000,000 Philadelphia Academy of Natural Science
2,700,000 University of Minnesota
2,000,000 University of Idaho
2,000,000 Utah State University
1,750,000 University of Nebraska
1,650,000 Brigham Young University
1,500,000 Purdue University
1,200,000 Northern Arizona University
1,200,000 Colorado State University
1,100,000 Yale University
1,100,000 University of Missouri
1,030,000 Iowa State University
1,000,000 Texas A & M University
1,000,000 Washington State University

ALABAMA

Auburn

AUBURN UNIVERSITY ENTOMOLOGICAL MUSEUM, DEPARTMENT OF ENTOMOLOGY, AUBURN UNIVERSITY, AUBURN, AL 36849. [AUEM]

Director: Dr. Wayne E. Clark. Phone: (205) 844-2565. Professional staff: Dr. Michael L. Williams. The collection is a part of the research projects of the Alabama Agricultural Experiment Station. It was initiated in 1924 with the donation of about 80 Schmitt boxes of insect specimens, mostly Coleoptera, by Dr. Henry Good. It is housed in drawers in cabinets, and on slides. There are in excess of 245,000 pinned, alcohol pre served, and slide mounted specimens, mostly North American, but recent specimens from the Neotropical Region have been added. The collection is particularly strong in Coleoptera, Orthoptera, Odonata, Homoptera, especially Coccoidea, and Siphonaptera. It incorporates the R. F. Wilkey Coccoidea collection. [1992]

Tuscaloosa

INSECT COLLECTION, ALABAMA MUSEUM OF NATURAL HISTORY, UNIVERSITY OF ALABAMA, P. O. BOX 870340, TUSCALOOSA, AL 35487-0340. [UANH]

Assistant Director: Dr. John C. Hall. Phone: (205) 348-7550; FAX: (205) 348-9292. This collection is maintained by the Department of Biology, University of Alabama [*see below*.]

INSECT COLLECTION, DEPARTMENT OF BIOLOGICAL SCIENCES, UNIVERSITY OF ALABAMA, TUSCALOOSA, AL 35487. [UABD]

Director: Dr. G. Milton Ward. Phone: (205) 348-1798; FAX: (205) 348-1786. This collection is maintained in support of the research and teaching efforts of the Aquatic Biology Program in the Department of Biological Sciences. The collection specializes in state and regional fauna, and consists of both terrestrial and aquatic specimens, with an emphasis on the immature and adult forms of aquatic species. The pinned collection is maintained in approximately 120 Cornell drawers, and the alcohol preserved specimens in approximately 10,000 vials. Among aquatic taxa, the U.A. insect collection has a large number of Trichoptera, especially Hydroptilidae. The terrestrial collection is strongest in Coleoptera as a result of the maintenance of the Loding beetle collection. The Loding collection, owned by the Alabama Museum of Natural History [*ee above*], contains mostly Alabama species, but also contains many other specimens from around the U.S.A. Most of the collection dates from the 1920s and 1930s, with some specimens collected around 1900. [1992]

THE GEOLOGICAL SURVEY OF ALABAMA, P. O. BOX 0, TUSCALOOSA, AL 35486. [GSAT]

Director: Dr. Steve Harris. This is a collection of aquatic insects. [*Too late to receive reply in time for this edition.*]

ALASKA

[*No known collection of insects in Alaska.*]

ARIZONA

Flagstaff

COLLECTION OF INSECTS, BOX 5640, BIOLOGY, NORTHERN ARIZONA UNIVERSITY, FLAGSTAFF, AZ 86011. [NAUF]

Director: Dr. C. Dan Johnson. Phone: (602) 523-2505. Professional staff: Dr. R. S. Beal, Dr. C. N. Slobodchikoff. About 1.2 million specimens are housed in 110 cabinets with 1,320 drawers. Coleoptera are best represented with excellent ecological supporting data for the Bruchidae. Insects are especially well represented from the Southwest, Mexico, Central and South America. [1986]

Grand Canyon

SCIENTIFIC STUDY COLLECTION, GRAND CANYON NATIONAL PARK, GRAND CANYON, AZ 86023. [GCNP]

Curator: Carolyn Richard. Phone (602) 638-7769. There are approximately 6,000 insects in the collection, including numerous butterflies, moths, and beetles. There are a few type specimens from Grand Canyon subspecies. [1992]

Portal

SOUTHWESTERN RESEARCH STATION OF THE AMERICAN MUSEUM OF NATURAL HISTORY, PORTAL, AZ 85632. [SWRS]

Director: Dr. Wade C. Sgerbrooke. Phone: (602) 558-2396. The collection consists of about 14,000 specimens (no long series) of the insects of the Chiricahua Mountains and surrounding valleys. It is housed in drawers in cabinets. Most of the material is identified at least to genus. There is a good collection of arachnids from the area, all identified to species [1992].

Tempe

FRANK M. HASBROUCK INSECT COLLECTION, DEPARTMENT OF ZOOLOGY, ARIZONA STATE UNIVERSITY, TEMPE, AZ 85281-3571. [ASUT]

Director: Dr. Michael E. Douglas. Phone: (602) 965-3571. Curator Emeritus: Dr. Mont A. Cazier. The collection contains approximately 400,000 specimens of insects (no spiders or other arthropods). The major portion of the collection consists of Coleoptera, Lepidoptera, Hymenoptera, and Diptera. Smaller orders are represented but have not been strongly developed. There is a large, well developed collection of Arizona cicadas, robber flies, butterflies, skippers, tiger beetles, and a slide collection of Arizona mosquitoes. The collection is housed in 112 metal cabinets with 1,344 drawers. [1992]

Tucson

DEPARTMENT OF ENTOMOLOGY COLLECTION, UNIVERSITY OF ARIZONA, TUCSON, AZ 85721. [UAIC]

Director: *Unknown*. Phone: (602) 621-1635. Professional staff: Carl A. Olson. North American species of Meloidae, Anthicidae, Buprestidae, Chrysomelidae, Scarabaeidae, Cerambycidae, Curculionidae, plus Arizona species of Orthoptera, aquatic and semiaquatic Hemiptera and Coleoptera, Trichoptera, and Ephemeroptera are well represented. The collection is housed in 1,600 drawers, plus vial storage of larvae, many of the aquatics, and assorted arthropod groups such as Chilopoda, Diplopoda, and Arachnida. Major emphasis is on Arizona and adjacent areas. A fair amount of ecological supporting data is included. About 90% of the data concerning Arizona species is computerized, using 42 geographical divisions of the state (no site specific data included). Material is available for loans; usual loan rules apply. No holotypes are kept in the collection. [1986]

ARKANSAS

Fayetteville

DEPARTMENT OF ENTOMOLOGY COLLECTION, UNIVERSITY OF ARKANSAS, FAYETTEVILLE, AR 72701. [UADE]

Director: Dr. J. B. Whitfield. Phone: (501) 575-2482. Professional staff: Mr. C. E. Carlton. The collection consists of about 20,000 species of insects, with over 500,000 pinned specimens housed in drawers, 100,000 alcohol preserved specimans in vials, and 15,000 slides (mostly Protura). The collection is strong in Carabidae (worldwide), Cerambycidae, Staphylinidae, Scarabaeidae, and Braconidae (Worldwide). [1992]

CALIFORNIA

Baker

DESERT STUDIES CENTER, P. O. Box 490, Baker, CA 92309. [CSDS]

Curator of Insects: Dr. Gary Shook, CSU, San Bernardino. Phone: (714) 773-2428. This desert field station, a segment of the California State University, provides housing, laboratory, and equipment support to research workers in several disciplines as well as college and university courses. The insect collection focuses on Mojave Desert species. [1992]

Berkeley

ESSIG MUSEUM OF ENTOMOLOGY, DEPARTMENT OF ENTOMOLOGICAL SCIENCES, UNIVERSITY OF CALIFORNIA, BERKELEY, CA 94720. [EMEC] [=CISC and CIS]

Director: Dr. John A. Chemsak. Phone: (415) 642-4779. Professional staff: J. T. Doyen, H. V. Daly, J. A. Powell, L. Caltagirone, and W. W. Middlekauff. The collection consists of about 4.5 million specimens on pins plus uncounted numbers in fluid. It is housed in Wellman Hall, occupying three levels of the rotunda, filling more than 340 CAS type cases (Coleoptera, 100; Lepidoptera, 65; Hymenoptera 63; Diptera, 39; Hemiptera, 30; miscellaneous orders, 14; storage and fluid, 29). The collection, originally started as the California Insect Survey, has a very strong representation of material from Western United States, Mexico (especially strong because of almost annual expeditions from Berkeley to Mexico since the early 1950's), and Central America. Strengths of the collection are the Coleoptera (especially Cerambycidae, Tenebrionidae, Chrysomelidae, Cleridae, Buprestidae, and Meloidae), Lepidoptera (especially Microlepidoptera), Hymenoptera (especially Apoidea), and Hemiptera. Present in the Essig Museum are the collections of Essig (aphids), Usinger (Hemiptera), Jensen (psyllids), Kellogg (Mallophaga and Anoplura) and Cott (Thysanoptera). The Arachnida collection is one of the best in the country according to contemporary workers. Also well represented are the Acarina especially from the Furman and Pritchard collections. Except for a few specialized collections, only Western Hemisphere material is retained. Other material is accessioned to the California Academy of Sciences in San Francisco. With the exception of some slide collections, and Lepidoptera, types are placed on indefinite loan to the California Academy of Sciences. Essig Museum primary types total about 1,300. Tens of thousands of paratypes have not been cataloged. Publications: *Bulletin of the California Insect Survey*. [1992]

Davis

THE BOHART MUSEUM OF ENTOMOLOGY, UNIVERSITY OF CALIFORNIA, DAVIS, CA 95616. [UCDC]

Collection Manager: Prof. Lynn S. Kimsey. Phone: (916) 752-0493. Professional staff: Prof. Robert K. Washino; Prof. Emeritus Richard M. Bohart; Prof. Robin W. Thorp; Prof. Philip S. Ward; Prof. Arthur M. Shapiro; Dr. Robert B. Kimsey; Prof. Emeritus Albert A. Grigarick. The research collection was established in 1946, and has since become a widely recognized resource in systematic entomology. It contains about 6 million specimens which include 1,500 primary types with a slide collection of about 241,000 specimens. The collections of Coccoidea and Hymenoptera are among the most significant in North America. It includes the only comprehensive material of some mite families and the largest world representation of the phylum Tardigrada. Pinned specimens and the alcohol collection are housed in drawers and mobile compactors. Slide mounted specimens are stored in 100 capacity slide boxes on bookshelves. Among the special collections are: Hymenoptera: 1,000,000 specimens, including the most comprehensive collection of aculeate wasps (R. M. Bohart); a large and rapidly growing collection of Chalcidoidea and Tiphioidea, Chrysididae of the world (R. M. Bohart and L. S. Kimsey), and bees with voucher specimens from pollination ecology studies from R. W. Thorp; Diptera: about 200,000 specimens including an important collection of mosquitoes from R. M. Bohart and R.

K. Washino, and voucher specimens from mosquito transmitted virus disease studies from W. C. Reeves; Lepidoptera: Buckett and Bauer collection of about 150,000 spread and identified of most North American macro-moths; Hamilton Tyler papilionid collection; voucher specimens of seasonal polyphenism and ecological studies of butterflies from A. M. Shapiro; a considerable amount of unidentified material from Mexico and Central America; Strepsiptera: The R. M. Bohart slide collection which is perhaps the best of this small insect order; Coleoptera: 700,000 specimens representing particularly the fauna of western North America, Mexico, and the neotropics; Thysanoptera: 16,500 slides of a cosmopolitan collection of thrips made by S. F. Bailey; Coccoidea: 80,000 slides and 1,000,000 specimens on habitats, mainly from G. F. Ferris and H. L. McKenzie; Pseudoscorpionida: 15,000 slides including examples of all western North America species from R. O. Schuster; Tardigrada: a cosmopolitan collection determined at least to genus consisting of about 45,000 slides from A. G. Grigarick, D. S. Horning, and R. O. Schuster; Acarina: representative material of most families, with excellent representation of the families Canestriniidae, Cheyletidae, Heterocheylidae, Phytoseiidae, Stigmaeidae, and Trombiculidae from studies of R. O. Schuster, F. M. Summers, M. M. J. Lavoipierre, and E. W. Jameson. [1992]

Death Valley

DEATH VALLEY MUSEUM, DEATH VALLEY NATIONAL MONUMENT, DEATH VALLEY, CA 92328. [DVNM]

Director: Mr. Virgil J. Olson. Phone: (619) 786-2331. Professional staff: Dr. Charles Douglas, Peter Sanchez, and Shirley Harding. This collection is pinned and labeled but not identified to species. It consists of about 1,500 representative specimens in two 12 drawer insect cabinets. [1986]

Fresno

FRESNO COUNTY DEPARTMENT OF AGRICULTURE, 1730 SOUTH MAPLE, FRESNO, CA 93702. [FCDA]

Director: Dr. Norman J. Smith. Phone: (209) 488-3510. The collection contains approximately 70,000 specimens which includes 66,000 pinned and labeled specimens, 3,000 alcohol vials of specimens, and 1,000 slide mounted specimens. Majority of pinned specimens are noneconomic native Fresno County insects, many of which are expertly identified. Many economic speciemens found in Fresno County are also included. Emphasis is mainly on Hymenoptera and Lepidoptera, but all orders are represented. The collection is housed in drawers in seven cabinets and in alcohol vials. [1992]

Long Beach

ENTOMOLOGICAL COLLECTIONS, CALIFORNIA STATE UNIVERSITY AT LONG BEACH, LONG BEACH, CA 90840. [CSLB]

Curator: E. L. Sleeper. All of North America and all families of Coleoptera are represented in the collection. Emphasis is on Carabidae, Scarabaeidae, Meloidae, Tenebrionidae, Nitidulidae, Cerambycidae, and Curculionidae. The collection consists of 530 CAS and USNM drawers and 20 Schmitt boxes [1968]. [*It is not certain whether this collection is to be maintained by the College, or whether it is the private collection of Dr. Sleeper. If so, it is being deposited in the California Academy of Sciences. 1992.*]

Los Angeles

INSECT COLLECTION, LOS ANGELES COUNTY MUSEUM OF NATURAL HISTORY, 900 EXPOSITION BLVD., LOS ANGELES, CA 90007. [LACM]

Acting Head: Julian P. Donahue. Phone: (213) 744-3364. FAX: (213) 746-2999. Professional staff: Roy R. Snelling, Brian P. Harris. Since its founding in 1913 the collection has grown as a result of field work by a succession of curatorial staff, the donations and bequests of numerous private collections, and the absorption of several institutional collections (University of California at Los Angeles, Stanford University, and the Allan Hancock Foundation/University of Southern California). By 1992 the collection contained over 5 million specimens: a pinned collection housed in nearly 11,000 Cornell-type drawers in a large electrical compactor (capacity of 17,010 drawers); a wet collection stored in about 2,300 jars (4-, 8-, and 16-oz. size); and a slide collection stored in about 450 slide boxes (100 slide capacity each). There is a large collection of primary and secondary types. The collection emphasizes the fauna of western North America, but has considerable Neotropical representation, particularly in the orders Hymenoptera and Lepidoptera, reflecting the interests and expertise of past and current staff. Hymenoptera account for about 45% of the total collection. The collection of Formicidae is especially good and World-wide; it includes the collections of W. S. Creighton, A. C. Cole, substantial parts received from G. C. and J. Wheeler, and the D. H. Janzen collection of *Acacia*-ants. Non-aculeate representation is only fair, except in Cynipidae and Proctotrupoidea. The Lepidoptera collection is particularly strong in Nearctic Geometridae, all families of southwestern moths and butterflies, and Neotropical moths (especially Arctiidae and Cossidae), including a large moth collection from the peninsula of Baja California, Mexico. Extensive special collections include the G. F. Augustsiv collection of fleas (Siphonaptera), insects from eastern Pacific islands (southern California Channel Islands, Revillagigedos, Cocos, Galapagos), and insects from many endangered or sensitive habitats in southern California [1992].

Affiliated Collections:

Cicero, Joseph M., 740 NE 23rd Ave., #42D, Gainesville, FL 32609 USA. [JMCC] Phone: (904) 377-6796 [permanent address: 13641 Terrabella, Arleta, CA 91331. Phone: (818) 899-8959]. This is a collection of about 25,000 beetles and a small collection of miscellaneous orders, with emphasis on larviform females of Coleoptera [1992]. (Registered with LACM.)

Dimock, Thomas E., 111 Stevens Circle, Ventura, CA 93003 USA. [TEDC]

Phone: (805) 650-6420; 647-5931. This is a collection of about 3,000 specimens of Lepidoptera, including 2,000 Nymphalidae (of which 550 are *Vanessa*); other families, 800, and other orders, 200. Paratypes of one species are held in the collection, which is stored in drawers in cabinets. Specimens may be borrowed; usual loan rules apply [1992]. (Registered with LACM.)

Dobry, Keith, 7837 Melba Ave., Canoga Park, CA 91304 USA. [KDIC] Phone: (818) 883-8606; 883-8606. The collection contains about 10,000 specimens, 5,000 in alcohol, remainder either pinned or papered. Specimens may be borrowed for study; usual loan rules apply [1992]. (Registered with LACM.)

Evans, Dr. Arthur V., 595 West Sierra Madre Blvd., #B, Sierra Madre, CA 91024 USA. [AVEC] Phone: (818) 355-3138. This is a collection of about 25,000 specimens of Coleoptera, almost all Scarabaeidae, in drawers, boxes, vials, and slide boxes. Material may be borrowed; usual loan rules apply [1992]. (Registered with LACM.)

Gorelick, Glenn A., 360 Toyon Road, Sierra Madre, CA 91024 USA. [GAGC] Phone: (818) 914-8634; 355-2837. This is a general collection consisting of about 3,000 specimens, mostly Lepidoptera (9,000 donated to LACM 1982, 1986), Coleoptera, Diptera, and Hymenoptera, including paratypes of subspecies of Callophrys. Specimens are pinned, papered, and in alcohol vials. Material may be borrowed for study; usual loan rules apply [1992]. (Registered with LACM.)

Griffin, Richard, 8519 Chester St., Paramount, CA 90723 USA. [RGIC] Phone: (310) 531-4320; 633-4391. About 1,000 specimens, mostly beetles are represented in this collection [1992]. (Registered with LACM.)

Gudehus, Dr. Donald H., Randall Laboratory, University of Michigan, Ann Arbor, MI 48109 USA. [DHGC] Phone: (313) 764-7561; 995-3119. About 1,400 butterflies, 1,400 moths, and 200 other insects are included in this collection, half pinned and half papered. Material may be borrowed for study, but one must supply shipping container, assume costs, and responsibility [1986]. (Registered with LACM.)

Hair, Dr. Christopher, 6134 Shady Lane Glade Ave., North Hollywood, CA 91606 USA. [CHIC] Phone: (818) 995-0170; 763-4718. This is a collection of about 3,000 specimens of butterflies, including Neotropical species, pinned, in Riker mounts, and some papered. Specimens may be studied only at above address [1992]. (Registered with LACM.)

Konopka, Ronald, 430 South Santa Anita Avenue, Pasadena, CA 91107 USA. [RKIC] Phone: (818) 793-0532. The collection consists of about 400 specimens of Lepidoptera [1992]. (Registered with LACM.)

Kristensen, Charles, [*Address unknown*] [CKSF] This is a collection of about 5,000 spiders stored in alcohol vials, *location unknown* [1992]. Registered with LACM.)

Landing, Dr. Benjamin, 4513 Deanwood Drive, Woodland Hills, CA 91364 USA. [BHLC] Phone: (818) 609-2426; 345-4427. This is a collection of about 3,500 butterflies, 1,500 moths, mostly from North America, but some from Scandinavia, Yugoslavia, Italy, Israel, India, Australia, Papua New Guinea, Mexico, Japan, Costa Rica, Colombia, Ecuador, and Switzerland. Specimens may be borrowed for study; usual loan rules apply [1986]. (Registered with LACM.)

Larsson, Sean G., 23711 Pastiempo Lane, Harbor City, CA 90710 USA. [SGLC] Phone: (310) 539-2539. This is a collection of 250 specimens representing most orders of insects, pinned, in alcohol, and some papered [1986]. (Registered with LACM.)

Leuschner, Ron, 1900 John St., Manhattan Beach, CA 90266 USA. [RLIC] Phone: (310) 545-9415. This is a collection of over 100,000 specimens of Lepidoptera. No primary types are kept. It is stored in cabinets. Material may be borrowed for study; usual loan rules apply [1986]. (Registered with LACM.)

Lowe, Graeme, Caltech 216-76, Pasadena, CA 91125 USA. [GLIC] Phone: (310) 356-6838; 449-2665. This is a collection of about 2,000 specimens, mostly

Arachnida, some Coleoptera and other orders of insects. Specimens are stored in alcohol, or pinned. Material may be borrowed; usual loan rules apply [1986]. (Registered with LACM.)

Ludtke, Alvin F., 6524 Stoneman Drive, North Highlands, CA 95660 USA. [AFLC] Phone: (916) 344-1626. This is a collection of over 30,000 specimens, mostly Lepidoptera, adults and immatures, and also some Hymenoptera and Diptera parasitic on Lepidoptera. Most specimens are from western U.S.A., but some are from Puerto Rico, Peru, Ecuador, Mexico, and Venezuela. Material may be borrowed for study; usual loan rules apply [1986]. (Registered with LACM.)

Mordaigle, R. C. de, 927 Hartzell St., Pacific Palisades, CA 90272 USA. [RCDM] Phone:(310) 466-4741; 454-8871. This is a collection of over 10,000 specimens of Lepidoptera, mostly butterflies, from Mexico and Central America, stored in 50 drawers in cabinets, or papered (90%). Specimens may be borrowed for study; usual loan rules apply [1986]. (Registered with LACM.)

Oppewall, J., P. O. Box 5051, Santa Monica, CA 90405 USA. [JOIC] Phone: (310) 399-7548. This is a collection of about 3,000 butterflies of southern California, with some from Japan and Europe. Material may be borrowed for study; usual loan rules apply [1986]. (Registered with LACM.)

Robertson, Jim, 10839 Kelmore St., Culver City, CA 90230 USA. [JRCC] Phone: (310) 839-7021. The collection currently contains over 23,000 pinned and labelled specimens, mainly Coleoptera, particularly Cerambycidae, Scarabaeidae, Curculionidae, and Buprestidae. Material of 19 other orders is represented. Most of the material is from U.S.A., but there are 2500 specimens of Cerambycidae (60% of the Japanese longhorn fauna), and 800 specimens of Scarabaeidae and Lucanidae, 50% of which are identified. Cerambycidae, Scarabaeidae, Curculionidae, and miscellaneous others families have been collected in Costa Rica, Panama, Venezuela, Colombia, Peru, and Brazil. The collection is housed in drawers and Schmitt boxes [1986]. (Registered with LACM.)

Snider, John M., 3520 Mulldae Ave., San Pedro, CA 90732 USA. [JMSC] Phone: (310) 832-2387. This is a collection of about 5,000 specimens of butterflies (Lepidoptera), pinned and papered. Material may be borrowed for study; usual loan rules apply [1986]. (Registered with LACM.)

Streit, Barney D., *Address and location of collection unknown.* This collection contains about 5,000 specimens of insects, with some concentration of *Pleocoma* (Coleoptera: Scarabae idae), housed in drawers. [1992]. (Registered with LACM.)

Strong, Arthur R., 4549 Pedley Road, Glen Avon, CA 92509-3241, USA. [ARSC] This is a collection of butterflies of western U.S.A., approximately 380 unit trays in drawers. Material may be borrowed for study; usual loan rules apply [1992]. (Registered with LACM.)

Taylor, David C., 17372 Mira Loma Circle, Huntington Beach, CA 92647 USA. [DCTC] Phone: (714) 546-5570; 536-0805. This is a general collection of about 3,000 insects and spiders, pinned, papered, and in alcohol vials. Material may be borrowed for study; usual loan rules apply [1986]. (Registered with LACM.)

Verity, David S., 7346 West 83rd St., Los Angeles, CA 90045 USA. [DSVC] Phone: (213) 825-2714; 641-2630. This is a general collection of over 31,000 specimens, the majority of which are Coleoptera with emphasis on Buprestidae (15,000 specimens) from western U.S.A. and Mexico, including many paratypes, host records, and distribution records. There are also 4,000 specimens of Histeridae, largely Saprininae from southwestern North America. The collection is housed in drawers and Schmitt type boxes [1986]. (Registered with LACM.)

Walters, George C., 3650 Watseka, Apt. 3, Los Angeles, CA 90034 USA. [GCWC] Phone: (213) 837-0638. This collection of Coleoptera contains about 25,000 specimens, with 15,000 of these Buprestidae, 3,000 Cerambycidae, and 1,500 Scarabaeidae. Material may be examined only on premises [1986]. (Registered with LACM.)

Wilson, Donald A., *Address and location of collection unknown.* [DAWC]. This is a collection of over 10,000 speci mens of Coleoptera, including 4,000 specimens of Cerambycidae and 3,350 specimens of Cicindelidae. Material may not be borrowed for study [1992]. (Registered with LACM.)

DEPARTMENT OF ZOOLOGY COLLECTION, UNIVERSITY OF CALIFORNIA, LOS ANGELES, CA 90024. (*Collection moved to LACM.*) [1992]

Mineral

MUSEUM COLLECTIONS OF LASSEN VOLCANIC NATIONAL PARK, LASSEN PARK, MINERAL, CA 96063. [LVNP]

Director: Mr. Richard Vance. Phone: (916) 595-444. Professional staff: Ellis Richard. The collection is restricted to the insects of Lassen Volcanic National Park. It is housed in 11 drawers in a cabinet. [1986]

Northridge

INSECT COLLECTION, DEPARTMENT OF BIOLOGY, SAN FERNANDO VALLEY STATE UNIVERSITY, NORTHRIDGE, CA 91330. [SFVS]

Director: Dr. Peter F. Bellinger. Phone: (213) 885-3359. The collection is housed in 144 drawers and in alcohol vials. It is a general collection from which specimens may be borrowed for study. [1986]

Paicines

MUSEUM COLLECTIONS OF PINNACLES NATIONAL MONUMENT, PAICINES, CA 95043. [PINN]

Curator: Steve De Benedetti. Phone: (408) 389-4485. Approximately 1,100 mounted specimens in 11 drawers, mostly unidentified to species. [1992]

Pleasant Hill

DIABLO VALLEY COLLEGE, 321 GOLF CLUB ROAD, PLEASANT HILL, CA 94533. [DVCC]

Director: David Cox. Phone: (415) 685-1230. The collection consists of 600 species of economic significance in northern California, housed in drawers and boxes [1986].

Pomona

ENTOMOLOGY DEPARTMENT, CALIFORNIA POLYTECHNIC UNIVERSITY, 3801 TEMPLE AVENUE, POMONA, CA 91768. [CPUP]

[Contact: Dr. Donald Force. Eds. 1992]

Riverside

UCR ENTOMOLOGICAL TEACHING AND RESEARCH COLLECTION, UNIVERSITY OF CALIFORNIA, RIVERSIDE, CA 92521. [UCRC]

Curator: Saul I. Frommer. Phone: (714) 787-4315. Professional staff: Dr. John D. Pinto. The collection contains a large representation of specimens from the southwestern United States and Mexico. Pinned material is housed in over 2,100 drawers in cabinets, plus a large slide mounted collection and specimens in alcohol vials. Strong representation of the following is present: Chalcidoidea, Apoidea, Meloidae, Staphylinidae, Bombyliidae, Sciomyzidae, Thysanoptera, and Aphidae. Strong representation of entomophagous insects from all over the world utilized in biological control and associated insectary and quarantine facilities are present. Non-insectan arthropod collections are maintained. Of particular interest is that dealing with predaceous Acarina. A fledgling collection of Araneae exists. A good collection of immature insects has been developed. Currently the collection is engaged in an arthropod survey of the university maintained P. L. Boyd Desert Research Center. The University has attempted to develop a collection of arthropods associated with the plant, jojoba (*Simmondsia chinensis* (Link) Schneider) and encourages contributions to this effort by other individuals and institutions. [1992]

Sacramento

CALIFORNIA STATE COLLECTION OF ARTHROPODS, ANALYSIS AND IDENTIFICATION UNIT, CALIFORNIA DEPARTMENT OF FOOD & AGRICULTURE, 1220 N ST., RM. 340, SACRAMENTO, CA 95814. [CDAE]

Director: Dr. Marius S. Wasbauer. Phone: (916) 445-4521. Professional staff: Dr. F. G. Andrews (Coleoptera); Ms. K. S. Corwin (Diptera); Dr. T. D. Eichlin (Lepidoptera); Dr. E. M. Fisher (Diptera); Mr. R. J. Gill (Homoptera); Dr. A. R. Hardy (Hemiptera, Orthoptera, small orders); Mr. T. Kono (Homoptera, Thysanoptera, Acarina); Ms. M. Moody (Arachnida); Mr. T. N. Seeno (Coleoptera); Dr. J. Sorensen (Homoptera); Dr. R. E. Somerby (Lepidoptera); and Dr. M. S. Wasbauer (Hymenoptera). The collection is maintained as an adjunct to the arthropod identification services provided by the California Department of Food and Agriculture. The collection contains approximately 900,000 pinned, labeled specimens, most of which are identified; 120,000 vials of alcohol preserved specimens, 30,000 slides; and 5,000 envelopes. Specimens from general collections and from the specialized research activities of the systematists on the staff are added to the collection at the rate of about 50,000 per year. There is not a large backlog of unmounted material. The collection contains several thousand paratypes. Holotypes are not retained. The collection is housed in 2,400 drawers. There are many associated collections (see affiliated collection section). The geographical scope of the collection is New World. At present, its composition is primarily Nearctic, but in a number of groups, there is good Neotropical representation.

[1986]

Affiliated Collections:

Hardy, Alan R., Insect Identification Lab., Room 340, 1220 "N" St., Sacramento, CA 95814 USA. [ARHC] This collection is restricted to Scarabaeidae. The southwestern U.S.A. and western Mexico are well represented. The 40,000 specimens are housed in drawers. There are some ecological data. The collection is especially rich in sand dune and desert species. Material is available for exchange and loan; usual loan rules apply [1986]. (Registered with CDAE).

Penrose, Richard L., 27116 Vista Delgado St., Valencia, CA 91355 USA. [RLPC] Phone: (213) 575-5471; 254-4177. This collection is restricted to Cerambycidae, with about 15,000 specimens representing more than 1,000 species and subspecies, mostly from U.S.A., Mexico (primarily Baja California), Costa Rica, and Panama. Many U.S.A. specimens are host associated. It is housed in 36 drawers. Specimens may be borrowed for study; usual loan rules apply [1986]. (Registered with CDAE.)

Tyson, W. H., 1240 W. Pico St., Fresno, CA 93705 USA. [WHTC] Phone: (209) 224-7103. This collection is specialized in North American Cerambycidae (Coleoptera), with over 15,000 specimens. Also well represented are North American Cleridae, Elateridae, and Buprestidae. Many specimens have host data. There is also some material from Central and South America, Asia, and a few other places. The collection is housed in 54 drawers and over 40 Schmitt boxes. Material is available for exchange and for loan; usual loan rules apply [1986]. (Registered with CDAE.)

Santa Barbara

DEPARTMENT OF INVERTEBRATE ZOOLOGY, SANTA BARBARA MUSEUM OF NATURAL HISTORY, 2559 PUESTA DEL SOL RD., SANTA BARBARA, CA 93105. [SBMN]

Curator: Dr. F. G. Hochberg. Phone: (805) 682-4711. Professional staff: P. H. Scott, L. P. Marx, Dr. A. M. Wenner, Dr. L. J. Friesen, Dr. H. W. Chaney, and research associates. The collection consists of 30,000 insects (mostly pinned), 1,500 arachnids (in alcohol vials), all North American. Strongest holdings are mostly local, with a large collection of Channel Islands insects and arachnids, identified and curated. [1992]

San Diego

ENTOMOLOGY DEPARTMENT, SAN DIEGO NATURAL HISTORY MUSEUM, BALBOA PARK, P.O. BOX 1390, SAN DIEGO, CA 92112. [SDMC]

Director: Dr. Mick Hager. Phone: (619) 232-3821, ext. 223. Professional Staff: David K. Faulkner. The Entomology Department is the oldest research department of the Museum, located on the third floor of the Museum building in Balboa Park. It houses one of the best representations available of the insects occurring in southern California and Baja California, Mexico. There are over 865,000 mounted and labelled specimens, 70% of which are identified to species. The number of specimens incorporated each year is about 15,000. Holdings of non-insect terrestrial invertebrates, such as arachnids, are small, with no more than 10,000 undetermined specimens stored in alcohol and housed in three storage cabinets. A type collection is also maintained by the department. Most holotypes and paratypes represent species described from southern California or Baja California, with the majority being either Lepidoptera or Coleoptera. Traditionally, the collections have reflected a strong research emphasis in the southwestern U.S.A. and northwestern Mexico, especially the peninsula and islands of Baja California. Other geographic areas from which insect material has been obtained are the Nearctic, Neotropics, and Southern Africa. The primary emphasis of research has been with the orders Coleoptera and Lepidoptera, these two groups accounting for 60% of the department's specimens. In addition, there is a fairly large collection of Nearctic Neuroptera and Odonata. Many loans are made each year. The collection is housed in 68 cabinets containing over 2,000 drawers. Publications sponsored: "Proceedings of the San Diego Natural History Museum," and "Occassional Papers of the San Diego Natural History Museum." [1992]

San Francisco

DEPARTMENT OF ENTOMOLOGY, CALIFORNIA ACADEMY OF SCIENCES, GOLDEN GATE PARK, SAN FRANCISCO, CA 94118. [CASC] [=CAS]

Director: Dr. Wojciech J. Pulawski. Phone: (415) 221-5100. Professional staff: Dr. Paul H. Arnaud, Dr. David H. Kavanaugh, Dr. Edward S. Ross, also Research Associates, Associates, and Curatorial Assistants. The collection is World-wide in scope, with over 7 million processed specimens including over 15,000 primary types. It is without equal in western North America and is one of the four largest in the U.S.A. Its collections of Coleoptera (and especially aquatic beetles) of western North America and the desert areas of South America are unsurpassed. Outstanding collections of other orders include Embioptera: 200,000 specimens including over 95% of known species; Thysanoptera: one of the World's most important with primary types of about one-fifth of the known species; its collection of the dipterous family Platypezidae is the world's largest; its collection of the parasitic fly family Tachiniidae is only surpassed by that of the Smithsonian Institution. Regionally, the collection is also strong in insects from western South America, Australia, India, the Congo area of Africa, China, the Philippine Islands, and Papua New Guinea. It is unsurpassed in its collections of insects from the Galapagos Islands and Baja California, Mexico. The collection occupies a total of 8,360 square feet, including offices, laboratories, and library. The collection is housed in drawers in compactor cabinets, and the alcohol collection in cases. A Hitachi Scanning Electron Microscope has recently been acquired. Publications sponsored: The "Pan-Pacific Entomologist," published by the Pacific Coast Entomological Society, has its headquarters in the Entomology Department. [1986]

Affiliated Collections:

Brown, Dr. Kirby W., P. O. Box 1809, Stockton, CA 95201 USA. [KWBC] Phone: (209) 468-3300. This is a specialized collection of Tenebrionidae (Coleoptera), including Alleculidae, Lagriidae, and Nilionidae, consisting of about 12,000 specimens representing over 1,000 species, including immatures (about 1,000 specimens in alcohol), mimics of Tenebrionidae (about 100 specimens), photographs of about 300 type specimens of tribe Asidini. Specimens are from Western Hemisphere, especially U.S.A., Mexico, Brazil, and Chile. There is a fair representation of specimens from the Old World, including Europe, Vietnam, Australia, and South Africa. The collection is housed in 66 drawers and 16 boxes. Specimens of many other groups are available for exchange for Tenebrionidae. An extensive research library has been developed (about 80% complete for New World taxonomic literature). Photocopies of hard-to-get papers will be provided for researchers. Also, copies are wanted of papers lacking in the library [1992]. (Registered with CASC.)

Nutting, Dr. Willard H., [WHNC] [*Deceased; collection deposited in CASC.*] [1992]

O'Brien, Dr. Lois B., 3009 Brookmont Drive, Tallahassee, FL 32312-2406 USA. [LBOB] Phone: (904) 385-7267. This world collection is restricted to Fulgoroidea. The U.S.A., Mexico, Central America, and Chile are especially well represented. The 100,000 specimens, including types, are housed in unit trays in 150 drawers in cabinets. Material is available for exchange or loan; usual loan rules apply [1986]. (Registered with CASC.)

O'Brien, Dr. Charles W., 3009 Brookmont Drive, Tallahassee, FL 32312-2406 USA. [CWOB] Phone: (904) 385-7267. This world collection is restricted to the Curculionidae, Brenthidae, Anthribidae, and Scolytidae (Coleoptera). The Curculionidae of the U.S.A., Mexico, Central America, and Chile are especially well represented. The owner's study interest is the Curculi onidae. The 500,000 specimens are housed in 709 drawers, including 76 holotypes, 62 allotypes, and several thousand paratypes. Host plant data are extensive. Material is available for exchange and for loan; usual loan rules apply [1986]. (Registered with CASC.)

Roth, Vincent D., Box G, Portal, AZ 85632 USA. [VDRC] Phone: (602) 558-2396. This is a collection of arachnids, mainly Agelenidae spiders with representatives of all spiders of the Chiricahua Mountains, Arizona. It is housed in alcohol jars [1986]. (Registered with CASC.)

Rumpp, Norman L., 3511 Blackstone St., Las Vegas, NV 89121-3723 USA). [NLRC] Phone: (702) 731-5626. This is a collection of Coleoptera, 30,000 specimens of Cicindelidae, and 18,000 specimens of other families, housed in drawers in cabinets [1986]. (Registered with CASC.)

Tilden, J. W., [JWTC] [*Deceased; collection deposited in CASC.* 1992]

Valentine, Dr. Barry D., Museum of Zoology, Ohio State University, Columbus, OH 43210. [BDVC] Phone: (614) 457-0973. World Anthribidae and other Coleoptera; anthribids stored in 45 Cornell-type drawers, about 20,000 specimens representing 325 genera, 2,000 species. Of these 220 species are represented by 400 paratypes and 850 species have been compared with holotypes. The entire geographic range of the family is well represented. Biological and distributional data are better than in most collections. Exchanges with museums and collectors continue. All available beetle families and reprints are offered, especially from Nearctic, Neotropical, and east African areas, in exchange for *any* anthribids. (Registered with CASC). [1992]

San Jose

J. GORDON EDWARDS MUSEUM OF ENTOMOLOGY, DEPARTMENT OF BIOLOGICAL SCIENCES, SAN JOSE STATE UNIVERSITY, ONE WASHINGTON SQUARE, SAN JOSE, CA 95192-0100. [SJSC]

Director: Dr. J. Gordon Edwards. Phone: (408) 924-4876. Professional staff: Curtis Takahashi, Curator, Dr. Ronald E. Stecker, and Richard Worth. The collection consists of about 700,000 pinned, labelled, and identified specimens, over 6,000 processed vial specimens, and approximately 2,000 slide mounts. All of North America, including most families, is represented in the collection with special emphasis on California, Arizona, and the Pacific Northwest. About 15 drawers of African, 27 drawers of South American, 11 drawers of Japanese, 6 drawers of Vietnamese, 15 drawers of Pacific island, 16 drawers of European, and 10 drawers of Mexican insects are representated in the exotic collection. Among the special collections are: Hugo H. Huntzinger Collection of Korean insects (24 drawers); Benjamin H. Pickel Collection of Midwestern U.S.A. insects; collections from Costa Rica, Quintana Roo and Yucatan made by Richard E. Main and J. Gordon Edwards (approximately 20 drawers); collections from Death Valley National Monument, Mexico, and Sequoia/Kings Canyon National Park made by R. Stecker; and collections from Glacier National Park made by J. Gordon Edwards. [1992]

Affiliated Collections:

Cope, James S., 6689 Mt. Holly Dr., San Jose, CA 95120 USA. [CJSC] Phone: (408) 578-9100, ext. 71; 268-5492. This collection of about 25,000 specimens (about 6,500 species), housed in drawers, is limited to Cerambycidae of the world with emphasis on North America. Specimens are available for exchange. [1992] (Registered with SJSC.)

Edwards, Dr. J. Gordon, Department of Biological Sciences, San Jose State University, San Jose, CA 95192 USA. [JGEC] The collection contains 120,000 specimens, with most families of North American beetles are well represented, with Carabidae, Buprestidae, Cerambycidae, Chrysomelidae, Dytiscidae, Hydrophilidae, and Tenebrionidae best represented. Specialized collections include alpine material from Glacier National Park, the Grand Teton Range. Alpine Coleoptera are among the special study interests of the owner. The collection is stored in 92 drawers, 210 Schmitt boxes, and 8,500 larvae in vials. Material is available for exchange and loan for study; usual loan rules apply [1992]. (Registered with SJSC.)

Stockton

SAN JOAQUIN COUNTY AGRICULTURAL COMMISSIONER, P. O. BOX 1809, STOCKTON, CA 95201. [SJAC]

Director: Dr. Kirby W. Brown. Phone: (209) 468-3300. This is a collection of terrestrial arthropods and molluscs of San Joaquin County, California, and of exotic species intercepted in quarantine inspections. It represents extensive surveys of a local fauna largely ignored by collectors. It consists of approximately 1,300 identified species, represented by 22,000 specimens stored in 96 drawers in cabinets, 12 vial drawers, and 5 boxes of microscope slides. About 3/4 of the specimens are fully curated and identified at least to genus. Most specimens are fully documented [1992].

Twentynine Palms

MUSEUM COLLECTIONS OF JOSHUA TREE NATIONAL MONUMENT, 74485 NATIONAL MOUNMENT DR., TWENTYNINE PALMS, CA 92277. [JTNM]

Superintendent: David E. Moore. Chief curator: Rosie Pepito. Museum Registrar: Melanie Yeager. The Monument has a small insect collection restricted to species occurring within the Monument. It consists mostly of representative families of Coleoptera, some determined to species, others to genus only. The orders Diptera, Lepidoptera, Neuroptera, Hemiptera, and Hymenoptera are also represented, but there are few specimens and these are not identified. The total collection consists of about 700 specimens housed in drawers. [1992]

COLORADO

Boulder

UNIVERSITY OF COLORADO MUSEUM, CAMPUS BOX 218, BOULDER, CO 80309. [UCMC]

Director: Urless N. Lanham. This collection of about 350,000 pinned specimens is housed in 2,000 drawers in cabinets. About 10% of the collection is North American butterflies; of the rest, over half are bees, mainly from the Rocky Mountain area and with a specialty in the genus *Andrena*. The Gordon Alexander Orthoptera collection consists of 25,000 Colorado specimens. Specimens of all orders are available for loan. The collection contains over 100 types, mostly of bees. [1986]

J. T. Polhemus Collection, 3115 South York St., Englewood, CO 80110. [JTPC] Phone: (303) 781-6190. Curators: J. T. Polhemus and Dan A. Polhemus. Approximately 350,000 specimens of aquatic Heteroptera, with World-wide coverage, containing about 85% of described species, 95% of described genera, and about 15% of World genera and species undescribed. The mounted collection is stored in 125 CASC drawers, 600 Schmitt boxes, and 7,000 alcohol vials. In addition there is a collection of Miridae mainly from western U. S. A., housed in 75 drawers. [1992] (Registered UCMC.)

Colorado Springs

MAY NATURAL HISTORY MUSEUM, 710 ROCK CREEK CANYON RD., COLORADO SPRINGS, CO 80926. [MCSC]

Director: Mr. John M. May. Phone: (719) 576-0450. Professional staff: Carla Bucy, Louise Steer, and Lynda Senko. Approximately 100,000 specimens of arthropods from all over the world are housed in 550 show case type drawers. Most of the collection is of large, spectacular Lepidoptera, Coleoptera, Orthoptera, and most of the related orders from tropical and subtropical countries. Library contains about 2,000 volumes. This collection is available for viewing by the general public. [1992]

Denver

DENVER MUSEUM OF NATURAL HISTORY, 2001 COLORADO BLVD. DENVER, CO 80205-5798 [DNHC]

Dr. Richard S. Peigler. Phone: (303) 370-6353. The collection consists of more than 150,000 pinned insects, most of which are butterflies and moths, but there are hundreds of Coleoptera, such as especially Cicindellidae. A few primary types of Lepidoptera are held. Publication: "Proceedings of Denver Museum of Natural History." [1992]

Fort Collins

INSECT COLLECTION, ROCKY MOUNTAIN FOREST AND RANGE EXPERIMENT STATION, 240 WEST PROSPECT, FORT COLLINS, CO 80521. [RMSC]

Director: R. E. Stevens. This collection is specialized, mostly composed of forest insects from the Rocky Mountains, especially Colorado. The collection contains species injurious to trees and others associated with them. The collection is housed in 25 drawers. Some ecological data are available. Material is not available for exchange, but loans may be made; usual arrangements apply. [1986]

C. P. GILLETTE ARTHROPOD BIODIVERSITY MUSEUM, DEPARTMENT OF ENTOMOLOGY, COLORADO STATE UNIVERSITY, FORT COLLINS, CO 80523. [CSUC]

Director: Dr. B. C. Kondratieff. Phone: (303) 491-7314. Professional staff: Dr. Howard E. Evans, Dr. P. A. Opler, and David A. Leatherman. The Colorado State University Insect Collection, established in 1870 by C. P. Gillette, is the most comprehensive and expanding collection of the Montana, Wyoming, Colorado, and New Mexico region. The collection has an excellent representation of 12 orders of insects of national importance, especially strong in the coverage of Rocky Mountain species. The 16,250 slide Palmer aphid collection and the holdings of 30 holotypes and over 2,500 secondary types are especially notable resources.

The Natural Resource Ecology Laboratory Soil Arthropod Collection is one of the most extensive soil arthropod collections in the western United States, and includes material from three Long-term Ecological Research Sites. This collection consists of approximately 9,200 slides (including secondary type series and voucher specimens), 60 alcoholic series from cultures, and 100 residual alcohol Berleseates of soil arthropods from Colorado, Wyoming, Montana, Nebraska, Kansas, New Mexico, Arizona, Utah, and Texas.

The insect collection is house in 105 cabinets of 12 drawers each and 20 slide cabinets, containing an estimated 1.2 million specimens, chiefly southern Rocky Mountains and southwest U. S. A. material. The Palmer aphid collection, Gillette leafhopper collection, Evans Hymenoptera collection, Natural Resource Ecology Laboratory Soil Arthropod collections, with 30 holotypes, and over 2,500 secondary types, are especially notable resources. Additionally, a comprehensive alcohol collection of many North American species of Ephemeroptera, Plecoptera, and Trichoptera are represented. [1992]

Fruita

MUSEUM COLLECTIONS, COLORADO NATIONAL MONU-MENT, FRUITA, CO 81521. [CNMC]

Director: Henry A. Schoch. Phone: (303) 858-3617. The collection consists of 220 insect specimens, somewhat representative of populations occurring within the monument boundries, but by no means complete. The collection is housed in 6 drawers. [1986]

Pueblo

DEPARTMENT OF RESEARCH AND GRADUATE STUDIES, UNIVERSITY OF SOUTHERN COLORADO, 2200 BONFORTE BLVD., PUEBLO, CO 81001-4901. [USCC] [*No reply*]

CONNECTICUT

Mystic

DENISON PEQUOTSEPOS NATURE CENTRE, P. O. Box 122, PEQUOTSEPOS RD., MYSTIC, CT 06355. [DPNC]

Director: Margarett J. Philbrick. Phone (203) 536-1216. Twelve drawer cabinet of Connecticut beetles collected and prepared by Roy W. Johnson. [1992]

New Haven

DEPARTMENT OF ENTOMOLOGY COLLECTION, CONNECTICUT AGRICULTURE EXPERIMENT STATION, BOX 1106, NEW HAVEN, CT 06504. [CAES]

Chief, Department of Entomology: Louis A. Magnarelli. Phone: (203) 789-7241. Professional staff: Theodore G. Andreadis, Carol Lemmon, Chris T. Maier, Mark S. McClure, Kirby C. Stafford, III, Kimberly A. Stoner, Kenneth A. Welch, Ronald M. Weseloh. The collection contains about 80,000 pinned and mostly identified specimens, housed in 360 Cornell drawers in cabinets, and over 9,000 slidemounts. Several thousand specimens in alcohol. Types, with the exception of 13 mite species, have been deposited in the U. S. National Museum of Natural History. In addition, the Townsend collection of eastern U.S.A. beetles is housed in 65 Schmitt boxes. Specimens are used for the identification of state insects and for revisionary taxonomic work. [1992]

DIVISION OF ENTOMOLOGY, PEABODY MUSEUM OF NATURAL HISTORY, YALE UNIVERSITY, NEW HAVEN, CT 06520. [PMNH]

Director: Prof. Charles L. Remington. Phone: (203) 432-3900; 432-5001. The collection consists of about 1.1 million specimens, with 260 primary types of Lepidoptera, Coleoptera, Hymenoptera, Diptera, Hemiptera, Orthoptera, Neuroptera, Apterygota, and Arachnida, and about 900 secondary types. Included in the collection are the following: Petrunkevitch's Arachnida; Gerould *Colias* sex-linkage collection; miscellaneous genetic collections, *e.g.*, gynandromorphism, hybridization, polymorphism, mutation, melanism, and mimicry; California offshore islands; Rapa Island and other Pacific Islands; selected New England bogs and islands; demonstration/teaching collections (*e.g.*, speciation, biogeography, *etc.*); fossils (amber deposits, Florissant, shale, *etc.*). As a part of the general collections there is an extensive amount of material preserved in alcohol, especially immature stages of Lepidoptera. The Division also maintains a general entomological library including sets of periodicals, books, and an extensive reprint collection. In addition, there is a separate unique Arachnology library originating from the A. Petrunkevitch library. Publicaions sponsored: 'Discovery"; "Postilla"; "Peabody Museum Bulletin." [1986]

Storrs

DEPARTMENT OF ECOLOGY AND EVOLUTIONARY BIOLOGY, BOX U-43, UNIVERSITY OF CONECTICUT, STORRS, CT 06269-3043. [UCMS] [=UCSE]

Collection Manager: Dr. Jane O'Donnell. Phone: (203) 486-4451). The collection consists of 115,000 pinned specimens and 40,000 vials of alcohol vials, mostly from northeastern U.S.A., but with scattered holdings from South Africa and Australia. There are good representations of Hemiptera, and some families of Coleoptera. The collection is housed in drawers in cabinets. [1992]

Affiliated Collections:

Slater, Dr. James A., Department of Ecology and Evolutionary Biology, Box U-43, University of Connecticut, Storrs, CT 06269-3043 USA. [JASC] Phone: (203) 486-2227. This is a collection of 170,000 pinned, 1 million specimens in alcohol, and 8,000 alcohol vials of nymphs. This is a world-wide collection with many paratypes, especially in the family Lygaeidae, housed in drawers in cabinets [1992]. (Registered with UCMS.)

Rettenmeyer, Dr. Carl W., Department of Ecology and Evolutionary Biology, Box U-43, University of Connecticut, Stores, CT 06269-3043 U.S.A. [CWRC] Phone: (203) 486-4460. This collection of 20,000 pinned Hymenoptera and 20,500 alcohol vials, is supplemented by 5,500 microscope slides and 30,000 photographs. The pinned insects are stored in drawers and cabinets. [1992] (Registered with UCMS.)

DELAWARE

Kirkwood

INSECTS OF DELAWARE, LUMUS POND STATE PARK, WHALE WALLOW NATURE CENTER, RT. 71, BEAR, DE 19701. [LPSP]

Director: Mr. Robert Ernst. Phone: (302) 836-1724. This collection consists of 1,560 specimens, most of them collected in the 1920's and 1930's, housed at a nature center. The collection is for public viewing during the summer [1986].

Newark

DEPARTMENT OF ENTOMOLOGY AND APPLIED ECOLOGY COLLECTION, UNIVERSITY OF DELAWARE, NEWARK, DE 19711. [UDCC]

Director: Dr. Robert T. Allen. Phone: (302) 831-2526. The beetle collection consists of about 19,000 specimens. A representative collection of aquatic insects is included. About 6,000 specimens of identified

Delaware spiders in vials are included. The entire collection is housed in 20 Cornell drawer cabinets. It consists of mostly local material from the state. [1992]

DISTRICT OF COLUMBIA

UNITED STATES NATIONAL ENTOMOLOGICAL COLLECTION, DEPARTMENT OF ENTOMOLOGY, U.S. NATIONAL MUSEUM OF NATURAL HISTORY, WASHINGTON, DC 20560. [USNM]

Director: Dr. Jonathan A. Coddington. Phone: (202) 357-2317. FAX: (202) 786-2894. Professional staff (Smithsonian Institution personnel): Dr. John M. Burns (Curator of Lepidoptera); Dr. Jonathan A. Coddington (Curator of Arachnida and Myriapoda); Dr. Donald R. Davis (Curator of Lepidoptera); Dr. Terry L. Erwin (Curator of Coleoptera); Dr. Oliver S. Flint, Jr. (Curator of Neuropteroids); Dr. Richard C. Froeschner (Curator of Hemiptera); Dr. Karl V. Krombein (Senior Entomologist, Hymenoptera); Dr. Ronald J. McGinley (Curator of Apoidea); Dr. Wayne N. Mathis (Curator of Diptera); Dr. Robert K. Robbins (Curator of Lepidoptera), and Dr. Paul J. Spangler (Curator of Coleoptera). Professional staff (Systematic Entomology Laboratory, U. S. Department of Agriculture; those research entomologists marked with an * are at USDA, Bldg. 004, BARC-West, Beltsville, MD 20705): Dr. David Adanski (Lepidoptera); Dr. Donald M. Anderson (Coleoptera); *Dr. Edward W. Baker (Acarina); Dr. Susanne W. T. Batra (Hymenoptera); Dr. R. W. Carlson (Hymenoptera); Dr. Douglas C. Ferguson (Lepidoptera); Dr. Raymond J. Gagné (Diptera); Dr. Robert D. Gordon (Coleoptera); Dr. E. Eric Grissell (Hymenoptera); Mr. Thomas J. Henry (Hemiptera); Dr. Ronald W. Hodges (Lepidoptera); Dr. James P. Kramer (Hemiptera); Dr. Paul M. Marsh (Hymenoptera); Dr. Arnold S. Menke (Hymenoptera), *Dr. Douglass R. Miller (Homoptera); *Dr. Gary Miller (Homoptera); *Mr. Sueo Nakahara (Thysanoptera); Dr. David A. Nickle (Orthoptera); Dr. Allen L. Norrbom (Diptera); Dr. James Pakaluk (Coleoptera); Dr. Robert V. Peterson (Diptera); Dr. Robert W. Poole (Lepidoptera); *Miss Louise M. Russell (Homoptera); Dr. Michael E. Schauff (Hymenoptera); *Mr. Robert L. Smiley (Acarina); Dr. Curtis C. Sabrosky (Retired, Diptera); Dr. David R. Smith (Hymenoptera); Dr. M. Alma Solis (Lepidoptera); Mr. Theodore J. Spilman (Retired, Coleoptera); *Dr. Manya B. Stoetzel (Homoptera); Dr. F. Christian Thompson (Diptera); Dr. Natalia J. Vandenberg (Coleoptera); Dr. Richard E. White (Coleoptera); Dr. Norman E. Woodley (Diptera). Professional staff (Walter Reed Biosystematics Unit): Dr. Ralph E. Harbach (Research Entomologist, Diptera), Dr. Jayson Glick (Research Entomologist, Diptera). Dr. Yiau-min Huang (Research Entomologist, Diptera); Mr. E. L. Peyton (Research Entomologist, Diptera); Dr. Richard C. Wilkerson (Research Entomologist, Diptera). Professional staff (Siphonaptera Project): Dr. Robert Traub (Research Entomologist, Siphonaptera);

The collection is the second largest in the world, with about 30 million specimens. Type material is represented by about 100,000 primary types and by thousands of secondary types. Worldwide in scope, the collections are the largest in coverage of the Nearctic and Neotropical Regions. Outside of these geographical areas, the Department and associates have large study collections from other areas, especially Sri Lanka, the Philippines, Micronesia, and Egypt. The Coleoptera collection is worldwide, both adults and immatures, exceeding 7 million specimens.

It is the largest beetle collection in the Western Hemisphere. Nearly 20,000 types reside in the collection. The Lepidoptera collection contains the most complete representation of both larvae and adults in the Western Hemisphere and is second only to that of the British Museum in its total coverage. The total number of Lepidoptera is estimated to be nearly 4 million specimens. The Hemiptera and Homoptera collection occupies about 4,000 drawers and contains over 7,000 holotypes. More than 200,000 slides are included in the collection. The Hymenoptera collection comprises about 15% of the total collection and it is rich in material of aculeates and entomophagous parasites from most parts of the world. There are about 2 million pinned specimens in 4,500 drawers, including more than 15,000 holotypes. The Diptera collection ranks among the most extensive in the world, containing more than 5,500 drawers of pinned material, 470,000 slide mounted specimens, and specimens in alcohol. Approximately 20,000 primary types are included. Important holdings of other orders of insects, including Trichoptera, Thysanoptera, Plecoptera, Neuroptera, Isoptera, Mecoptera, Mallophaga, Anoplura, Siphonaptera, Odonata, Orthoptera, Embioptera, Zoraptera, and Psocoptera, are also maintained by the Department. The collections also include the classes Chilopoda, Diplopoda, Arachnida, Symphyla, and Pauropoda. Pertinent to the arachnid orders, the largest and most important holdings are represented in the Acarina and Araneida. The spider collection, considered to be a prominent resource for taxonomic studies, is currently being actively curated under the auspices of a curator in this group. The largest arachnid collection is that of the order Acarina. In terms of its holdings of mites parasitizing man, animals, and plants, it is the finest in existence. Excluding oribatid mites, it is the foremost collection in the Western Hemisphere; in global purview it ranks second only to that of the British Museum. Only two collections are currently maintained separately within the Department: the Casey collection of Coleoptera is one of the most significant historical accessions at the Museum, and added nearly 177,000 beetle specimens, including 9,200 types and about 20,000 species. The Drake collection of Hemiptera was bequeathed to the Department in 1957, and includes over 100,000 specimens of Hemiptera, among which there are more than 1,000 holotypes. Material is available for loan for taxonomic research by qualified students and professionals. [1986]

Affiliated Collections:

Dressler, Dr. Robert, Apartado 2072, Balboa, R. de Panama. [RBDC]. Phone: 52-2485. This is a reference collection of orchid pollinators [1986]. (Registered with USNM.)

Engleman, Dr. Dodge, PSC Box 806, APO Miami, FL 34005 USA. [HDEC] Phone: Panama 46-5601; 45-0259. This is a collection of 10,000 specimens of Hemiptera, especially Neotropical species, mainly Pentatomidae, stored in drawers in cabinets, and in Schmitt boxes [1986]. (Registered with USNM.)

Erwin, Dr. Terry L., Department of Entomology, Smithsonian Institution, Washington, DC 20560 USA. [TLEC]. This collection is deposited in the USNM collection.

Garrison, Dr. Rosser W., 1030 Fondale St., Azusa, CA 91702 USA. [RWGC] Phone: (818) 575-5469; 966-2889. About 20,000 New World and 1,000 Old World specimens of Odonata are contained in this collection, including all species found in Alaska, Canada, U.S.A., and Antilles, except for 21 species (two of which are undescribed), representatives of every genus found in North America and northern Neotropical Region, and every extant family in the world. The entire collection is cataloged and housed in clear envelopes, each specimen with data cards [1986]. (Registered with USNM.)

Hespenheide, Dr. Henry A., Department of Biology, University of California, Los Angeles, CA 90024 USA. [CHAH] Phone: (213) 825-3170; 649-0904. The collection specializes in Nearctic and Neotropical Buprestidae and Curculionidae, including Asian material (Baudon, Buprestidae) and secondary types of about 40 species from all groups/areas, all housed in 105 drawers. Special collections include Buprestidae, Agrilinae and Trachyinae; Curculionidae, Zygopinae and Tachygoninae; Chrysomelidae, Hispinae; Diptera, Asilidae; Hymenoptera, Chalcididae. Material is available for exchange and for loans; usual rules apply. Field work has been conducted in Costa Rica and Panama. [1986] (Registered with USNM.)

Lenczy, Dr. Rudolph, Deceased. [RLIE] Collection now at the USNM.

Steyskal, George C., Florida State Collection of Arthropods, P. O. Box 147100, Gainesville, FL 32614 USA. [GCSC] Phone: (904) 372-3505. Entire collection deposited at USNM. [1992]

Surdick, Rebecca F., Route 2, Box 1072, Front Royal, VA 22630 USA. [RFSC] Phone: (703) 635-8553. This collection of 30,000 specimens is mainly aquatic insects with emphasis on Nearctic Plecoptera. These are stored in alcohol vials in cabinets [1986]. (Registered with USNM.)

Ward, Dr. Robert D., [RDWC]. *Address and location of collection unknown.* This is a collection of about 27,000 specimens of Coleoptera, with emphasis on Carabidae, especially Cicindelinae (16,000 specimens). It also contains 800 wing slides of Adephaga, Archostemata, and Neuropteroidea. The collection is housed in 54 drawers and 50 Schmitt boxes [1986]. (Registered with USNM.)

FLORIDA

Gainesville

AMERICAN ENTOMOLOGICAL INSTITUTE, 3005 SW 56TH AVE., GAINESVILLE, FL 32608. [AEIC]

President: Dr. Dale Habeck. Curator: Dr. David B. Wahl. Phone: (904) 377-6458. This institute is an independent research foundation dedicated to the study of Hymenoptera systematics, with emphasis on the Ichneumonoidea; its collection consists of about 589,000 specimens, the collection of Ichneumonidae is the most complete and best arranged for this group. Holdings of Braconidae are significant as well comprising about 174,000 specimens. The remainder of the collection consists of about 19,000 Symphyta, 94,000 "microhymenoptera" and small apocritan families, and 105,000 Aculeata. The type collection numbers about 3,400 holotypes. In addition, there are about 57,000 specimens belonging to other insect orders. The Hymenoptera material is available for loan to qualified students and professionals for systematic research. Also housed at the Institute is the Lloyd collection of Lampyridae. [1992]

Affiliated Collections:

Gupta, Dr. Virendra, 2716 N.W. 37th Terrace, Gainesville, FL 32605 USA. [GPTA] Phone: (904) 392-9279; 371-4071. Approximately 50,000 specimens of Ichneumonidae and some Braconidae from India, including many identified reference specimens and some types of species described by Gupta and colleagues in "Ichneumonologia Orientalis," part 1-9, are included in this collection. Also in the identified reference material are specimens from other parts of the Orient obtained by exchange, including paratypes. The collection is housed in boxes [1986]. (Registered with AEIC.)

FLORIDA STATE COLLECTION OF ARTHROPODS, DIVISION OF PLANT INDUSTRY, 1911 34TH ST. SW, P.O. BOX 147100, GAINESVILLE, FL 32614. [FSCA]

Director: Dr. A. B. Hamon (acting). Curator: Dr. M. C. Thomas. Phone: (904) 372-3505. Professional staff: Mr. Harold A. Denmark (retired), Dr. G. B. Edwards, Dr. A. B. Hamon, Dr. John B. Heppner, Dr. Frank W. Mead, Dr. Lionel A. Stange, Dr. Gary J. Steck, Dr. Howard V. Weems, Jr. (retired), Dr. Robert E. Woodruff (retired), and seven technicians. In addition there are many Research Associates located in Gainesville and elsewhere that take part in the identification and curation of the collection. The prime area of interest encompasses Florida, the Southeastern United States, Greater and Lesser Antilles, and land areas in and bordering the Gulf of Mexico, and the Caribbean Sea. Collections from other parts of the World, especially tropical and subtropical areas of South America, Africa, and Indo-Australia, are being developed at a rapid rate. A special effort is being made to obtain worldwide representation of insects and other arthropods of known economic importance.

The FSCA has grown rapidly in recent years. It is the largest arthropod collection in the Southeastern United States and the seventh largest in the country. It includes the main collection located in the Doyle Conner Building, some collections curated by taxonomists in the Department of Entomology and Nematology, Department of Zoology, and the Florida State Museum (see separate entry, below), University of Florida, Gainesville, and the collections of the Laboratory of Aquatic Entomology, Florida A. & M. University, Tallahassee (see separate entry, below).

The FSCA consists of about 6.5 million curated arthropod specimens, including more than 24,000 type specimens, more than 6,300 of which are primary types. Included in the FSCA but housed elsewhere are 450,000 aquatics at Florida A & M University, 20,000 Arachnida at the University of Florida's Department of Zoology, and 216,000 immatures in the University of Florida's Department of Entomology and Nematology. The FSCA also houses 5,000 biological control voucher specimens. In addition, there are over 30,000 bulk samples of arthropods from many parts of the World preserved in 75% isopropyl alcohol consisting of many millions of specimens, mostly unsorted and unidentified.

The library of the FSCA comprises over 12,500 volumes and 472 current journal subscriptions, primarily on arthropod taxonomy, plus extensive reprint files. The University of Florida libraries house another

5.8 million books and archives of all subjects, including about 800,000 scientific books in the Hume Science Library.

The FSCA maintains a unique program of Research Associates currently with more than 300 members worldwide. This program is coordinated by Dr. Gary Steck.

The collection is also a member institution of the Center for Systematic Entomology which has its headquarters in Gainesville. In addition to the FSCA, collections of the Center include 800,000 Lepidoptera, primarily Rhopalocera, of the Allyn Museum of Entomology (located in Sarasota, Florida), a branch of the Florida State Museum. The combined holdings of these various collections exceed 7 million processed specimens. The "Societas Internationalis Odonatologica," at its last annual meeting (1985) held in Paris, formalized its International Odonata Research Institute, and designated FSCA the site for the new institute where office and workers are provided. Loans are made; usual rules apply. Publications sponsored: "Arthropods of Florida and Neighboring Land Areas," the "Entomology Circular" series, and the "Occasional Papers of the Florida State Collection of Arthropods." [1992]

Affiliated Collections:

Arbogast, Richard T., 114 Monica Blvd., Savannah, GA 31419 USA. [RTAC] Phone: (912) 925-6270. This Lepidoptera collection contains over 2,500 specimens, both pinned and papered. Two paratypes are kept in the collection [1986]. (Registered with FSCA.)

Atkinson, Dr. Thomas H., Department of Entomology, University of California, Riverside, CA 92521 U.S.A. [THAC] Phone: (714) 787-9344. This collection of Coleoptera contains over 10,000 specimens containing in order of importance, Scolytidae, Platypodidae, Cerambycidae, and Buprestidae. It is particularly strong in species from Mexico and Southeastern United States. Paratypes of approximately 40 species are included. [1992]. (Registered with FSCA.)

Baggett, Howard David (Dave), 403 Oleander Dr., Palatka, FL 32177 U.S.A. [HDBC]. This is a collection of 800 specimens of Lepidoptera, mainly from Florida, with all families represented, especially Noctuidae, Geometridae, and Pyralidae, including one paratype. Material may be borrowed; usual loan rules apply [1992]. (Registered with FSCA.)

Baranowski, Dr. R. M., Agricultural Research and Education Center, 18905 S.W. 280th St., Homstead, FL 33031 USA. [RMBB] Phone: (305) 247-4624; 247-7494. This is a collection of 40,000 specimens of Hemiptera, including over 400 paratypes. Specimens may be borrowed for study; usual loan rules apply [1986]. (Registered with FSCA.)

Beck, Dr. Andrew F., Navy Disease, Vector Ecology, and Control Center, Box 43, N. A. S., Jacksonville, FL 32210 USA. [AFBC] Phone: (904) 772-2424. About 8,000 specimens are housed in this collection of Lycaenidae (Lepidoptera). Material may be borrowed for study; usual loan rules apply [1986]. (Registered with FSCA.)

Belmont, Robert A., P. O. Box 2626, Naples, FL 33939 USA. [RABC] Phone: (813) 574-5352. This Lepidoptera and Coleoptera collection, stored in 2 cabinets, contains about 4,000 specimens, mainly Geometridae. Specimens may be borrowed, in person, with a guarantee to return them [1992]. (Registered with FSCA.)

Benton, Dr. Allen H., 292 Water St., Fredonia, NY 14063 USA. Phone: (716) 679-0462. [AHBS]. This is a collection of over 5,000 slides of Siphonaptera; no holotypes are retained. Specimens may be borrowed; usual loan rules apply [1992]. (Registered with FSCA.)

Berner, Dr. Lewis, Department of Zoology, University of Florida, Gainesville, FL 32611 USA. [LBIC] Phone: (904) 392-1568; 373-9629. This is a collection of 250,000 specimens of Ephemeroptera, including some holotypes and paratypes [1986]. (Registered with FSCA.)

Bick, George H., 1928 SW 48th Ave., Gainesville, FL 32608 USA. [GHBC] Phone: (904) 372-4805. Entire collection deposited in FSCA [1992].

Boscoe, Richard W., 150 Ridge Pike, Apt. 201, Lafayette Hill, PA 19444 USA. [RWBC] This is a collection of 3,000 specimens of Lepidoptera housed in cabinets. Specimens may not be borrowed for study [1986]. (Registered with FSCA.)

Brattain, R. Michael, 3206 Longlois Drive, Lafayette, IN 47904 USA. [RMBC] Phone: (317) 448-7123; 447-3627. This is a collection of about 40,000 Coleoptera housed in drawers and Schmitt boxes [1986]. (Registered with FSCA.)

Brower, Dr. Lincoln P., Department of Zoology, University of Florida, Gainesville, FL 32611 USA. [LPBC] Phone: (904) 392-1107. This is a collection of 3,000 Lepidoptera, and some mimetic Diptera and Hymenoptera, pinned and papered. These are mostly research specimens and teaching exhibits. [1986] (Registered with FSCA.)

Burris, Daniel L., 16208 September Dr., Lutz, FL 32459. This is primarily a butterfly collection, with 15,000 specimens, 1,000 pinned, remainder papered. Specimens may be borrowed; usual loan rules apply. [1992] (Registered with FSCA.)

Carlson, Dr. Paul H., [*Address and location of collection unknown.*] [PHCC] [1992]. (Registered with FSCA.)

Cavanaugh, Robert William, Jr., [*Address and location of collection unknown.*] [1992]. (Registered with FSCA.)

Coffman, Mr. John M., Route 1, Box 331, Timberville, VA 22853 USA. [JMCI] Phone: (703) 896-8149. This is a collection of 10,000 Lepidoptera with some Hymenoptera and Coleoptera. Specimens may be borrowed for study; usual loan rules apply [1986]. (Registered with FSCA.)

Cokendolpher, James C., 2007 29th St., Lubbock, TX 79411 USA. [JCCC] Phone: (806) 744-0318. This is primarily a reference collection composed entirely of Opiliones and is worldwide with emphasis on Phalangioidea of North America, northern Europe, and Japan. The collection, about 4,000 small alcohol jars and vials, contains no primary types, but some paratypes are present. Some material is available for trade. Loans are made to specialists. The collection also contains examples of most ants (Formicidae) from western Texas and common spiders (Araneae) from western Texas. The library has good coverage of taxonomic literature. Extensive holdings are also present on Opiliones in general as well as almost complete holdings of literature on Schizomida. Included with the collection is about 800 catalogued microscope slide preparations of dissected parts of Opiliones, and of all developmental stages of over 1,000 individuals from western U.S.A. [1992]. (Registered with FSCA.)

Daigle, Jerrell James, Water Quality Analysis Section, FPER, 2600 Blair Stone Rd., Tallahassee, FL 32301 USA. [JJDC] Phone: (904) 488-0780; 878-8787. This is a collection of 1,200 Odonata in cellophane envelopes. Specimens may be borrowed by qualified Odonatologists [19926]. (Registered with FSCA.)

Deonier, D. L., Department of Zoology, Miami University, Oxford, OH 45056 USA. [DLDC] Phone: (513) 529-5454; 523-7701. The collection is primiarly Diptera, but all orders are represented to a great or lesser extent, with 87,000 specimens, including 2 primary types, 700 slides, and 1,500 vials of alcohol material. Material may be borrowed for study; usual loan rules apply [1986]. (Registered with FSCA.)

Döberl, Manfred, Seweg 34, D-8423, Abensberg, Germany. [MDIC]

Donnelly, Dr. Thomas W., Department of Geological Sciences and Environmental Studies, State University of New York, Binghamton, NY 13901 USA.

[TWDC] Phone: (607) 777-2264; FAX: (607) 777-2288. This is a collection of Odonata, particularly from Central America, India, Thailand, Fiji, northern South America, and the U.S.A., including over 20,000 specimens, mainly papered. Specimens may be borrowed; usual loan rules apply [1986]. (Registered with FSCA.)

Dow, Linwood C., 230 Emerald Lane, Largo, FL 32640 USA. [LCWC] Phone: (813) 855-8850. This is a North American collection of Lepidoptera of about 18,000 specimens with emphasis on Florida species. It includes four paratypes. It is stored in drawers. Specimens may be borrowed; usual loan rules apply [1992]. (Registered with FSCA.)

Dozier, Annie, 297 Dovehaven Drive, Pickens, SC 29671 USA. [ANDC] This is a collection of Mutillidae, primarily South Carolina specimens, plus material from other areas of U.S.A. and foreign countries, stored in drawers. There is also a general collection of local insects containing miscellaneous groups, but primarily Coleoptera [1992]. (Registered with FSCA.)

Dozier, Byrd K., P. O. Box 699, Keystone Heights, FL 32656 USA. [BKDC] Phone: (619) 421-6095. This is a collection of about 15,000 specimens of Buprestidae, worldwide, housed in 60 drawers and 20 boxes [1992]. (Registered with FSCA.)

Dozier, Herbert L., 297 Dovehaven Drive, Pickens, SC 29671 USA. [HLDC] This is a collection of Coleoptera, primarily from Maryland, plus material from other areas of the U.S.A. and foreign countries. Most families represented to some extent; Coccinellidae fairly well represented. The collection is stored in 30 drawers, 50 Schmitt and Mason boxes. Most species in the collection are represented by only a few specimens. The collection includes 225 paratypes of 89 species of Coccinellidae [1992]. (Registered with FSCA.)

Drees, Bastiaan M., P. O. Box 2150, Bryan, TX 77806-2150 USA. [BMDC] Phone: (409) 845-6800; 846-0524. This collection includes 500 Tabanidae from West Virginia, Ohio, and Texas as a reference collection. The Rhopalocera reference collection is from West Virginia, Ohio, Texas, and Mexico (200 specimens). There is a general collection, especially Coleoptera, Neuroptera (1,000 specimens), and 2,000 alcohol vials of immature and soft bodied insects. The pinned collection is housed in 59 Schmitt boxes [1986]. (Registered with FSCA.)

Dunkle, Dr. Sidney W., Biology Department, Collin County Community College, Spring Creek Campus, Plano, TX 75074 USA. [SWDC] Phone: (214) 881-5989. This a collection of over 1,900 species of Odonata. Specimens may be borrowed for study; usual loan rules apply [1992]. (Registered with FSCA.)

Dunn, Gary A., Young Entomology Society, Inc., 1915 Peggy Place, Lansing, MI 48910-2553 USA. [GADC] Phone: (517) 887-0499. This collection consists of Carabidae and Cicindelidae (Worldwide), with best representation from North America (especially New Hampshire, Michigan, Florida, and Texas). There are about 35,000 specimens stored in 95 drawers and 15 Schmitt boxes. Material may be borrowed for study; usual loan rules apply. (Registered with FSCA.) [1992]

Fairchild, Dr. G. B., 16 NW 22 Drive, Gainesville, FL 32601 USA. [GBFC] Phone: (904) 372-3505; 378-7180. This is a collection of over 17,000 specimens representing 1,741 species of Tabanidae and Psychodidae, including 41 holotypes and over 1,000 secondary types, housed in drawers in cabinets. Material may be borrowed for study; usual loan rules apply [1992]. (Registered with FSCA.)

Farnworth, Dr. Edward G., [*Address and location of collection unknown.*] [EGFC]. [1992]. (Registered with FSCA.)

Flowers, Dr. R. Willis, 1208 Victory Garden Dr., Tallahassee, FL 32301 USA. [RWFC] This is a collection of 5,000 specimens, mostly Coleoptera and Lepidoptera, from Central America and Africa [1986]. (Registered with FSCA.)

Foote, Dr. Benjamin A., Department of Biological Sciences, Kent State University, Kent, OH 44240 USA. [BAFC] Phone: (216) 672-2817; 673-1520. This is a collection of 5,000 Diptera, including 2 paratypes. Material may be borrowed for study; usual loan rules apply [1986]. (Registered with FSCA.)

Fragoso, Dr. Sergio Augusto, EMBRAPA/Museu Nacional, Quinta Boa Vista, Rio de Janerio, 20.942 Brazil. [SAFC] Phone: 225-7113. This is a collection of 20,000 specimens of worldwide Cerambycidae, with about 15 holotypes [1986]. (Registered with FSCA.)

Frank, Dr. J. Howard, Department of Entomology and Nematology, University of Florida, Gainesville, FL 32611-0620 USA. [JHFC] Phone: (904) 392-1901. This is a collection of 100,000 specimens of Staphylinidae, primarily of Florida, other southern states, and the circumcaribbean region. Stored in unit trays in 92 drawers on mounted and labelled adults, plus some stored in boxes, one cabinet of alcohol vials with unsorted adults and immature stages. Holotypes are not held, but deposited in other public collections. Material may be borrowed for study; usual loan rules apply [1992]. (Registered with FSCA.)

Fritz, Manfredo, Instituto Investigaciones Entomologicas Salta, Casilla Correo 539, 4400 Salta, Argentina. [MAFC] This is a collection of 50,000 specimens of Hymenoptera including about 100 primary types. Material may be borrowed for study; usual loan rules apply [1986]. (Registered with FSCA.)

Gage, Edward V., P. O. Box 380622, San Antonio, TX 78280 USA. [EGCC] Phone: (512) 684-7873. This is a worldwide collection of Lepidoptera and Coleoptera, the latter with emphasis on Scarabaeidae, Cicindelidae, and Tenebrionidae. The Lepidoptera collection consists of 5,000 pinned and 10,000 papered specimens, including allotypes, paratypes, and topotypes, housed in 96 drawers and numerous boxes. Specimens may be borrowed for study; usual loan rules apply, except that 30 days is the maximum loan limit [1986]. (Registered with FSCA.)

Gall, Lawrence F., Department of Biology, 257 OML, Yale University, New Haven, CT 06520 USA. [LFGC] Phone: (203) 436-8847; 562-1986. This is a collection of 3,000 adult and 2,000 immature Lepidoptera, especially Rhophalocera from North America, Holarctic Catocala and relatives, and a synoptic collection of northeastern U.S.A. moths. Specimens may be borrowed for study; usual loan rules apply [1986]. (Registered with FSCA.)

Gatrelle, Ron, 126 Wells Rd., Goose Creek, SC 29601. This is a collection of Lepidoptera, especially butterflies, with complete coverage of eastern butterflies. (Registered with FSCA.) [1992]

Gerberg, Dr. Eugene J. [EJGC]. 5819 NW 57th Way, Gainesville, FL 32606. Phone: (904) 373-7384. Lyctidae and Bostrichidae (Coleoptera) of the world are represented in this specialized collection, along with a large collection of butterflies from many parts of the world. In addition there is a miscellaneous collection of Coleoptera, Lepidoptera, and Culicidae, particularly *Stegomyia* of east Africa. Material is available for exchange but not for student loan. Excellent library resources are available. The collection is housed in drawers in cabinets [1992]. (Registered with FSCA.)

Giesbert, Edmund F., 9780 Drake Lane, Beverly Hills, CA 90210 USA. [GIES] Phone: (213) 276-7691. This is a collection of Coleoptera, mostly Cerambycidae, from North and Central America, and the West Indies. All material is personally collected in the field, uniformly curated, and identified by the collector. The primary collection contains about 30,000 cerambycids representing about 2,000 species, and approximately 2,500 Cleridae of 350 species, housed in 62 drawers and a number of boxes. Also a secondary collection consists of about 5,000 other Neotropical Coleoptera (Scarabaeidae, Buprestidae, Cicindelidae, Elateridae, and Curculionidae) in boxes. The working collection contains paratypes of about 50 species, and over 200 specimens of undescribed cerambycid species [1986]. (Registered with FSCA.)

Goodwin, Dr. James T., International Agriculture Programs, P. O. Box 2477, Texas A & M University, College Station, TX 77843 USA. [JTGC] Phone: (409) 822-7717; 775-7697. This is a collection of 12,000 specimens of Tabanidae (Diptera), and Odonata, including about 100 paratypes, pinned and in alcohol vials. Material may be borrowed for study and exchange; usual loan rules apply [1986]. (Registered with FSCA.)

Gross, Scott W., [*Deceased; location of collection unknown.*] [SWGC]. (Registered with FSCA.)

Hall, Dr. David G., 1200 Pinewood St., Clewiston, FL 33440. Phone: (813) 983-5341. Insects associate with sugarcane in Florida form a collection of 4,500 pinned specimens and 350 vials of alcohol specimens. (Registered with FSCA.) [1992]

Heitzman, Roger L., P. O. Box 38261, Baltimore, MD 21231 USA. [RLHC] Phone: (202) 454-5872; 422-6819. The collection of 30,000 specimens is mainly Lepidoptera, with specimens of Coleoptera, Diptera, Hymenoptera, Neuroptera, and a few specimens of other orders. Material may be borrowed for study; usual loan rules apply [1986]. (Registered with FSCA.)

Henry, Parker R., 10960 SW 89th Terrace, Miami, FL 33176 USA. [PRHC] Phone: (305) 279-3005. This is a collection of over 15,000 specimens of Lepidoptera, mainly butterflies. Material may be borrowed; usual loan rules apply [1986]. (Registered with FSCA.)

Heppner, Dr. John B., Florida State Collection of Arthropods, Division of Plant Industry, P. O. Box 147100, Gainesville, FL 32614 USA. [JBHC] Phone: (904) 372-3505; 373-5630. About 95,000 Lepidoptera, especially micros, including about 1,200 paratypes, and 1,200 genitalia slides, mostly Nearctic, particularly southwestern USA and Florida, with some families with worldwide representation compose this collection which may be borrowed; usual rules apply [1992]. (Registered with FSCA.)

Hovore, Frank T., c/o Placerita Canyon Nature Center, 19152 W. Placerita Canyon Road, Newhall, CA 91321 USA. [FTHC] Phone: (805) 259-7721; 259-7920. This is a collection of Coleoptera; synoptic collections of Buprestidae, Cleridae, and Scarabaeidae, one CASC drawers each; Scarabaeidae, genus *Pleocoma*, 15 drawers, about 4,000 specimens representing all species and subspecies, males and females, with paratypes of most species. Holotypes are deposited with CASC and LACM; Cerambycidae in about 80 drawers including field pinned specimens to be curated and integrated; 20 redwood boxes of duplicate or exchange specimens. North America, Mexico, and Central America well represented. Material may be borrowed for study; usual loan rules apply [1992]. (Registered with FSCA.)

Jenkins, Dr. Dale W. and Joanne F., 3028 Tanglewood Drive, Sarasota, FL 33579 USA. [DWJC] Phone: (813) 921-2627. This is a collection of about 50,000 specimens of Lepidoptera, including paratypes of some butterflies, mostly pinned, some in papers. Material may be borrowed; usual loan rules apply [1986]. (Registered with FSCA.)

Johnson, James W., Department of Entomology, Michigan State University, East Lansing, MI 48824 USA. [JWJC] This collection contains Aranae, Hymenoptera, Coleoptera, Neuroptera, Lepidoptera, Diptera, Hemiptera, Homoptera, and Apterygota, over 3,000 specimens, pinned, in alcohol, and on slides. Material may be borrowed for study; usual loan rules apply [1992]. (Registered with FSCA.)

Jordan-Soto, Pablo E., Apartado Postal 2067, San Pedro Sula, Honduras. [PEJS] Phone: 56-2078; 56-2153. This collection contains over 16,000 specimens, mostly identified, stored in 82 boxes in cabinets [1986]. (Registered with FSCA.)

Kendall, Roy O., 5598 Mt. McKinley Drive, San Antonio, TX 78251 USA. [ROKC] Phone: (512) 684-2518. This collection contains over 100,000 pinned adults and 2,000 vials of immature Lepidoptera, and some Coleoptera, Diptera, Hymenoptera, and a few specimens of other orders. Paratypes of 5 species are included in the collection. Material may be borrowed on short term loan; usual rules apply. (Registered with FSCA.) [1992]

Knopf, Dr. Kenneth W., 7303 NW 47th Court, Gainesville, FL 32606 USA. [KWKC] Phone: (904) 377-1705; 373-5947. This is a collection of about 10,000 specimens of Odonata, representing about 1,000 species, primarily from the New World, but with some world representation. The collection is housed in cellophane envelopes [1992]. (Registered with FSCA.)

Kutash, Marc Roger, 4314 South Anita Boulevard, Tampa, FL 33611 USA. [MRKC] Phone: (813) 238-0397; 238-8811; 837-5377; 876-9471. This is a collection of over 1,000 specimens of Rhopalocera, Heterocera, Orthoptera, Odonata, and Coleoptera, pinned, and stored in drawers [1986]. (Registered with FSCA.)

Lampert, Col. Lester L., Jr., [*Address and location of collection unknown.*] [LLLJ] [1992]. (Registered with FSCA.)

Landolt, Peter J., 2247 SW 43rd Place, Gainesville, FL 32608 USA. [PJLC] Phone: (904) 378-8389. This is a collection of about 15,000 specimens of Lepidoptera, Hymenoptera (Vespidae, Sphecidae), and Coleoptera, mostly pinned, some in alcohol vials. Material may be borrowed for study; usual loan rules apply [1986]. (Registered with FSCA.)

Lehman, Robert D., Casa G 12, Colonia El Sauce, Segunda Etapa, La Ceiba, Honduras. [RDLC] This is a collection of over 3,000 specimens of moths and butterflies of Honduras, with over 1,000 species, mostly named, except for some of the butterflies. It is housed in drawers in cabinets [1986]. (Registered with FSCA.)

Lloyd, Dr. James E., 915 NW 40th Terrace, Gainesville, FL 32605 USA. [JELC] Phone: (904) 392-1901; 378-3819. This is a collection of about 8,000 specimens of Coleoptera, mostly Lampyridae, worldwide. It contains fewer than 30 primary types. They are housed in unit trays in drawers. Material may be borrowed for study; usual loan rules apply [1986]. (Registered with FSCA.)

MacRae, Mr. Ted C., 7500 Greenhaven Dr., Sacramento, CA 95831 U.S.A. [TCMC] Phone: 916-757-4700; 916-393-3700. Over 40,000 specimens, mostly Coleoptera, are housed in 60 Cornell drawers and 36 Schmitt boxes in cabinets. Emphasis is on Nearctic Buprestidae, most with host labels and genitalia extracted; some paratypes. The remainder of the collection is mostly Cerambycidae, Chrysomelidae, Scarabaeidae, some other beetles, and U.S.A. Lepidoptera. Specimens are available for borrowing or exchanging; usual loan rules apply [1992]. (Registered with FSCA.)

Martinez, Antonio, Juan Carreras 508-C.C. 21, Rosario de Lerma, 4405 Salta, Argentina. [AMIC] This is a collection of about 44,000 specimens, mostly Scarabaeidae, some Meloidae, Carabidae, and Staphylinidae, and some Hemiptera, Reduvoidea, stored in boxes and on slides [1986]. (Registered with FSCA.)

Mead, Dr. Frank W., Division of Plant Industry, P. O. Box 147100, Gainesville, FL 32614 USA. [FWMC] Phone: (904) 372-3505; 376-7285. This is a collection of Culicidae (Diptera), mostly from Ohio and Florida, which includes nearly 4,000 specimens, representing 49 species, and some additional unidentified specimens stored in 25 Schmitt boxes [1986]. (Registered with FSCA.)

Minno, Marc C., 3033-18 Diamond Village, Gainesville, FL 32603 USA. [MCMF] This is a collection of over 12,000 specimens of Lepidoptera, with good representation from U.S.A., and about 3,500 specimens from western Colombia, some French, and some Indian material. Specimens are pinned in drawers stored in cabinets, with some alcohol preserved specimens. Material may be borrowed for study; usual loan rules apply [1986]. (Registered with FSCA.)

Muchmore, Dr. William B., Department of Biology, University of Rochester, Rochester, NY 14627 USA. [WBMC] Phone: (716) 275-3844. This is a collection of over 20,000 specimens of Pseudoscorpionida (Arachnida). The types are deposited in the FSCA. It is stored in alcohol and on slides [1992]. (Registered with FSCA.)

Muma, Dr. Martin H., [MHMC] [*Deceased; collection at FSCA.*] [1992].

Neal, Thomas M., Jr., 1705 NW 23rd St., Gainesville, FL 32605 USA. [TMNJ] Phone: (904) 378-3550; 375-1916. About 5,000 specimens of Lepidoptera are represented in this collection which is stored in drawers. Material may be borrowed for study; usual loan rules apply [1986]. (Registered with FSCA.)

Nelson, Dr. Gayle H., College of Osteopathic Medicine of the Pacific, 309 Pomona Mall, Pomona, CA 91766-1889 USA. [GHNC] Phone: (714) 623-6166; 980-4313. Over 150,000 specimens of Coleoptera, pinned, papered, and in alcohol, stored in 60 drawers and 270 Schmitt boxes, in cabinets, comprise this collection. Material may be borrowed for study; usual rules apply [1986]. (Registered with FSCA.)

Nordin, Dr. John S., 2217 Skyview Ave., Laramie, WY 82070 USA. [JSNC] Phone: (606) 223-1580. This collection is mostly Lepidoptera, with 40,000 pinned and 5,000 papered specimens, mostly Hesperiidae, Geometridae, Noctuidae, and micros from U.S.A. and Canada, stored in drawers. Material may be borrowed for study; usual loan rules apply [1986]. (Registered with FSCA.)

Passoa, Steven,USDA, APHIS, PPQ, Bldg 3, Rm 109, 8995 E. Main St., Reynoldsburg, OH 43068 USA. [SPIC] All orders are represented in this collection of over 10,000 specimens, but emphasis is on the Lepidoptera. The collection is strongest in Nearctic Lepidoptera and insects of Honduras. Preserved larvae are included. Slides of larval mouthparts and adult genitalia are stored in boxes, and immatures are preserved in alcohol. Host plant samples are pressed and mounted on herbarium sheets. Material may be borrowed for study; usual loan rules apply [1986]. (Registered with FSCA.)

Porter, Dr. Charles C., Department of Biological Sciences, Fordham Univer sity, Bronx, NY 10458 USA. [CCPC] Phone: (212) 579-2571, 579-2557; 798-8535. This is a collection of over 100,000 specimens of Hymenoptera, with emphasis on Ichneumonidae, Eumenidae, Vespidae, Sphecidae, and Apoidea. In addition, there are some specimens of Lepidoptera, Odonata, Coleoptera, Neuroptera, and a few other orders. Material may be borrowed for study; usual loan rules apply [1986]. (Registered with FSCA.)

Preston, Floyd W. and June D., 832 Sunset Drive, Lawrence, KS 66044 USA. [FWJP] Phone: (913) 843-6212. Over 16,000 pinned and 40,000 papered specimens of Rhopalocera (Lepidoptera) are contained in this collection. Specimens are primarily from North America north of Mexico. A few paratypes are included. The collection is housed in drawers and cabinets. Specimens may not be borrowed for study [1986]. (Registered with FSCA.)

Paulson, Dennis R., Washington State Museum, University of Washington, Seattle, WA 98195 USA. [DRPC] Phone: (206) 543-4486; 632-7953. This collection of 40,000 specimens of Odonata is stored in plastic envelopes, in vertebrate specimen cases. Material may be borrowed for study; usual loan rules apply [1986]. (Registered with FSCA.)

Plomley, John M., 7531 Lincoln St., Hollywood, FL 33024. [JMPC] Phone: (305) 987-1118. This is a Lepidoptera collection of 10,000 specimens from eastern U.S.A. and Canada. Specimens may be borrowed for study; usual loan rules apply [1986]. (Registered with FSCA.)

Reiskind, Dr. Jonathan, Department of Zoology, University of Florida, Gainesville, FL 32611 USA. [JRIC] Phone: (904) 392-1187; 378-8290. About 3,000 Arachnida from Central and South America, and Borneo are contained in this collection, stored in alcohol vials. Material may be borrowed for study; usual loan rules apply [1986]. (Registered with FSCA.)

Richman, Dr. David B., Department of Entomology and Plant Pathology, Box 3BE, New Mexico State University, Las Cruces, NM 88003-0057 USA. [DBRC] This is a collection of 500 specimens, 250 species of Arachnida, mostly Scorpionida and Araneae, stored in alcohol vials. Specimens may be borrowed only if returned within one month because the collection is synoptic and in constant use [1986]. (Registered with FSCA.)

Riley, Dr. Edward G., Department of Entomology, 402 Life Sciences Bldg., Louisiana State University, Baton Rouge, LA 70803 USA. [EGRC] Phone: (504) 388-1634; 766-2717. This is a collection of nearly 100,000 specimens, over half of which are Chrysomelidae, and the remainder of other families of Coleoptera, stored in 80 drawers and 161 Schmitt boxes. Material may be borrowed for study; usual loan rules apply [1986]. (Registered with FSCA.)

Schuster, Dr. Jack C., Universidad del Valle de Guatemala, Guatemala City, Guatemala. [JCSC] Phone: 690-791. This is a collection of about 5,000 specimens of Passalidae, Syrphidae, and Apoidea [1986]. (Registered with FSCA.)

Staines, C. L., Jr., 3302 Decker Place, Edgewater, MD 21037. [CLSJ] Phone: (410) 956-2174. Seven thousand specimens of Coleoptera in Cornell drawers and Schmitt boxes, strong in Chrysomelidae, Scarabaeidae, and aquatic families, including some secondary types [1992]. (Registered with FSCA.)

Stephan, Karl H., Route 1, Box 913, Red Oak, OK 74563 USA. [KHSC] Phone: (918) 465-5201. This is a collection of about 30,000 specimens of Coleoptera from Latimer County, Oklahoma. At present about 97 families, 1,097 genera, and 2,851 species are represented. An estimated 70% of these have been identified by specialists. Specimens are housed in 140 Schmitt boxes. Material is available for study; usual loan rules apply [1986]. (Registered with FSCA.)

Tennessen, Ken, 1949 Hickory Ave., Florence, AL 35630. This is a collection of over 10,000 specimens of Odonata, plus alcohol preserved larvae of Odonata. (Registered with FSCA.) [1992]

Thomas, Dr. M. C., Division of Plant Industry, P. O. Box 147100, Gainesville, 32614. [MCTC] Phone: (904) 372-3505. This is a collection of about 25,000 specimens of Florida Coleoptera, pinned, and arranged in drawers. Material may be borrowed for study; usual loan rules apply [1986]. (Registered with FSCA.)

Tidwell, M. A., Department of Environmental Health, College of Health, University of South Carolina, Columbia, SC 29208 USA. [MATC] This is a collection of 20,000 Diptera, particularly Simuliidae, Culicidae, Tabanidae, and Culicoides. Material may be borrowed for study; usual rules apply [1986]. (Registered with FSCA.)

Vernon, John B., 1135 McClelland Drive, Novato, CA 94947 USA. [JBVC] Phone: (415) 541-1508; 897-1081. This is a collection of 8,000 butterflies (Lepidoptera) from U.S.A. and Mexico, with exchanges from Canada, England, and Europe. Material may be borrowed for study; usual loan rules apply [1986]. (Registered with FSCA.)

Vick, Dr. Kenneth W., 10104 E. Franklin Ave., Glendale, MD 20769 USA. [KWVC] Phone: (904) 374-5772; 472-3059. This collection specializes in Cicindelidae (Coleoptera), with 4,000 specimens representing the majority of U.S.A. species, some foreign species, and a few paratypes are included; and Scarabaeidae (Coleoptera), with 5,000 specimens with emphasis on species from southeastern U.S.A. [1986]. (Registered with FSCA.)

Wappes, James E., Rt. 2, Box 16BB, Atwater Road, Chadds Ford, PA 19317 USA. [JEWC] This is a collection of about 60,000 Coleoptera with emphasis on Cerambycidae (25,000 specimens), including numerous paratypes. Material may be borrowed for study; usual loan rules apply [1992]. (Registered with FSCA.)

Wilkerson, Dr. Richard C., P. O. Box 217, Welcome, MD 20693 USA. [RCWC] Phone: (202) 357-1856. This is a collection of 12,000 specimens of Tabanidae (Diptera). Specimens may be borrowed for study; usual loan rules apply [1986]. (Registered with FSCA.)

Woodruff, Dr. Robert E.,Florida Department of Agriculture, Division of Plant Industry, Gainesville, FL 32602 USA. [REWC] Phone: (904) 372-3505; 376-1914. This is a worldwide collection of over 100,000 Scarabaeidae (Coleoptera), with greatest strength in the New World, in the subfamilies Coprinae, Aphodiinae, and Geotrupinae. Most material is pinned, with good ecological data and most with genitalia extracted. In addition to the pinned collection (stored in unit trays in drawers) there are several hundred thousand specimens stored in alcohol, much of which is identified. Specimens may be borrowed; usual rules apply. The collection is in the process of being donated and incorporated into FSCA [1986].

Young, Dr. Frank N., Department of Zoology, Indiana University, Bloomington, IN 47401 USA. [FNYC] Phone: (812) 335-3985; 332-1224. This collection contains about 100,000 pinned specimens of water beetles (Coleoptera), including many paratypes. Specimens are from North America, particularly Florida, Mexico, as well as South America. The collection is housed in drawers and Schmitt boxes. The collection is in the process of being dispersed to FSCA, UMMZ, and FMNH [1986]. (Registered with FSCA.)

Zieger, Charles F., 3751 Sommers St., Jacksonville, FL 32205 USA. [CFZC] Phone: (904) 384-2679. This is a collection of 3,000 specimens of butter flies and moths (Lepidoptera). Material may be borrowed for study; usual loan rules apply [1986]. (Registered with FSCA.)

Zoebisch, Tomas G., CATIE Project MID, 7170 Turrialba, Costa Rica [TGZC] This is a collection of 2,000 specimens of Coleoptera, mainly Scarabaeidae from Central and Southern Mexico, with a few from Africa, Europe, and Asia. Specimens may be borrowed for study; usual loan rules apply. [1992]. (Registered with FSCA.)

FLORIDA STATE MUSEUM, UNIVERSITY OF FLORIDA, GAINESVILLE, FL 32611. [FSMC]

This collection contains only Lepidoptera, no other insects and spiders, (see under Sarasota, below).

UNIVERSITY OF FLORIDA AGRICULTURE EXPERIMENT STATION [SEE FLORIDA STATE COLLECTION OF ARTHROPODS, FSCA.]

UNIVERSITY OF FLORIDA DEPARTMENT OF ENTOMOLOGY [SEE FLORIDA STATE COLLECTION OF ARTHROPODS, FSCA.]

Homestead

MUSEUM COLLECTIONS OF EVERGLADES NATIONAL PARK, PARK HEADQUARTERS, (Attention: South Florida Research Center) P. O. BOX 279, HOMESTEAD, FL 33030. [EGNP]

Museum Curator: Mr. Daniel E. Foxen. Phone: (305) 242-7826. Professional staff: One Museum Technician and one Museum Clerk. The insect collection contains about 2,500 specimens collected throughout the Everglades National Park, Fort Jefferson National Monument/Dry Tortugas, and south Florida in general. Ninety percent of the collection is pinned and labelled. [1992]

Lake Placid

ARCHBOLD BIOLOGICAL STATION COLLECTION, P. O. BOX 2057, LAKE PLACID, FL 32852-2057. [ABSC]

Director: Dr. Mark Deyrup. Phone: (813) 465-2571; FAX (813) 699-1927. About 240 families of insects and about 15 families of spiders are represented by 70,000 specimens in this collection. Ten species of primary types in 3 orders are included. Specimens may be borrowed for study; usual loan rules apply. [1986]

Leesburg

CLEMENTS' MUSEUM OF EXOTIC INSECTS, WILLIAM B. CLEMENTS, 3004 Casteen Road, LEESBURG, FL 32748-8549. [CMEI]

Although this collection exists, no information is available about it after several requests [1992]

Orlando

ARTHROPOD COLLECTION, BIOLOGY DEPARTMENT, UNIVERSITY OF CENTRAL FLORIDA, P. O. BOX 25000, ORLANDO, FL 32816-2368. [UCFC]

Curator: Stuart M. Fullerton (retired), 469 S. Central Ave., Oviedo, FL 32765. Phone: (407) 365-5279. This collection consists of about 29,800 pinned specimens of insects housed in three cabinets of 25 drawers each, and numerous Schmitt boxes. There are 2,400 4-dram vials of specimens in alcohol housed in two special cabinets. Most of the collection is from the five major eco-zones of the 1,100 acre campus in East Orange County, Florida. The remainder is from the surrounding areas. The collection is strong in Hymenoptera (52%), Coleoptera (20%) and Diptera (12%). The collection is relatively new and is growing. There is much work to be done in identification to a generic and specific level. Volunteer help is welcome. [1992]

Tallahassee

CENTER FOR STUDIES IN ENTOMOLOGY, FLORIDA A. & M. UNIVERSITY, TALLAHASSEE, FL 32307. [FAMU]

Director: Dr. William L. Peters. Phone: (904) 599-3912. Professional staff: Dr. R. W. Flowers, Dr. M. D. Hubbard, Dr. C. W. O'Brien, Dr. M. L. Pescador, Dr. A. R. Soponis, Mr. J. Jones, Mr. G. B. Marshall, Mrs. J. G. Peters. The Florida A. & M. Aquatic Insect Collection contains the following: Ephemeroptera collection, 232 genera, about 130,000 specimens, type specimens in 54 genera; 83,000 specimens of Chironomidae, mostly on slides; 150,000 specimens of aquatic insects other than Ephemeroptera and Chironomidae; general collection of 37,300 pinned insects in drawers. Publications sponsored: "Eatonia." This collection has been developed in conjunction with the Florida State Collection of Arthropods (see above). [1992]

Affiliated Collections:

Pescador, Dr. Manuel L., 5208 Touraine Dr., Tallahassee, FL 32303 USA. [MLPC] This collection consists of 300 Philippine Ephemeroptera [1992]. (Registered with FAMU.)

Peters, Dr. William, 1803 Chuli Nene, Tallahassee, FL 32301 USA. [WPEC] This is a collection of 30,000 Ephemeroptera, worldwide, especially New Caledonia. [1992] (Registered with FAMU.)

Sarasota

ALLYN MUSEUM OF ENTOMOLOGY, 3707 BAYSHORE ROAD, SARASOTA, FL 33580. [FSMC]

Director: Dr. Lee D. Miller. Phone: (813) 355-8475. Curator: Dr. Jacqueline Y. Miller. This large collection of butterflies is a part of the Florida State Museum (see above). [1986]

GEORGIA

Athens

MUSEUM OF NATURAL HISTORY, ENTOMOLOGY COLLECTION, UNIVERSITY OF GEORGIA, ATHENS, GA 30602. [UGCA]

Director: Dr. Warren T. Atyeo. Phone: (706) 542-2816. Professional staff: Cecil L. Smith (Hemiptera); Robert W. Matthews (Hymenoptera); J. Bruce Wallace (Trichoptera). The collection is housed in 4,150 drawers. Total holdings include about 550,000 pinned specimens, primarily from southeastern USA, with Georgia well represented. Special collections include: the P. W. Fattig collection; 70,000 slides and 300,000 specimens of feather mites; and 5,000 slides of free-living laelapids. Slides are housed in 3,000 100-slide capacity boxes. Primary types are deposited in USNM, AMNH, and FMNH. Material is available for exchange and loans including student loans; loans are made on understanding that primary and half the secondary types are to be returned. [1992]

Affiliated collection:

Starr, Dr. Christopher K., Department of Zoology, University of the West Indies, St. Augustine, Trinidad. [CKSC] Phone: (809) 663-1334, ext. 2046. FAX: (809) 663-9684. Several thousand aculeate Hymenoptera, primarily social wasps are housed at the Museum of Natural History, Athens, GA. Regular loans may be arranged through the curator of entomology at that museum, subject to usual loan rules, except permission from owner must be obtained prior to destructive study [1992]. (Registered with UGCA.)

HOOGSTRAAL CENTER FOR TICK RESEARCH, DEPARTMENT OF ENTOMOLOGY, GEORGIA SOUTHERN UNIVERSITY, STATESBORO, GA 30460 [HCTR].

[*The following information was obtained from a brochure; no reply to questionnaire.*] Director: Dr. James E. Keirans. Phone: (912) 681-5497. Dr. Harry Hoogstraal (1917-1986) was among the most honored biologists and medical entomologists of our time. During his 37 years of service as director of the Medical Zoology Department of Naval Medical Unit No. 3 in Cairo, Egypt, Dr. Hoogstraal assembled the world's largest tick specimen collection and reference library. The author of more than 500 publications, Dr. Hoogstraal was recognized as the world's foremost authority on ticks and tick-borne diseases. At the time of his death, the collection contained 740 species (of the 850 described species), including about 300 types, obtained from 167 different countries [1992].

HAWAII

Honolulu

DIVISION OF PLANT INDUSTRY, HAWAII DEPARTMENT OF AGRICULTURE, 1428 S. KING ST. (P. O. BOX 22159), HONOLULU, HI 96823-2159. [HDOA]

Head curator: Mr. Bernarr Kumashiro. Phone: (808) 973-9534. The collection consists of: 140,000 pinned specimens; 2,200 alcohol vials; 8,500 specimens on slides; 2,200 specimens awaiting preparation; and 125 primary types, housed in 500 drawers and 80 slide boxes. About 10,000 specimens are received annually. It contains a part of the Hawaiian Sugar Planters' Association collection, native cerambycids and scolytids, biocontrol agents of plants and insect pests, obtained through worldwide exploratory programs. [1992]

J. LINSLEY GRESSITT CENTER FOR RESEARCH IN ENTOMOLOGY, DEPARTMENT OF ENTOMOLOGY, BERNICE P. BISHOP MUSEUM, P. O. BOX 19000A, 1525 BERNICE STREET, HONOLULU, HI 96819. [BPBM]

Director: Dr. Scott E. Miller. Phone: (808) 848-4194; FAX (808) 841-8968. Professional staff: Dr. Neal L. Evenhuis, Dr. Francis G. Howarth, Dr. John T. Medler, Mr. Gordon M. Nishida, Dr. Dan A. Polhemus, and Dr. G. Allan Samuelson. This collection is the largest of Pacific insects and arachnids in the world. It also encompasses arthropods other than Crustacea and the entirely marine groups. There are over 13 million specimens including more than 15,000 primary types. The majority of specimens are pinned, though many are stored mounted on slides, in alcohol in vials, or in transparent envelopes. The collection is generally most strongly representative of the Pacific Basin and surrounding land masses, viz., Hawaii and other Oceanic island groups, the Papuan Subregion, southern and eastern Asia, Australia, subantarctic islands, and far southern South America. Synoptic specimens are included from other areas, and some families are well represented from many regions (*e.g.*, Coleoptera: Chrysomelidae, Cerambycidae; Diptera: Culicidae; Acari: several families of parasitic Gamasida). Incorporated collections of special interest include those from the "Fauna Hawaiiensis" survey; Hawaiian Sugar Planters' Association Collection (part); Pacific Entomological Survey plus many others covering south and central Pacific islands; "Insects of Micronesia" project; surveys in Antarctica and subantarctic islands; ship- and plane-trapping collections for airborne dispersal studies; surveys of parasitic arthropods of the Pacific, Southeast Asia; and cave-associated arthropods from the Pacific, Southeast Asia, and Australia. The Diptera: Bombyliidae collection of the National Museum of Natural History (=USNM) is on long term loan through a site enhancement agreement. The collection occupies about 10,000 sq. ft., with compacted storage systems containing over 20,000 drawers, 7,000 jars, and 5,000 wooden slide boxes. Sponsored publications: "Insects of Micronesia," "Bishop Museum Bulletin in Entomology," and Bishop Museum Occasional Papers." "International Journal of Entomology" (formerly

"Pacific Insects") and "Pacific Insects Monographs" are no longer published. The "Journal of Medical Entomology" is currently published by the Entomological Society of America. [1992]

Affiliated Collection:

Evenhuis, Neal L., Department of Entomology, Bishop Museum, P. O. Box 19000-A, Honolulu, HI 96817 USA. [NLEC] Phone: (808) 848-4138; 524-6203. This is a collection of adult Bombyliidae from mainly southwestern U.S.A. and northern Mexico, with synoptic World collection, including most known genera, housed in 20 drawers [1992]. (Registered with BPBM.)

DEPARTMENT OF ENTOMOLOGY, COLLEGE OF TROPICAL AGRICULTURE, UNIVERSITY OF HAWAII, MANOA, 3050 MAILE WAY, HONOLULU, HI 96822. [CTAM]

Director: Dr. J. W. Beardsley. Phone: (808) 948-6737. Professional staff: Dr. Kenneth Y. Kaneshiro, Dr. M. Lee Goff, Dr. D. Elmo Hardy, Mr. John S. Strazanac, Mr. Dick Tsuda. The collections' main emphasis is the insect fauna of the Hawaiian Islands, with especially strong holdings in Diptera, Hymenoptera, and Coccoidea. The approximately 200,000 pinned specimens are housed in 1,200 drawers in cabinets. The more than 15,000 immature specimens in the alcoholic collection are housed in 10 cabinets. The collection includes the most complete collection of Hawaiian Diptera, the largest collection of Australasian Tephritidae, and world-wide collections of Bibionidae and Pipunculidae [transferred to BPBM, Eds.]. The Diptera holdings were amassed and sustained mainly through the efforts of Dr. Hardy. The very large Coccoidea collection (*ca.* 8,000 slide mounts) was built up by Dr. Beardsley [transferred to BPBM, Eds.]. Three staff members are engaged in systematic studies. Dr. Hardy works on Diptera and specializes in Bibionidae, Drosophilidae, Pipunculidae, and Tephritidae. Dr. Beardsley works on parasitoid Hymenoptera and also specializes on Coccoidea. Dr. Goff works on Hawaiian Acari and specializes in Trombiculidae and Paramegistidae. Other staff members include Mr. Tsuda (assistant curator), Dr. Kaneshiro (Coordinator of the Hawaiian Drosophila Research Stock Center), and Mr. Strazanac (collection manager). [1986]

HAWAII VOLCANOES NATIONAL PARK, U. S. DEPARTMENT OF THE INTERIOR, NATIONAL PARK SERVICE, P. O. BOX 52, HAWAII NATIONAL PARK, HI 96718. [HVNP]

Director: Dr. Charles P. Stone. Phone: (808) 967-8211. Professional staff: Mr. Clifton J. Davis, State Entomologist (Retired); Dr. David Foote, Research Associate. The collection consists of about 4,000 pinned specimens and about 150 in alcohol. The material is the result of faunal surveys primarily in National Parks on the Island of Hawaii. It is organized in a specimen-level computer database and the collection is housed at the Hawaii Field Research Center. [1992]

IDAHO

Boise

DEPARTMENT OF BIOLOGY, BOISE STATE UNIVERSITY, BOISE, ID 83725. [BIDA]

Director: Dr. Charles W. Baker. Phone (208) 385-3499. The collection consists of more than 20,000 local specimens, largely student collected and housed in 5 drawers and vials of alcohol. [1986]

Caldwell

ALBERTSON COLLEGE OF IDAHO COLLECTION, ORMA J. SMITH MUSEUM OF NATURAL HISTORY, ALBERTSON COLLEGE OF IDAHO, CALDWELL, ID 83605. [CIDA]

Director: Mr. William H. Clark. Phone: (208) 375-8605. Professional staff: Dr. Eric Yensen (Museum Director), Paul E. Blom, Robert L. Cheley, Sean D. Farley, Dr. Donald R. Frohlich, George Stephens, and Dr. David M. Ward, Jr. The collection is housed in 50 CAS cabinets, with over 1,200 drawers and in 3 large cabinets of alcohol vials. It contains mainly western U. S. A. and Mexican material plus worldwide reference material. It is strong in Coleoptera, Hymenoptera (Formicidae), and Lepidoptera, with special collections from Baja California, Mexico, including vouchers of Long Term Ecological Research sites and projects. It includes the Robert L. Chehy Lepidoptera collection. Some secondary types are housed here but Holotypes are deposited in CASC. Sponsors "Idaho Entomology Group Newsletter." [1992]

Affiliated Collections:

Blom, Paul E. 4130 Hwy. 8, Troy, ID 83871 USA. [PEBC] This is a general synoptic collection of insects housed in 48 drawers in cabinets [1992]. (Registered with CIDA.)

Clark, Mr. William H., 6305 Kirkwood Road, Boise, ID 83709 USA. [WHCC] Phone: (208) 375-8605. This is a general insect and arthropod collection in four CAS cabinets and in alcohol, with about 50% of the specimens Coleoptera. Specimens are from western U.S.A. and Mexico. A special collection of ants (Hymenoptera: Formicidae) from Idaho, Mexico (especially Baja California) and a general worldwide reference collection is included. About 10,000 alcohol vials, 4 CAS cabinets (96 drawers), and numerous Schmitt boxes comprise the storage facilities. Large library of Formicidae and Baja California [1992]. (Registered with CIDA.)

Shook, Gary, 5298 W. Boone St., Boise, ID 83705. [OGAS] This is a collection of 3,400 identified Cicindelidae and 1234 identified Cerambycidae (Coleoptera) from U.S.A., Europe, southeast Asia, and Africa, stored in drawers and cabinets [1992]. (Registered with CIDA.)

Moscow

W. F. BARR ENTOMOLOGICAL COLLECTION, ENTOMOLOGY DIVISION, DEPARTMENT OF PLANT, SOIL, AND ENVIRONMENTAL SCIENCES, UNIVERSITY OF IDAHO, MOSCOW, ID 83843. [WFBM]

Director: Dr. James B. Johnson. Phone: (208) 885-7543. Professional staff: Mr. Frank W. Merickel. The collection includes about 2 million specimens housed in 2,088 drawers, 1,200 museum jars and 9,500 slides. About 670,000 specimens have been identified, about 200,000 have been curated, but not identified and 130,000 remain to be curated. Most of the material is from North America, with an emphasis on the western U.S.A. and Baja California, Mexico. Supporting ecological data is good for specimens associated with desert shrubs of southern Idaho. The collection has good taxonomic coverage of Coleoptera, Acrididae, Sphecidae, Asilidae, Bombyliidae and many aquatic groups. Material is available for exchange or loan; usual rules apply. Holotypes are placed in the CASC on an indefinite loan basis. The museum also houses the Melville H. Hatch library which is particularly strong in works on Coleoptera taxonomy. Included is the special collection of William F. Barr Coleoptera, excluding his Buprestidae and Cleridae; Frank M. Beer collection of Buprestidae, Cicindelidae, Carabidae (*Scaphinotus*); Richard G. Dahl, Cicindelidae, and Robert E. Miller, Noctuidae. [1992]

Affiliated Collections:

Barr, Dr. William F., 1415 Borsh Ave., Moscow, ID 83843. Phone: (208) 882-2886. Phone: (208) 882-2886. [WFBC] Buprestidae housed in 41 CAS drawers from North and Central America; 53 CAS drawers of World Cleridae, with a very large species representation from North and Central America (including the Caribbean area); a moderate representation of South American and Australian species. Primary types are deposited in the CAS collection, but many paratypes are included in each family [1992]. (Registered with WFBM.)

Johnson, Dr. James B., 3318 Highway 8 East, Moscow, ID 83843. Phone: (208) 883-0543. [JBJC] About 50,000 specimens in 96 CAS drawers, *ca.* 50 Schmitt boxes and *ca.* 1,000 vials, mostly adults, some larvae, with emphasis on Neuroptera, Megaloptera, Raphidioptera, Coleoptera, Mecoptera, and Hymenoptera, especially Chrysopidae, Carabidae, Parasitica, excluding Ichneumonoidea, and Aculeata, excluding Formicidae and Apoidea; collection is basically synoptic, with small series of many taxa, mostly from North America. About 50 paratypes of Neuroptera, Coleoptera, Mecoptera, and Hymenoptera are included [1992]. (Registered with WFBM.)

Westcott, Richard L., Plant Division, Oregon Department of Agriculture, 635 Capitol NE, Salem, OR 97310-0110 USA. [RLWE] Phone: (503) 378-6458; (Home) 362-1595. The collection consists almost entirely of Coleoptera, mostly from the western and southwestern U.S.A., and Mexico. There are about 30,000 specimens, mostly Buprestidae, Cerambycidae, and Scarabaeidae. The Buprestidae constitute a research collection of 10,000 specimens of 700 species from North America (incl. Mexico) and a synoptic collection of exotic genera [1992]. (Registered with WFBM.)

Pocatello

IDAHO MUSEUM OF NATURAL HISTORY, CAMPUS BOX 8096, IDAHO STATE UNIVERSITY, POCATELLO, ID 83209. [ICIS]

Curator of Insects: Dr. Stefan Sommer. Phone: (208) 236-2335; C. Anderson. The collection currently contains about 4,416 insects and arachnid specimens from southern Idaho, northern Utah, the southwestern U.S.A., Mexico, Australia, and a small number of specimens from Russia. The largest portion of this collection is of insects from the introduced tumbleweed, *Salsola Kali* L. Periodicals sponsored: "Tebiwa: The Journal of the Idaho Museum of Natural History." This is an annual publication printed by the Idaho State University Press. [1992]

ILLINOIS

Carbondale

SOUTHERN ILLINOIS UNIVERSITY ENTOMOLOGY COLLECTION, AND INVERTEBRATE COLLECTION, RESEARCH MUSEUM OF ZOOLOGY, DEPARTMENT OF ZOOLOGY COLLECTION, SOUTHERN ILLINOIS UNIVERSITY, CARBONDALE, IL 62901. [SIUC]

Director: Dr. J. E. McPherson. Phone: (618) 536-2314. This is a general collection of mostly southern Illinois specimens with a scattering of material from various parts of the New World. It contains about 125,000 pinned specimens housed in more than 700 drawers in cabinets, and about 50,000 specimens in alcohol. Many have been identified by specialists. Material for exchange and loan if the borrower will accept miscellaneous unsorted material. The invertebrate collection is curated by Dr. Joseph A. Beatty. It contains about 75,000 specimens of Arachnida including spiders from islands of western basin of Lake Erie, various Pacific islands, and *Arrenurus* Southern Illinois water mites, stored in vials. [1986]

Champaign

ILLINOIS NATURAL HISTORY SURVEY INSECT COLLECTION, 607 E. PEABODY DRIVE, CHAMPAIGN, IL 61820. [INHS]

Director: Dr. Lawrence M. Page. Phone: (217) 333-6846. Professional staff: W. E. LaBerge, D. W. Webb; D. J. Voegtlin; M. E. Irwin, W. Brigham, J. Bouseman, A. Brigham; Contract person, Collection Manager: K. R. Methven. Phone: (217) 244-2149. Established in 1858, and located on the campus of the University of Illinois, the insect and related arthropod collection contains over 6 million specimens, with nearly 4,000 primary types, and uncounted secondary types. Approximately 8,000 drawers and 1/4th of the alcohol material are housed in compactors. The remaining alcohol material is housed in cabinets. Slide mounted specimens are stored in 100 capacity slide boxes on bookshelves. A good deal of the collection dates prior to the 1900's. Although the collection's primary representation is Illinois and the rest of the Midwestern states, it is worldwide in scope. All orders are represented and it is strong in Coleoptera, Lepidoptera, Hymenoptera, Collembola, Trichoptera, Ephemeroptera, Plecoptera, Homoptera, and Thysanoptera. The collection is also a repository for the International Soybean Collection (INTSOY). Publications sponsored: "Bulletin of the Illinois Natural History Survey," "Illinois Natural History Survey Biological Notes," "Illinois Natural History Survey Biological Monographs." [1992]

Chicago

INSECT COLLECTION, FIELD MUSEUM OF NATURAL HISTORY, ROOSEVELT ROAD AND LAKE SHORE DRIVE, CHICAGO, IL 60605. [FMNH]

Director: Dr. A. F. Newton. Phone: (312) 922-9410, ext. 354. Professional staff: Dr. J. Kethley, Mr. D. A. Summers, Mr. P. P. Parrillo, Mr. H. G. Nelson, Dr. R. L. Wenzel (Emeritus). The arthopod (exclusive of Crustacea) collection, worldwide in scope, contains about 9 million fluid preserved and dry prepared specimens, over 4 million of which are pinned specimens. This includes over 205,000 identified species and an estimated 9,200 primary types.

The Coleoptera are especially well represented and include many notable World collections. For example, collections of the family Staphylinidae represent one of the two ranking collections in the World and include more than 500,000 pinned, 300,000 unprepared specimens, representing 30,000 identified species and more than 4,000 types. In this collection are those of M. Bernhauer, A. Bierig, C. Seevers, and L. Benick. The Ptiliidae with about 1 million specimens, mostly in alcohol, represent the most extensive material of its kind in existence. Other exceptional strengths in the Coleoptera collection include the families Histeridae, Tenebrionidae (including the L. Peña collection with over 200 holotypes), Scarabaeidae (esp. Cetoniinae), Lucanidae (including E. Knirsch and B. Benesh collections), Cerambycidae, Buprestidae, Cleridae, Elateridae (J. Knull collection), Leiodidae (Knirsch collection), and Pselaphidae (O. Park collection). The recently added N. M. Downie collection [NMDC] (88,000 specimens) includes over half of the named North American beetle species.

Other strengths include the Lepidoptera (H. Strecker butterflies returned to FMNH from Allyn Museum), the K. Brancsik general collections, Isoptera (identified representatives of nearly half of the world's species), and Diplopoda and Chilopoda (considerable number of identified New World and Africa species, including about 200 holotypes). The Arachnida-Acarina holdings include the water mite collection (of Ruth Marshall and David Cook), which contains primary types of 765 or more species, representing more than 90% of those described from North America making this the most important type collection of water mites in the New World. Exotic water mite material includes 200 identified species (146 holotypes) from the Ethiopian Region and 173 species from India (143 holotypes). The R. Loomis collection of Trombiculidae contains about 150,000 slides as well as many specimens in alcohol.

Collections of ectoparasitic insects are extensive and include ticks (50% of the described species for the World); batflies (40,000 specimens representing 70% of the World's species of which 40% are represented by types); fleas (about 1/3 of the described species and types of 25% of the Neotropical species); sucking lice (about 1/3 of the described species); and many species of parasitic mites.

Collections of bulk samples include 10,000+ partially sorted Berlese samples and 3,000 bait-trap samples from throughout the World, many with full ecological data. These have yielded about 4 million specimens of little studied groups. For example, the representation of Coleoptera from the bulk samples is extremely comprehensive at the family level - about 80% of described families represented in about 1 million specimens.

Other specialized collections include the Baltic amber collection which is second only to that at the MCZC in the New World (3,500+ specimens), and the spider collection which contains over 300,000 specimens. [1992]

Affiliated Collections:

Baumgartner, Donald L., 150 S. Walnut St., Palatine, IL 60067 USA. [DLBC] Phone: (312) 359-1855; 359-5767. This is a collection of about 7,700 Diptera from Illinois, and Peru, including Culicidae of Illinois, and Calliphoridae from Illinois, Arizona, Hawaii, Bermuda, and Peru. It is housed in 25 drawers [1990]. (Registered with FMNH.)

Downie, N. M., 505 Lingle Terrace, Lafayette, IN 47901 USA. [NMDC] Phone: (317) 423-1048. The entire collection has been moved to the FMNH [1992].

Hamilton, Dr. Robert W., Department of Biology, Loyola University of Chicago, 6525 North Sheriden Road, Chicago, IL 60626 USA. [RWHC] Phone: (312) 508-3628. This is a teaching and research collection housed at the above address, consisting of 225 redwood boxes containing about 50,000 pinned specimens, 800 vials containing about 10,000 alcohol preserved specimens. The collection is strongest in Coleoptera, especially Curculionidae of North America, and a special collection of 20 boxes of parasitic Hymenoptera. The collection also contains voucher material for major taxonomic paper on arthropods of three Illinois prairie remants and paratypes of new species of Curculionoidea. [1992] (Registered with FMNH.)

Peña G., Luis E., Casilla 2974, Santiago, Chile. [LEPG] Phone: 56-2-8441374. The collection is composed of the following: Diptera of diverse groups, specially Nemestrinidae and Asilidae. Lepidoptera: Rhopalocera from Chile and Bolivia. Coleoptera: several families, particularly Tenebrionidae, Cicindelidae, Buprestidae, Carabidae, Scarabaeidae, Lampyridae, Alleculidae, and Lucanidae. The Tenebrionidae includes specimens from the Andes Range, High Plateau of Bolivia, Chile, Peru, and Argentina, Patagonian steppe, deserts of Peru and Chile, the nothofagus forests, subtropical areas of Bolivia and Argentina, pampas of Argentina. There is a total of 713 species, 148 holotypes, 92 allotypes, 2,416 paratypes, 2,761 topotypes, with a total of about 50,000 specimens. The collection of Scarabaeidae includes 733 species; Cerambycidae, 665 species, and many specimens compared with types. The collection contains a grand total of 193 holotypes, 2,842 allotypes, 2,781 topotypes, with a total for all collections of nearly 64,000 pinned specimens, and over 1 million papered specimens. The collection is housed in over 200 drawers in cabinets, and nearly 200 boxes. New collection: Lepidoptera from Chile; Tenebrionidae from Chile, Perú, Argentina, Bolivia, including certain groups (Nycteliini, Thinohathini, etc.) Cerambycidae, Lucanidae, Scarabaeidae, and Carabidae, all from Chile. Additions total about 12,000 pinned and labeled specimens, including approximately 200 secondary types, in 80 boxes. (Collection registered with FMNH.) [1992]

CHICAGO ACADEMY OF SCIENCES, MUSEUM OF NATURAL HISTORY, LINCOLN PARK, 2001 N. CLARK ST., CHICAGO, IL 60614. [CASM]

Collection manager: Mr. Ron Vasile. Phone: (312) 549-0606, ext. 2030. About 30,000 specimens, most from the Midwest area, 1880-1910. Lepidoptera and Coleoptera comprise the bulk of the collection. Approximately 500 spiders collected by Donald Lowrie in the Midwest and Arizona are stored here. (See News of the Lepidopterists Society,

May/June 1990, p. 50-51.) Publication sponsored: "Bulletin of the Chicago Academy of Sciences." [1992]

Macomb

DEPARTMENT OF BIOLOGICAL SCIENCES COLLECTION, WESTERN ILLINOIS UNIVERSITY, MACOMB, IL 61455. [WIUC]

Director: Dr. Y. Sedman. Phone: (309) 298-1366. This is a general collection; no arachnids, but with moths and butterflies of west central Illinois. The collection is housed in drawers. [1986]

Normal

DEPARTMENT OF BIOLOGICAL SCIENCES, ILLINOIS STATE UNIVERSITY, NORMAL, IL 61761. [ISUC]

Curator: Dr. Douglas Whitman. Phone: (309) 438-5123. Collection contains 24,000 insect specimens mainly from Illinois. [1992]

Springfield

ILLINOIS STATE MUSEUM, SPRING AND EDWARDS STREETS, SPRINGFIELD, IL 62706. [ISMS]

Director: Dr. R. Bruce McMillan. Phone: (217) 782-7386. Professional staff: Dr. Everett D. Cashatt. Phone: (217) 782-6689. There are about 55,000 insects in drawers or alcohol vials, including about 32,000 Lepidoptera, 8,000 Odonata, and 3,000 spiders [1992].

INDIANA

Bloomington

DEPARTMENT OF BIOLOGY COLLECTION, INDIANA UNIVERSITY, BLOOMINGTON, IN 47401. [IUIC]

Director: Dr. Frank N. Young. Most of the collection, other than teaching material is aquatic Coleoptera. Indiana and vicinity are covered with special emphasis on Scarabaeidae in addition to water beetles. The collection contains some W. S. Blatchley material, but none of the Blatchley types. It is housed in 15 drawers and about 100 Schmitt boxes. No holotypes are kept in the collection. Material is available for exchange and loan for study. [1986] [Note: at least some of the collection has been dispersed in recent years. A number of collections, including BPBM and FSCA, have received parts of it. Eds. 1992]

Indianapolis

INDIANA DEPARTMENT OF NATURAL RESOURCES, DIVISION OF ENTOMOLOGY, 402 W. Washington St., Rm. W-290, INDIANAPOLIS, IN 46204. [ISMC]

Director: Dr. Robert D. Waltz. Phone: (317) 232-4120. No type specimens, but W. S. Blatchley material (in part) is included in the collection which is housed in 48 Cornell drawers, including Coleoptera, Lepidoptera, Homoptera (esp. leafhoppers), and an extensive collection of subterranean aphids and ant associates with host data [1992].

Lafayette

ENTOMOLOGY RESEARCH COLLECTION, DEPARTMENT OF ENTOMOLOGY, PURDUE UNIVERSITY, WEST LAFAYETTE, IN 47907. [PURC]

[Director: Dr. W. Patrick McCafferty. Phone: (317) 494-4598. Curator: Arwin V. Provonska. The collection contains approximately 1.5 million specimens, representing more than 100,000 species. Although taxa are broadly represented geographically, approximately 65% of the specimens are from Indiana, and it is by far the single most important resource center for the insects of Indiana. The majority of the holdings are pinned, but a sizable number of specimens (or parts thereof) are maintained on microscope slides. A rapidly growing collection of alcohol-preserved material is also housed. Voucher specimens from research conducted at Purdue are regularly maintianed. The collection is a primary resource in three areas: (1) the type collection of some 1,250 primary and secondary types; (2) the Blatchley collections of Coleoptera, Hemiptera, and Orthoptera; and (3) the aquatic insect collection (especially Ephemeroptera), which is becoming one of the important ones in North America. [1992]

IOWA

Ames

IOWA STATE UNIVERSITY INSECT COLLECTION, DEPARTMENT OF ENTOMOLOGY COLLECTION, IOWA STATE UNIVERSITY, AMES, IA 50010-3222. [ISUI]

Director: Dr. Robert E. Lewis. Phone: (515) 294-1815. This is a large general collection with particular emphasis on the fauna of Iowa. About 1,000,000 pinned or pointed adult insects are stored in 2,400 Cornell drawers in 55 cabinets, plus 2 cabinets of slide mounts. An additional 30,000 specimens (mainly larvae) are preserved in alcohol. Limited ecological data are available. Material is available for exchange or loan to qualified personnel. Loans are usually for two years, subject to renewal. Primary types are deposited in the USNM. [1992]

Affiliated Collection:

Lewis, Dr. R. E., Department of Entomology, Iowa State University, Ames, IA 50011 USA. [RELC] Phone: (515) 294-1815. This is a collection of about 8,500 accessions involving 50,000 slide mounted specimens of Siphonaptera, plus an unknown number in alcohol, representing more than 750 described species, worldwide (all known families and most genera), particularly strong in material from western U.S.A., Nepal, and the Middle East [1992]. (Registered with ISUI.)

Davenport

SAINT AMBROSE COLLEGE COLLECTION, DEPARTMENT OF BIOLOGY, ST. AMBROSE COLLEGE, DAVENPORT, IA 52803. [SACC]

Director: Dr. John Horn. Phone: (319) 383-8958. A local collection of about 4,000 specimens, including about 1,900 Lepidoptera (nearly 500 Nymphalidae), about 500 Coleoptera (Scarabaeidae and Carabidae are well represented), about 1,400 other insects, housed in drawers, and about 100 slide mounts, it is for teaching, and is not loaned. [1992]

KANSAS

Hays

COLLECTION OF INSECTS, STERNBERG MUSEUM, FORT HAYES STATE UNIVERSITY, HAYS, KS 67601. [FHKS]

Curator: Dr. Charles A. Ely. Phone: (316) 628-4214. This is a small regional collection of about 70,000 pinned and labelled specimens, largely from western Kansas. Specimens are sorted to family and butterflies are identified to species. This is a new collection with most growth since 1980. Exchange or loans can be arranged under the usual rules. Special collections include those resulting from research on geographical distribution of Kansas butterflies; abcut 3,000 voucher specimens (one specimen per species per county) from all 105 counties plus many duplicates. The collection is stored in 400 drawers. [1992]

Lawrence

SNOW ENTOMOLOGICAL MUSEUM, UNIVERSITY OF KANSAS, LAWRENCE, KS 66044. [SEMC]

Director: James S. Ashe. Phone: (913) 864-3065. Professional staff: Dr. Robert W. Brooks (Collection Manager), Dr. Charles D. Michener (retired), Dr. George W. Byers (retired). An estimated 3 million curated specimens comprise this collection; many more are not curated. It is well represented worldwide in Apoidea, aquatic Hemiptera, Cicadellidae (Homoptera), Diptera (especially Tipulidae), Mecoptera. It is strong in Diptera (especially Tipulidae), Coleoptera (especially Staphylinidae), and Hymenoptera. A good collection of Acarina is included; other non-insect arthropods are not well represented. The collection includes all insect orders. Over 8,200 holotypes, lectotypes, and paratypes are included. [1992]

STATE BIOLOGICAL SURVEY OF KANSAS INVERTEBRATE COLLECTION, 2045 CONSTANT AVE., CAMPUS WEST, UNIVERSITY OF KANSAS, LAWRENCE, KS 66044. [KSBS]

Director: Dr. Edward A. Martinko. Phone: (913) 864-4493. Professional staff: Dr. Leonard C. Ferrington, Jr., Mr. Paul Liechti, Dr. Dan Reinke, Mr. Ralph Brooks, and Mr. Don Huggins. The major emphasis of the collection is on aquatic invertebrates collected primarily from within the state of Kansas. Groups actively being worked on are: Chironomidae, in excess of 300 species, 600 alcohol collections identified to species, about 75 species, most of which are alcoholic and identified to genus or species; Plecoptera, about 40 species, alcoholic and identified to species. Inactive collection include over 5,000 vials of Coleoptera, Trichoptera, Hemiptera, Diptera, and Neuroptera. Hundreds of collections await processing, mostly light trap collections and benthic collections. Among the special collections are recently received collections of Chironomidae from the University of Alaska. This collection is from prepipeline survey conducted by the U. S. Environmental Protection Agency, and consists of about 1,800 vials of specimens in alcohol. Publications sponsored: "Technical Publications of the State Biological Survey of Kansas," "Bulletin of the State Biological Survey of Kansas," and "Reports of the State Biological Survey of Kansas." [1986]

Manhattan

DEPARTMENT OF ENTOMOLOGY COLLECTION, KANSAS STATE UNIVERSITY, MANHATTAN, KS 66502. [KSUC]

Curator: Dr. H. Derrick Blocker. Phone: (913) 532-6154; FAX (913) 532-6232. Approximately 750,000 specimens with strength in Coleoptera, Diptera, Homoptera, Hymenoptera, and Orthoptera. Small number of primary and secondary types are included. The collection is housed in 35 cabinets in modified Cornell drawers. Small alcohol-stored collection exists. Publication co-sponsored: "Journal of Kansas Entomological Society." [1992]

Curator of Ticks: Dr. Donald E. Mock. The tick collection currently includes 989 accession numbers, with ticks mostly in vials, but a few mounted on glass slides. Altogether there are >3,000 specimens. This special collection includes only ticks collected in Kansas. Thirty-three vials of voucher specimens are included. [1992]

KENTUCKY

Louisville

DEPARTMENT OF BIOLOGY INSECT COLLECTION, UNIVERSITY OF LOUISVILLE, LOUISVILLE, KY 40292-0001. [ULKY]

Curator: Dr. Charles V. Covell, Jr. Phone: (502) 588-6771. About 200,000 curated insect and spider specimens, mostly North American, but with significant recent collections of exotic Lepidoptera, housed in Cornell drawers and 4-dram alcohol vials. No primary types; less than 100 secondary types. Includes Covell's private collection of 15,000 Lepidoptera (butterflies, Geometridae). Publications sponsored: "Kentucky Lepidopterists." [1992]

LOUISIANA

Baton Rouge

LOUISIANA STATE UNIVERSITY INSECT COLLECTION, DEPARTMENT OF ENTOMOLOGY, LOUISIANA STATE UNIVERSITY, BATON ROUGE, LA 70803-1710. [LSUC]

Director: Dr. Joan B. Chapin. Phone: (504) 388-1834. Assistant curator: Vicky L. Moseley. Phone: (504) 388-1838. The collection is the principal repository in Louisiana for insects and related arthropods and contains 237,743 pinned specimens (housed in 65 Cornell cabinets containing 962 drawers), 3,933 vials and 8,830 slides, each containing one to many specimens. The collection is strongest in Coleoptera (51%) followed by Hemiptera (19%), including 18,000 specimens from H. M. Harris, Homoptera (9%0, Lepidoptera (6%), Diptera (6%), Hymenoptera (4%), and all other orders (5%). The majority of specimens are from Louisiana, and the remainder are mostly from North America, Mexico, Central and South America. Notable exceptions: 1,364 British moths. a worldwide synoptic collection of Hemiptera, and a reference collection of insects not known to occur in continental U.S.A. Families that are well represented in the LSUC are as follows: **Coleoptera**: Carabidae, Cerambycidae, Chrysomelidae, Coccinellidae, Curculionidae, Scarabaeidae, and aquatic Dryopoidea; **Diptera**: Tabanidae; **Hemiptera**: Coreidae, Lygaeidae, Miridae, Pentatomidae, and Reduvidae; **Lepidoptera**: butterflies and Noctuidae. The collection contains 744 paratypes, 1 syntype, 1 holotype, and 1 allotype, although primary types are usually deposited elsewhere. The insect collection from the University of Southwestern Louisiana [USWL] was transferred to LSUC in 1992. [1992]

Affiliated collection:

Brou, Vernon A., Jr. 74320 Jack Loyd Rd., Abita Springs, LA 70420 U.S.A. Phone: (504) 892-8732. [VABL] This collection consists of over 450 Cornell drawers, more than 300,000 insects of Louisiana, mostly Lepidoptera and Coleoptera; some secondary types of Lepidoptera. [1992]. (Registered with LSUC.)

MAINE

Augusta

ENTOMOLOGICAL LABORATORY COLLECTION, MAINE FOREST SERVICE, 50 HOSPITAL ST., AUGUSTA, ME 04330. [ELMF]

Curator: Richard G. Dearborn. Phone: (207) 287-2431. The collection which has been building since 1921 includes over 100,000 specimens in 300 Cornell drawers. The collection stresses Maine insects and in particular forest insects, parasites, and predators. Rearing data are available for many species. Over 4,000 species have been identified by specialists. Some paratypes are included. [1992]

Orono

DEPARTMENT OF ENTOMOLOGY COLLECTION, UNIVERSITY OF MAINE, ORONO, ME 04473. [UMDE]

Curator: Dr. Stephen A. Woods. Phone: (207) 581-2955. The collection of 108,545 insect specimens is housed in 611 drawers. About 5,000 specimens, most identified to species, of aquatic insects, is probably one of the best collections in the Northeast. Approximately 20,000 slides of aphids from the Edith Patch collection are included, along with Lepidoptera collections of Bruno M. Spies, Prof. Manton Copeland, and Charles Burton Hamilton. [1992]

MARYLAND

Baltimore

U.S. DEPARTMENT OF AGRICULTURE, ANIMAL AND PLANT HEALTH INSPECTION SERVICE, RM. 308, 40 SOUTH GAY ST., BALTIMORE, MD 21202. [APHI]

Curator: Mr. J. F. Cavey. Phone: (410) 962-4499; FAX (410) 783-9376. This collection contains in excess of 38,000 pinned insects and about 1,200 vials, each with one or more larval specimens totaling about 44,000 identified insects. The collection is housed in 96 drawers with unit trays, and 1,200 vials in half pint jars. About one third of this material is from many foreign countries and was collected by various U. S. Department of Agriculture personnel while inspecting foreign baggage, cargo, and means of transportation. Many of the specimens were identified by specialists at the Systematic Entomology Laboratory and the Smithsonian Institution. No types are kept in this collection. [1992]

College Park

DEPARTMENT OF ENTOMOLOGY COLLECTION, UNIVERSITY OF MARYLAND, COLLEGE PARK, MD 20742. [UMDC]

Curator: F. E. Wood. The collection has coverage for the Western Hemisphere, especially eastern North America, with Maryland well represented. Specimens are housed in 102 Cornell drawers. [*No further information available after several requests.*] [1968]

MASSACHUSETTS

Amherst

UNIVERSITY OF MASSACHUSETTS INSECT COLLECTION, DEPARTMENT OF ENTOMOLOGY, FERNALD HALL, UNIVERSITY OF MASSACHUSETTS, AMHERST, MA 01002. [UMEC]

Director: Dr. T. Michael Peters. Phone: (413) 545-2283. The collection is a comprehensive research collection of moderate size. Approximately 250,000 identified specimens are housed in drawers or Schmitt boxes. Slide mounted material is contained in 100 slide boxes. Several important private collections have been acquired by the Department. Among these are: The William B. Proctor Mount Desert Island Collection (about 15,000), The M. C. Lane Coleoptera (about 25,000), The H. B. Tietz Lepidoptera (about 4,000), and the Crampton exotic insects (250 Schmitt boxes). As with most other institutional collections, this is especially strong with the work of Alexander, Shaw, Smith, Peters, and Thompson. The Lepidoptera are also well represented due to McDonald, Smith, and Tietz. Coleoptera is the third of the heavily worked groups with Bill, Lane, and Frost specimens. A moderately extensive thrips collection is the result of the early work by Hines, plus exchange of paratypes with Hood. Approximately 1,000 types are housed here. [1986]

Cambridge

ENTOMOLOGY DEPARTMENT, MUSEUM OF COMPARATIVE ZOOLOGY, HARVARD UNIVERSITY, 26 OXFORD STREET, CAMBRIDGE, MA 02138. [MCZC]

Director: Dr. Edward O. Wilson. Phone: (617) 495-2464. Professional staff: Dr. Naomi Pierce, Dr. David G. Furth, and Dr. Frank M. Carpenter. Restricted to Hexopoda (spiders covered by MCZ Department of Invertebrates, see *addenda* below), this is the largest university collection in North America, with over 7 million prepared specimens, including over 5 million pinned specimens, 20,000 vials of specimens in alcohol, 20,000 microscope slides and 70,000 fossil insect specimens. Approximate percentages of each group are: Coleoptera (54%), Hymenoptera (25%), Lepidoptera (8%), Diptera (7%), other orders (5%), fossils (1%). Primary types of over 29,500 species are present, particularly in Coleoptera, Hymenoptera, and Diptera. The collection is worldwide in scope, with the U.S.A., West Indies (notably Cuba), Central and South America, Australia, Papua New Guinea, and the Philippines especially well represented. Exceptional strengths of the collection include Coleoptera: Carabidae, Chrysomelidae, Ciidae, Nitidulidae, and Staphylinidae; Diptera: Asilidae, Hippoboscoidea, Nemestrinidae, Phoridae, Tabanidae, and Tephritidae; Hymenoptera: Formicidae, Pompilidae, and Vespidae; Lepidoptera: Arctiidae, Catocalinae (with Erebinae), Geometridae, Sphingidae, and Tineidae; Neuroptera, Plecoptera, and Trichoptera; and the second largest fossil insect collection in the world. The pinned collection is housed in about 13,000 drawers in metal cabinets. Alcohol collection is stored in 1,400 canning jars and 5,000 neoprene stoppered vials. The fossils are in Agassiz-type drawers. Special collections include: General: N. Banks, C. T. Brues, G. C. Crampton, F. A. Eddy, G. B. Fairchild, H. Hagen, T. W. Harris, and R. Thaxter. Coleoptera: F. Blanchard, F. C. Bowditch, P. J. Darlington, Jr., W. G. Dietz, H. C. Fall, C. A. Frost, E. D. Garris, R. Hayward, G. H. Horn, J. F. Lawrence, J. L. LeConte, C. Liebeck, F. E. Melsheimer, C. T. Parsons, D. Ziegler, also, a synoptic tribal collection, larval, and host-determined fungicole collections. Diptera: M. Bates, C. W. Johnson, H. Loew, C. R. Osten-Sacken. Hymenoptera: J. Bequaert, H. E. Evans, A. Kinsey, C. Porter; Formicidae: W. L. Brown, Jr., J. W. Chapman, B. Finzi, W. Mann, N. Weber, G. C. and J. C. Wheeler (larvae), W. M. Wheeler, E. O. Wilson. Lepidoptera: M. Bates, C. Biezanko, J. Boll, S. A. Cassino, V. T. Chambers, S. A. Hessel, C. P. Kimball, C. Oliver, A. S. Packard, C. J. Paine, S. A. Scudder, L. Y. Swett. Fossil insects: F. M. Carpenter, H. Hagen, S. A. Scudder, C. T. Brues.

The director of the spider collection is Dr. Herbert W. Levi, phone (617) 495-2472 who reports the spider collection to contain 500,000 to 1 million spiders, worldwide, in alcohol.

Publications sponsored: "Bulletin of the Museum of Comparative Zoology," "Breviora," and "Psyche" (published by the affiliated Cambridge Entomological Club). [1992]

MICHIGAN

Ann Arbor

DIVISION OF INSECTS, MUSEUM OF ZOOLOGY, UNIVERSITY OF MICHIGAN, ANN ARBOR, MI 48109-1079. [UMMZ]

Collection Coordinator: Mr. Mark F. O'Brien. Phone: (313) 764-0471 or 747-2199. FAX (313) 763-4080. Email: userlps5-@um.cc.umich.edu, or ent-list@umichum.bitet. Professional staff: Mr. Mark F. O'Brien (Hymenoptera), Dr. Richard D. Alexander (Orthoptera), Dr. Thomas E. Moore (Cicadidae), Dr. Barry M. O'Connor (Acari); Adjunct Curators: H. Don Cameron (Arachnida), Theodore J. Cohn (Orthoptera), Paul B. Kannowski (Formicidae). This collection of 4,500,000 specimens contains outstanding collections of Orthoptera, Odonata, astigmatid Acari; very good collections of Coleoptera, Formicidae, Sphecidae, Lepidoptera, and Heteroptera. Emphasis is on the Nearctic and Neotropical faunas, especially Mexico, Honduras, Guatemala, and Costa Rica. Many new collections have been added from Malaysia and the Philippines. The collection includes 1012 primary types. [1992]

Bloomfield Hills

CRANBROOK INSTITUTE OF SCIENCE, BLOOMFIELD HILLS, MI 48013. [CISM]

Director: Dr. Dennis M. Wint. Phone: (313) 645-3260. Professional staff: Dr. Kathryn K. Matthew. The collection comprises 135 drawers of which 12 are for Hemiptera and Diptera, 12 for Hymenoptera, 13 for Ephemeroptera, Orthoptera, and Homoptera, 32 for Coleoptera, and 16 for Odonata, Trichoptera, Plecoptera, Mecoptera, Neuroptera, and Thysanoptera. Although these orders are represented primarily by specimens from Michigan and other U.S.A. localities, there is a significant number of specimens from tropical localities such as British Guiana and Costa Rica. The Lepidoptera comprise 50 drawers total, with the major families represented being Papilionidae, Nymphalidae, Sphingidae, Saturnidae, Noctuidae, and Geometridae. Neotropical Morphidae, Sphingidae, and Noctuidae are also represented. There are few specimens from Indonesia, Africa, China, India, and Australia. Although this collection has a broad scope for its size, emphasis is on use primarily for education and exhibits, although specimens are being added to the collection regularly. [1986]

East Lansing

DEPARTMENT OF ENTOMOLOGY COLLECTION, MICHIGAN STATE UNIVERSITY, EAST LANSING, MI 48824-1115. [MSUC]

Director and Curator: Dr. Frederick W. Stehr. Phone: (517) 353-8739 (office); 355-1803 (collection). Professional staff: M. C. Nielsen, Adjunct Curator, Lepidoptera; R. J. Snider, Adjunct Curator, Collembola, Araneida. The collection contains an estimated 2.5 million pinned insects, 10,000 vials of spiders, and a large collection of Collembola. Strengths are in Coleoptera, Lepidoptera, Hymenoptera, and Diptera, with lesser collections of other orders. About 250 primary types and 2,000 secondary types of insects and approximately 80 primary and 1,000 secondary types of Collembola. [1992]

Affiliated Collections:

Young, Dr. Daniel K., Department of Entomology, University of Wisconsin, Madison, WI 53706 USA. [DKYC] Phone: (608) 262-2078; 837-9787. This collection contains about 80,000 species of adult and larval Coleoptera, with worldwide representation, but with strength in U.S.A. (especially Michigan and California). Heteromerous families are particularly well represented, especially Pyrochroidae. The collection is stored in drawers and Schmitt boxes, and larvae in alcohol vials [1992]. (Registered with MSUC.)

Kalamazoo

KALAMAZOO PUBLIC MUSEUM, 315 SOUTH ROSE STREET, KALAMAZOO, MI 49001. [KPMC]

Director: Ms. Linda W. Hager. Phone: (616) 345-7092. Professional staff: Jean Stevens, Joel J. Orosz, Lynn S. Houghton. A small collection of insects and spiders are circulated to patrons holding a Kalamazoo Public Library card. The collection consists of about 550 specimens of which about 300 are butterflies and moths, 10 are spiders, and 240 are assorted insects. The collection is housed in Riker mounts. [1986]

MINNESOTA

St. Cloud

DEPARTMENT OF BIOLOGY COLLECTON, ST. CLOUD STATE COLLEGE, ST. CLOUD, MN 56301. [SCSC]

Director: Dr. Ralph Gundersen. Phone: (612) 255-4136. The collection emphasizes the fauna of Minnesota. It is housed in 450 drawers in 50 cabinets, containing over 150,000 specimens. Aquatic Coleoptera and Hemiptera (over 60,000 specimens fill 20 cabinets. Other insect groups are stored in 25 cabinets, and exotics, mainly from South America, are held in 5 cabinets. Ecological data are good for the Minnesota material. Locality data are present on all pinned material. Holdings of immature insects (except aquatics) and other Arthropoda groups are minimal and used mainly for teaching purposes. Loans will be made to recognized institutions or to those with adequate references; usual loan rules apply. [1986]

St. Paul

UNIVERSITY OF MINNESOTA INSECT COLLECTION, DEPARTMENT OF ENTOMOLOGY, 219 HODSON HALL, 1980 FOLWELL AVE., ST. PAUL, MN 55108. [UMSP]

Curator: Dr. Philip J. Clausen (specialty in Diptera and Coleoptera). Phone: (612) 624-1254, FAX: (612) 625-5299. Professional staff: Dr. Ralph W. Holzenthal, Director (Trichoptera), Dr. William E. Miller (Lepidoptera), Dr. John C. Luhman (Ichneumonoidea), Dr. Roger D. Price (Phthiraptera), Dr. Edwin F. Cook (Diptera, Aphididae). Contributions to the collection began in 1897 with specimens of insects and spiders from the North Shore of Lake Superior. During the next 95 years, the collection has grown in both quantity of material and quality of curation. Recent collecting efforts in Bolivia, Costa Rica, Ecuador, French Guiana, Peru, Trinidad, and Irian Jaya have added much exotic material to the collection. In the most recent survey (based on 1986 data on total holdings), the collection ranked as the 8th largest university affiliated insect collection in North America and 19th overall. The University of Minnesota Insect Collection is a charter member of the Association of Systematic Collections. The collection's 2,714,592 specimens, including 1,899 primary and 25,308 secondary types represent 39,526 described species. About 65% of the material is determined to species. The collection, located in room 341 Hodson, St. Paul campus, contains 363 12-drawer Cornell cabinets for pinned material, 18 steel cabinets for vials, and 1,030 linear feet of shelving for slide mounted material. Much of the museum's holdings is of special taxonomic or historical importance, including insects (especially Coleoptera) collected by Otto Luger; the C. E. Mickel Mutillidae collection, one of the best in the World, consisting of 27,000 specimens and 282 primary types; the Guthrie collection of Collembola; the Oestlund and Granovsky collection of Aphididae; Minnesota Trichoptera from the D. G. Denning collection; the Noctuidae collection of A. G. Richards; and the Price collection of Mallophaga, containing more than 220,000 slide mounted specimens. The entire collection is cataloged on a computerized inventory management system which provides information on the total number of specimens per family, percent of these determined to species, their storage method (pin, vial, or slide) and location in the collection, number and kinds of types, and grand totals of these categories for the entire collection. In addition, new material can be accessioned and merged into the system. Finally, loan information, loan letters, and renewal notices can be produced automatically. [1992]

SCIENCE MUSEUM OF MINNESOTA, 30 E. 10th ST., ST PAUL, MN 55101. [SMPM]

Curator: Dr. Frederick J. Jannett, Jr. Phone: (612) 221-9429. Professional staff, 2.5 (Biology). About 40,000 specimens, mostly pinned. No primary types. Voucher specimens of material published in series: "Scientific Publications of the Science Museum of Minnesota" and "Monographs, Science Museum of Minnesota." [1992]

MISSISSIPPI

Mississippi State

MISSISSIPPI ENTOMOLOGICAL MUSEUM, MISSISSIPPI STATE UNIVERSITY, DRAWER EM, MISSISSIPPI STATE, MS 39762. [MEMU]

Director: Dr. Richard L. Brown. Phone: (601) 325-2085. Professional staff: Mr. Terence Schiefer. The museum includes a research collection, a diagnostic laboratory, public exhibits, and library. The collection is adjacent to offices, greenhouses, growth chambers, and other rearing facilities, including an electron microscope center, microcomputer, and printing facilities. The arthropod collection contains over 750,000 pinned specimens, more than 30,000 vials, nearly 11,000 slides, over 2,800 cellophane envelopes of specimens, and 220 jars of unsorted bulk material. Five holotypes and 270 secondary types are housed in the museum; additional holotypes have been placed on indefinite loan to other collections. Most of the specimens have been collected in Mississippi; however, the collection also includes substantial material from other southeastern states and Central America. The pinned specimens are kept in over 1,400 drawers in 86 cabinets. The special collections include the Hepner Cicadellidae Collection that includes over 120,000 specimens, of which 80,000 are reared. The Charles Bryson Collection includes more than 12,000 butterflies and skippers. The Hoke-Lobdell collection of Nearctic scale insects includes over 6,000 slides and 1,100 envelopes of dried material. In addition, portions of the A. Balachowski European Collection, the Chermotheca Italica Collection, and the Kuwana Japanese Collection are present in the museum. It is nationally recognized for holdings of insects and other arthropods associated with cotton and other Malvaceae from the U.S.A. and Central America, originally developed as the USDA-ARS Cotton Insect Collection by William Cross. The Kislanko Aphid Collection of over 2,000 slides is of special importance because of his notebooks that provide detailed biological, behavioral, and descriptive information concerning the collected species. The museum includes over 18,000 vials of spiders and over 150,000 specimens of Coleoptera, of which Carabidae, Scarabaeidae, and Curculionidae are best represented. The John MacDonald Collection of Panamanian Lepidoptera is affiliated with the museum. [1986]

University

UNIVERSITY OF MISSISSIPPI INSECT COLLECTION, DEPARTMENT OF BIOLOGY, UNIVERSITY OF MISSISSIPPI, UNIVERSITY, MS 38677. [UMIC]

Director: Dr. Paul K. Lago. Phone: (601) 232-7203. This collection contains representatives of most families occurring in Mississippi and Alabama. Scarabaeidae, Cerambycidae, Chrysomelidae, aquatic Coleoptera and Trichoptera are best represented. The collection contains about 140,000 pinned specimens housed in 80 drawers, and about 7,000 alcohol vials of immatures. Supporting ecological data is moderate. Exchanges and loans are available; usual loan rules apply. No holotypes are present. Extensive collections of beetles have been made from wild carrot (*Daucus carota*), marijuana (*Cannabis sativa*), and magnolia (*Magnolia grandiflora*). [1992]

MISSOURI

Columbia

W. R. ENNS ENTOMOLOGY MUSEUM, 1-87 AGRICULTURE BLDG., UNIVERSITY OF MISSOURI, COLUMBIA, MO 65201. [UMRM]

Director: Dr. Robert W. Sites. Phone: (314) 882-2410 (Museum); (314) 882-2345 (Director). Professional staff: Ms. Kristin B. Simpson (Collection Manager). This collection contains 1.1 million specimens, 40,000 vials of specimens in alcohol, 1,600 drawers of pinned specimens, 20,000 microscope slides; 790 primary and secondary types. [1992]

St. Louis

ST. LOUIS SCIENCE CENTER, 5050 OAKLAND, ST. LOUIS, MO 63110. [SLSC]

Director: Dwight S. Crandell. Phone: (314) 652-5500. Professional staff: James G. Houser, Scott K. VanderHamm, Lynn Fendler. This collection contains the third largest collection of Lepidoptera in the state. The collection of 4,000 *Catocala* (underwing moths) ranks high among the N. A. collections of this group. Most of the spread specimens were collected in the early part of this century by well known entomologists of the period. The majority were collected by Ernst, Fred, and Herman Schwarz, along with numerous examples from C. L. Heink, E. D. Meiners, R. R. Rowley, H. McElhose, F. Malkmus, and H. I. O. Byne. All of the famous collecting localities at that period are well represented. Among the special collections are the O. C. K. Hutchinson collection of 5,000 specimens, most in envelopes, and others in glass mounts or Riker mounts; Bouton collection of 600 Denton mounts of Lepidoptera; and Dritschilo ground beetle collection, 825 specimens (Coleoptera, Carabidae). The majority of the collection is housed in 173 drawers in steel cabinets. [1986]

MONTANA

Bozeman

MONTANA STATE UNIVERSITY ENTOMOLOGY COLLECTION, ENTOMOLOGY RESEARCH LABORATORY, MONTANA STATE UNIVERSITY, BOZEMAN, MT 59717. [MTEC] [=DZEC]

Curator: Dr. Michael A. Ivie. Phone: (406) 994-4610. Professional staff: Associate curators: Dr. Kevin O'Neill, Dr. Richard S. Miller, Dr. Daniel F. Gustafson, Mr. Kelly Miller, and Mrs. LaDonna Ivie. Collection manager: Ms. Catherine Seibert. Phone: (406) 994-6995. The collection contains over 300,000 pinned specimens, 20,000 alcohol vials, and 10,000 microscope slides. Begun in 1899 by R. A. Cooley, the collection emphasizes Montana and the Northern Great Plains/Northern Rocky Mountain regions. Holdings are limited to the Nearctic region, plus Palaearctic material of pest or beneficial taxa. Holdings include: R. A. Cooley's Coccoidea, G. Roemhild's aquatic collection, the A. Buroughs flea collection, and the Elrod Lepidoptera (formerly at the University of Montana, Missoula). Taxonomic strengths include Montana Noctuidae (120 drawers), Montana butterflies (100 drawers), and Coccoidea (over 2,000 slides), and all acquatic orders for the Northern Rockies and Great Plains. A very large collection from the International Biosphere Reserves at Yellowstone and Glacier National Parks are included in with this collection, with acquatics, Coleoptera, and Hymenoptera particularly well represented. The collection is most actively growing in the Coleoptera and Hymenoptera, with Coleoptera holdings doubling since 1985. The pinned collection is housed in 60 California Academy cabinets, and the Ethanol material in five shelved metal cabinets. A supporting library of 72 shelf feet is maintained in the collection room. Primary and secondary type specimens are maintained separately. [1992]

Affiliated collection:

Ivie, Dr. Michael A., Entomological Research Laboratory, Montana State University, Bozeman, MT 59717. Phone: (406) 994-4610. [MAIC] This is a specialized collection of Coleoptera, with three main parts. A general world collection is aimed at representation of all major lineages of beetles; a systematic collection centers on worldwide species level research on bostrichids (*sensu lat.*) and the zopherid-monommid-colydiid lineage; and a West Indian collection supports species level faunistic and biogeographic work in that region. The collection contains minimal Nearctic material, but is rich in Neotropical, Palaearctic and Australasian specimens. A total of over 100,000 specimens includes approximately 40,000 pinned West Indian beetles, 40,000 pinned general beetles, 2,000 vials of beetle immatures, a disarticulation collection, and 10,000 uncurated specimens in ethanol. Pinned material is housed in Cornell drawers and unit trays; ethanol material in glass vials. Seventeen drawers of non-Coleoptera are also maintained. Much of the material is from specialized collecting methods, such as Berlese funnels, flight intercept, and Malaise traps. A series of lots in 95% ethanol is maintained for DNA studies. Holotypes are deposited in the USNM or other public collections, but several hundred paratypes are included in this collection. Material may be borrowed, usual rules apply. (Registered with MTEC.) [1992]

West Glacier

MUSEUM COLLECTIONS OF GLACIER NATIONAL PARK, WEST GLACIER, MT 59936. [GLNP]

Director: Ellen C. Seeley. Phone: (406) 888-5441, ext. 302. The collection is in 23 drawers in a cabinet: Coleoptera, 5 1/2 drawers; Diptera, 1/2 drawer; Lepidoptera, 15 drawers, and miscellaneous, 2 drawers. [1986]

NEBRASKA

Chadron

LABORATORY OF ARTHROPOD DIVERSITY, BIOLOGY DEPART-

MENT, CHADRON STATE COLLEGE, CHADRON, NE 69337. [CSCC]

Director: Dr. H. Randy Lawson. Phone: (308) 432-6298. The collection is a general regional collection containing about 15,000 specimens taken mostly within 200 miles of Chadron, NE. Ninety percent of the collection is from the panhandle counties of Nebraska, s.w. SD, and e. WY. It is housed in 80 drawers and 2,000 vials. There is a special collection of Nebraska Neuroptera, Odonata, and Lepidoptera (mainly butterflies). [1986]

Lincoln

UNIVERSITY OF NEBRASKA STATE MUSEUM, W436 NEBRASKA HALL, UNIVERSITY OF NEBRASKA, LINCOLN, NE 68588-05143. [UNSM]

Curator: Dr. Brett C. Ratcliffe. Phone: (402) 472-2614. Professional staff: Charles Messenger (Collection Manager). The collection began in the 1870s with the Orthoptera work of Lawrence Bruner. It now contains about 1.75 million pinned, papered, fluid preserved, and slide mounted specimens. Emphasis is on the Great Plains of U.S.A., followed by North America, and then the Neotropical Region. Specimens are about 8% Orthoptera, 7% Hemiptera, 25% Coleoptera, 7% Lepidoptera, 13% Diptera, 25% Hymenoptera, 3% smaller orders, and 13% Acarina. An entomological reprint collection and the University library are excellent in arthropod systematics. Primary types are housed separately, and many secondary types are contained in the collection. Thousands of specimens are on loan to specialists. The collections are used for study by visiting professionals as well as the faculty and students of the University. Special collections include: Lawrence Bruner collection and library of Orthoptera; Kenneth Fender synoptic collection of North American Cantharidae; and Worley and Pickwell spider collection. The collection is housed in 2,400 drawers in cabinets, 120 metal drawers of vials, and climate controlled room for microscope slide storage. Publications sponsored: "Bulletin of the University of Nebraska State Museum," and "Museum Notes." [1992]

Affiliated Collection:

Ratcliffe, Dr. Brett C., Systematics Research Collections, W436 Nebraska Hall, University of Nebraska, Lincoln, NE 68588-0514. [BCRC] Phone: (402) 472-2614. This collection consists of the Scarabaeidae of the World with strongest holdings from the Neotropical and Nearctic Regions. Primary types are deposited with the University of Nebraska State Museum as well as other institutions; secondary types are maintained in the collection. The collection is housed in unit trays in 120 USNM drawers. Specimens are available for exchange and loan. (Registered with UNSM.) [1992]

Hastings

HASTINGS COLLEGE COLLECTION OF ARTHROPODS, HASTINGS COLLEGE, 7TH AND TURNER AVE., HASTINGS, NE 68901. [HCCA]

Director: Dr. Charles Anthony Springer. Phone: (402) 463-2402. This is a general collection of arthropods from Nebraska and surrounding area. Coleoptera make up more than half of the 10,000 specimens, housed in 48 drawers. There is a special collection of world Byturidae (Coleoptera). [1986]

NEVADA

[The following collection is no longer taxonomic; returned form asking to be deleted: Department of Biology Collection, Nevada University, Las Vegas, NV 89109.] [UNLV]

Carson City

NEVADA STATE MUSEUM, 600 NORTH CARSON ST., CAPITOL COMPLEX, CARSON CITY, NV 89710. [NVMC]

Professional staff: Ann Pinzl. Phone: (702) 687-4810. The collection contains 50,000 specimens, 90% of which are butterflies from Nevada. They are housed in 75 drawers and 150 Schmitt boxes. [1992]

Reno

NEVADA STATE DEPARTMENT OF AGRICULTURE, P. O. BOX 11100, RENO, NV 89510. [NVDA]

Director: Mr. Jeff B. Knight. Phone: (702) 688-1180. This is the largest primary collection of insects and other arthropods in Nevada. The collection, comprised mostly of Nevada material, contains over 30,000 pinned, labelled, and identified specimens, over 1,700 vials containing labelled and identified specimens, and approximately 1,200 labelled and identified slide mounts. The pinned and vial collections are housed in 7 cabinets with drawers, and the slide collection is stored in slide boxes. Several thousand pinned specimens determined to family, subfamily, tribe, or genus are housed in Schmitt boxes or cigar boxes, and much unsorted material from mostly pan and pitfall traps is currently preserved in alcohol. Special collections include that of the late Ira LaRivers, a general collection containing over 16,000 pinned, labelled and identified specimens and numerous unidentified specimens in alcohol and the late P. C. Ting weevil collection containing approximately 4,000 pinned, labelled, and identified specimens. Both collections contain material primarily from North America, with a small amount from South America and the Old World. These collections are housed in 7 cabinets with drawers. Some holotypes, allotypes, and over 500 paratypes are represented in the main and special collections. [1992]

NEW HAMPSHIRE

Durham

UNIVERSITY OF NEW HAMPSHIRE INSECT COLLECTION, DEPARTMENT OF ENTOMOLOGY, UNIVERSITY OF NEW

HAMPSHIRE, DURHAM, NH 03824. [DENH]

Director: Dr. Donald S. Chandler. Phone: (603) 862-1707. Professional staff: Dr. John F. Burger, Dr. R. Marcel Reeves, and Dr. John S. Weaver. The collection contains 500,000 insect specimens in alcohol and pinned, 8,000 slides, and 22,000 other arthropods in alcohol. Strengths of the collection are the Lepidoptera, especially the Microlepidoptera, Odonata, Trichoptera, and Mallophaga, and Coleoptera. Coverage is very good for New England. There is a fair representation of world Lepidoptera and Tabanidae. Many parasites have been reared, and in certain groups, the host and ecology data are good. Collecting trips are occasionally sponsored by the Department. Material loaned; usual rules apply. Special collections include: R. L. Bickle collection of Trichoptera; J. F. Burger collection of Tabanidae; R. M. Reeves collection of Tetranychidae and Oribatoidea; D. S. Chandler collection of Pselaphidae and Anthicidae. The collection is housed in 12-drawer cabinets. [1992]

NEW JERSEY

New Brunswick

INSECT COLLECTION, RUTGERS-THE STATE UNIVERSITY, NEW BRUNSWICK, NJ 08903. [RUIC]

Director: Dr. Michael L. May. Phone: (201) 932-9459, 8872. The collection consists of over 300,000 specimens, with emphasis on Apoidea, Hydradephaga, and Odonata, stored pinned in drawers, Odonata in clear envelopes, and alcohol vials in a vial cabinet. The collection was initiated in 1879. The original nucleus was the Lepidoptera collections of G. D. Hulst and J. B. Smith. Later a Coleoptera collection (purchased from Mr. Boerner) and a general "State Experiment Station" collection were added. The collection is primarily representative of the fauna of New Jersey. [1992]

Princeton

INSECT COLLECTION, MUSEUM OF NATURAL HISTORY, PRINCETON UNIVERSITY, PRINCETON, NJ 08540. [MNHP]

Director: Dr. Donald Baird. Phone: (609) 452-4102. The collection consists of fossil insects and spiders, primarily from Florissant, CO (Oligocene) and Labrador (Cretaceous), including types, paratypes, and cited specimens, stored in drawers in steel cabinets. [1986]

Trenton

BUREAU OF NATURAL HISTORY, NEW JERSEY STATE MUSEUM, NEW JERSEY DEPARTMENT OF STATE, CN-530, TRENTON, NJ 08625-0530. [NJSM]

Curator: Mr. David C. Parris. Phone (609) 292-6330. Professional staff: Shirley Albright, Assistant Curator; Dr. William B. Gallagher, Registrar. This is a small collection of donated specimens, mostly from New Jersey; occupies one Lane cabinet, several hundred specimens.

[1992]

NEW MEXICO

Albuquerque

ARTHROPOD COLLECTION, DEPARTMENT OF BIOLOGY, UNIVERSITY OF NEW MEXICO, ALBUQUERQUE, NM 87131-1091. [UNMC]

Director: Dr. Clifford S. Crawford. Phone: (505) 277-3858. Professional staff: Carlos A. Blanco-Montero, Assistant curator. Phone: (505) 277-4225. This relatively new collection was developed over the past decade to support arthropod-related ecological research, mainly in Central and Northern New Mexico. The collection now houses over 22,000 pinned insects and many alcohol vials containing arachnids and myriapods. Large number of species represent the Biology Department's Long Term Ecological Research program at the Sevilleta National Wildlife Refuge as well as other collections representing montane, riparian, and urban environments. [1992]

Carlsbad

MUSEUM COLLECTIONS OF CARLSBAD CAVERNS NATIONAL PARK, 3225 NATIONAL PARK HIGHWAY, CARLSBAD, NM 88220. [CCNP]

Director: Franklin C. Walker. Professional staff: Douglas A. Buehler. This is a very general collection from within the park and its immediate vicinity. [No further details on the size or contents of the collection were provided. [1986]

Las Cruces

DEPARTMENT OF ENTOMOLOGY, PLANT PATHOLOGY AND WEED SCIENCE, BOX 30003, DEPT. 3BE, NEW MEXICO STATE UNIVERSITY, LAS CRUCES, NM 88003. [NMSU]

Director: Dr. David B. Richman. Phone: (505) 646-3542. This is a unified collection New Mexico Department of Agriculture, Extension, and this collection. The collection contains about 30,000 specimens in 15 cabinets with drawers. The Range Insect Collection contains 300 specimens in 12 drawers. No primary or secondary types are housed here. Voucher specimens from "Grasshoppers of New Mexico" project, Snakewood biological control project, and several graduate student papers are included. The private arachnid collection of David B. Richman is also housed here. [1992]

NEW YORK

Albany

INSECT COLLECTION, NEW YORK STATE MUSEUM, BIOLOGI-

CAL SURVEY, 3132 CULTURAL EDUCATION CENTER, ALBANY, NY 12230. [NYSM]

Director: Dr. Louis D. Levine. Phone: (518) 473-8496. Professional staff: Dr. Timothy McCabe, and Dr. Jeffery K. Barnes. This is a rapidly growing, general collection comprising nearly 500,000 pinned, 100,000 liquid preserved, and 7,500 slide mounted insect specimens, plus over 12,000 spiders. Most specimens are from New York State, but representation from other areas of North America, South America, and Europe is expanding. The collection was initiated in 1846 by Dr. Asa Fitch, who was employed by the Regents of the University of the State of New York to collect, arrange, and catalog insects for the State Cabinet of Natural History, lineal ancestor of the NYSM. Today the collection has over 200 primary types and 800 secondary types from all major orders. Special collections: Asa Fitch Homoptera types; Joseph Albert Lintner Lepidoptera Collection; Ephriam Porter Felt Gall Collection; the Felt Cecidomyiid Collection; W. W. Hill Lepidoptera Collection; University of Rochester Insect Collection (including the Moore and Wendt Collection). The collection is stored in unit trays arranged in 12-drawer cabinets. Publications sponsored: "New York State Museum Bulletin," and "New York State Museum Memoirs." [1992]

Buffalo

INSECT COLLECTION, BUFFALO MUSEUM OF SCIENCE, 1020 HUMBOLDT PARKWAY, BUFFALO, NY 14211-1293. [BMSC]

Director: Mr. Ernest E. Both. Phone: (716) 896-5200. Professional staff: Wayne K. Gall, Associate Curator (Ph.D. candidate, University of Toronto and Royal Ontario Museum) and Dr. Donald Anderson (retired USNM/USDA). The collection consists of over 300,000 insect specimens stored in 47 25-drawer cabinets and in Schmitt boxes. Previous entomologists at the museum include Augustus R. Grote, Coleman T. Robinson, Leon F. Harvey, David S. Kellicott, Ottomar Reinecke, Frank H. Zesch, Edward P. Van Duzee, Millard C. Van Duzee, and William Wild. Collections housed in the Museum include those of Augustus R. Grote (major portions of his collection were given to the Museum of Natural History (BMNH) and to the Museum of Natural History of Sao Paulo), Ottomar Reinecke, Edward P. Van Duzee, Mrs. George L. Squier, E. G. Love, Kirke B. Mathes, A. G. Dimond, and Millard C. Van Duzee (the majority of Edward P. Van Duzee collection went to the California Academy of Sciences). In the early fifties, the Ward's Natural Science Establishment of Rochester, New York, gave the Buffalo Society of Natural Sciences more than 100,000 insect specimens, mainly Lepidoptera, Coleoptera, and Orthoptera. Most of these specimens remain in their original papers. Special collections include: 2,060 specimens from the Mary and Harold Cohen 1979 collection from Rancho Grande, Venezuela, and 70,000 specimens of Lepidoptera collectect by the Cohens in Venezuela, Europe, and Mexico. Most of the collection is local, western New York with emphasis on Odonata, Orthoptera, Heteroptera, Hymenoptera, Coleoptera, and especially Lepidoptera. Publications sponsored: "Bulletin of the Buffalo Society of Natural Sciences," "Miscellaneous Contributions of the

Buffalo Society of Natural Sciences," and "Occasional Papers of the Buffalo Society of Natural Sciences." [1992]

Ithaca

CORNELL UNIVERSITY INSECT COLLECTION, DEPARTMENT OF ENTOMOLOGY, CORNELL UNIVERSITY, ITHACA, NY 14850. [CUIC]

Director: Dr. James K. Liebherr. Phone: (607) 256-4507. Professional staff: Dr. C. O. Berg (Diptera), Dr. W. L. Brown, Jr. (Hymenoptera), Dr. G. C. Eickwort (Hymenoptera), Dr. J. G. Franclemont (Lepidoptera), Dr. E. R. Hoebeke (Coleoptera), Dr. B. L. Peckarsky (aquatic insects), Dr. C. A. Tauber (Neuroptera), and Dr. Q. D. Wheeler (Coleoptera). This collection constitutes one of the major systematic resource centers in North America, situated at a university with a strong program, commitment, and history in insect systematics. The collection has been estimated to rank among the top two university maintained collections in North America and among the largest seven collections of all types. The Comstock Memorial Library of Entomology, one of the most outstanding entomological literary collections in the Western Hemisphere, combines with the insect collection to support systematic research in the Department. The collection is currently estimated to include in excess of 4 million pinned, labelled, and identified specimens, 40,000 vials of fluid preserved material, and 60,000 microscope slides. The collection represents more than 200,000 identified species of insects, and is a well balanced, world class collection of insects and related arthropods with depth and breadth in all major insect taxa. A collection of more than 20,000 vials of spiders is on indefinite loan to the American Museum of Natural History. There are also large holdings of unprocessed specimens, and a collection of residues from Berlese funnels, flight intercept traps and other mass collecting methods. More than 5,700 species are represented by primary or secondary types in the collection. Published lists of types are currently available for Coleoptera, Hymenoptera (in part), and some Diptera and Mallophaga taxa. The collection is housed in over 8,000 Cornell drawers. Alcohol preserved specimens are currently being transferred to a new storage system in glass museum jars, and microscope slides are stored in standard boxes and flat trays. Although the collection is global in scope, it has particular strengths in eastern North America, the Neotropical Region, Africa, and Eurasia. The collection has been moved into a new building, the new Comstock Hall with the drawers in a compactor system. [1986]

Affiliated Collections:

Dirig, Robert, The Liberty Hyde Bailey Hortorium Herbarium, 467 Mann Library Building, Cornell University, Ithaca, NY 14853 USA. [RDIC] Phone: (607) 256-7978, -7981 (days); 272-0313. The collection is primarily Lepidoptera of Northeastern North America, especially butterflies of New York. (Plus "exotic" Lepidoptera from England, Spain, Germany, Japan, Australia, New Zealand, Mexico, Guatemala, Bras\zil, Africa, Madagascar, and elsewhere; these are showy species, traded.) Regional emphasis is on New York State: Catskill Mountains, Karner Pine Bush, and associated sandy areas, Northern New York (greater Adirondack region, and Ithaca area. General insect collections also from sone of these area. There are about 5,500 specimens stored in drawers. Two butterfly paratypes are included (*Everes, Erynnis*). Specimens may be borrowed; usual regulations apply. [1992]

Franclemont, Dr. John G., Department of Entomology, Cornell University, Ithaca, NY 14853 USA. [JGFC]. This is a large, important collection of Lepidoptera, especially Noctuidae. No further information available [1986]. It is currently located at CUIC.

New York

DEPARTMENT OF ENTOMOLOGY COLLECTION, AMERICAN MUSEUM OF NATURAL HISTORY, CENTRAL PARK WEST AT 79TH ST., NEW YORK, NY 10024. [AMNH]

Director: Dr. Randall T. Schuh. Phone: (212) 873-1300. Professional staff: Dr. Lee H. Herman, Jr., Dr. Norman I. Platnick, Dr. Frederick H. Rindge, and Dr. J. G. Rozen, Jr. The collection contains about 16 million specimens in 45,000 drawers in 1,000 cabinets as follows: Arachnida, 1.2 million; Coleoptera, 1.7 million; Diptera, 830,000; Hemiptera, 425,000; Hymenoptera, 8.3 million; Lepidoptera, 2 million, and other groups, 1.7 million specimens. Greatest strengths in the collection are Isoptera, New World Rhopalocera, Nearctic Cynipoidea, Apoidea, Araneae, and others. Publications sponsored: "Bulletin of the American Museum of Natural History," and "American Museum Novitates." [1986]

Affiliated collection:

Mather, Bryant, 213 Mt. Salvs Rd., Clinton, MS 39056-5007 U.S.A. [MATH] Phone: (601) 924-6360. The collection includes 100,000 pinned specimens and about 100 genitalia slides of Lepidoptera; 5,000 pinned Neuroptera, Trichoptera, Mecoptera, and Plecoptera; 3,000 specimens of Orthoptera, Hymenoptera, Diptera, and Coleoptera. Notebook data on about 500,000 specimens, many of which have already been given to major museums. [1992]

Staten Island

STATEN ISLAND INSTITUTE OF ARTS AND SCIENCES, 75 STUYVESANT PLACE, STATEN ISLAND, NY 10301. [SIIS]

President and CEO: Hedy A. Hartman. Phone: (718) 727-1135. Curator of Science: Edward W. Johnson. The collection consists of about 500,000 specimens representing 16 orders, and is stored in approximately 3,500 boxes. Major portions of the collection are Homoptera (66,000 specimens), Coleoptera (314,000 specimens), Orthoptera (50,000 specimens), Lepidoptera (36,000 specimens) and Diptera (20,000 specimens). The several special collections are: William T. Davis Collection of Cicadas (60,000 specimens, world-wide in scope, with emphasis on North American species, including type specimens); Howard Notman Collection of Coleoptera (mostly Northeastern United States); and the William T. Davis Gall Collection (2,000 specimens of various orders). The collection contains about 100 primary types and numerous paratypes. The emphasis is on North American material, especially the Northeastern U.S.A. Publication sponsored: "Proceedings of the Staten Island Institute of Arts and Sciences." [1992]

Syracuse

COLLEGE OF ENVIRONMENTAL SCIENCE AND FORESTRY COLLECTION, STATE UNIVERSITY OF NEW YORK, SYRACUSE, NY 13210. [DFEC]

Director: Dr. Frank E. Kurczewski. Phone: (315) 470-6753. Professional staff: Dr. Roy A. Norton, Dr. Gerald N. Lanier. The collection consists of 450,000 pinned specimens housed in 1,200 drawers, including 10,000 Hemiptera, 260,000 Coleoptera, 100,000 Hymenoptera, and 70,000 Diptera from northeastern United States. In addition, 3,000 specimens of spiders, identified mostly to species, are housed in alcohol vials. The spiders are mostly from New York and Pennsylvania. Paratypes of Scolytidae, Braconidae, and Sphecidae are available by loan. Special collections: Many S. W. Blackman specimens of Scolytidae, predatory wasp association specimens (F. E. Kurczewski), and Tabanidae from the eastern Great Lakes identified to species by L. L. Pechuman. [1986]

NORTH CAROLINA

Raleigh

NORTH CAROLINA INSECT COLLECTION, PLANT INDUSTRY DIVISION, NORTH CAROLINA DEPARTMENT OF AGRICULTURE, P. O. BOX 27647, RALEIGH, NC 27611. [EDNC]

Director: Mr. James F. Greene. Phone: (919) 733-3610. Professional staff: Mr. Kenneth R. Ahlstrom, Dr. David L. Wray. This collection consists primarily of insects occurring in North Carolina. It was begun in the 1890's and contains many valuable specimens taken from ecosystems that no longer exist. It contains about 250,000 pinned, labelled, and identified specimens. It is associated with the North Carolina State Museum of Natural History collections and the North Carolina Biological Survey. It is housed in over 1,000 drawers in cabinets and 500 storage boxes. The primary area of geographic interest is North Carolina. Specimens will be loaned to specialists and those doing serious taxonomic work. The list, "Insects of North Carolina" and supplements is published in conjunction with the collection. [1986]

NORTH CAROLINA STATE UNIVERSITY INSECT COLLECTION, DEPARTMENT OF ENTOMOLOGY, BOX 5215, NORTH CAROLINA STATE UNIVERSITY, RALEIGH, NC 27607. [NCSU]

Director: Dr. Lewis L. Deitz. Phone: (919) 737-2833. Collection Manager: Mr. Robert L. Blinn. Phone: (919) 515-2833. Professional staff: Dr. James R. Baker, Dr. Maurice H. Farrier, Dr. Kenneth L. Knight, Dr. Herbert H. Neunzig, Dr. Clyde F. Smith, Dr. P. Sterling Southern, and Mr. David L. Stephan. The collection consists of more than 531,000 pinned specimens, 111,000 slides, and 22,000 vials. Formally initiated in 1952, it ranks among the largest arthropod collections of the Southeast. Though much of the material is from North Carolina, coverage of many groups spans the U.S.A. and beyond. Notable areas of specialization are:

Homoptera (382,000 specimens, world wide scope in Auchenorrhyncha and Aphididae, including major collections of Z. P. Metcalf, D. A. Young, W. Wagner, and C. F. Smith); Hymenoptera (71,000 specimens, including T. B. Mitchell's world collection of bees); Diptera (84,000 specimens, including K. L. Knight's mosquito collection and a sizable tabanid collection); Coleoptera (92,000 specimens, including over 13,000 elaterids and 11,000 scarabs); Lepidoptera (42,000 specimens, including valuable immature-adult associations, numerous noctuids, geometrids, sphingids, and over 8,000 pyralids); Acarina (33,000 slides, rich in Ascidae, Phytoseiidae, Rhodacaridae, Veigaiidae, and Zerconidae); Heteroptera (17,000 specimens, including many mirids and lygaeids); Orthoptera (12,000 specimens including the B. B. Fulton collection), and Siphonaptera (5,000 specimens, including the A. D. Shaftesbury collection). In addition to the collections above, notable voucher collections are those of soybean and grape arthropods from North Carolina. The collection serves as a repository for research vouchers and North Carolina reference material, including many specimens from the Plant Disease and Insect Clinic. More than 8,000 paratypes and other secondary types are incorporated, most holotypes (more than 500) are on indefinite loan to the USNM). The associated D. H. Hill Library includes three important literature collections: the Tippmann Collection (items on beetles and many old European serials), the Z. P. Metcalf Collection on auchenorrhynchous Homoptera, and the C. F. Smith collection on aphids. Pinned specimens occupy more than 2,000 museum drawers in 179 cabinets. [1992]

NORTH CAROLINA MOSQUITO AND MEDICALLY IMPORTANT ARTHROPOD COLLECTION, PHPM SECTION, NCDEHNR, P. O. Box 27687, RALEIGH, NC 27611-7687. [NCMA]

Director: Nolan H. Newton. Phone: (919) 733-6407. Professional staff: Dr. Barry Eugber. The collection contains about 1,000 labelled and identified adult mosquitoes, 500 larvae on slides, several thousand unmounted adult and larval mosquitoes with complete collection data, and several hundred other arthropods of medical importance (fleas, flies, ticks, *etc.*) with data. Most specimens are from North Carolina. A computerized data bank on mosquitoes with seasonal and distribution information is maintained and continually updated. The collection is stored in drawers, alcohol vials, and on slides. [1992]

Southern Pines

WEYMOUTH WOODS SANDHILLS NATURE PRESERVE, 400 NORTH FORT BRAGG ROAD, SOUTHERN PINES, NC 28387. [WWSP]

Director: L. M. Goodwin, Jr. Phone: (919) 692-2672. The collection consists of 1,200 insects and 200 spiders with representatives of most orders. The collection was begun in late 1978 with contributions from Thomas Howard, Richard Thomas, and Charlotte Gantz. Specimens may be borrowed for study. [1986]

NORTH DAKOTA

Fargo

NORTH DAKOTA STATE INSECT REFERENCE COLLECTION, ENTOMOLOGY DEPARTMENT COLLECTION, N. D. STATE UNIVERSITY, FARGO, ND 58102. [NDSU]

Director: Dr. David A. Rider (Hemiptera). Phone: (701) 237-7902. Professional staff: A. W. Anderson (immature insects); L. D. Charlet (Acari); G. Fauske (Lepidoptera); R. L. Post (Thysanoptera). Emphasis is primarily on Nearctic fauna, but includes some examples from all parts of the world. The collection contains approximately 900,000 specimens from the following orders (number of drawers): Coleoptera (450), Lepidoptera (300), Hemiptera (70), Homoptera (33), Diptera (80), Hymenoptera (75), Orthoptera (80), and miscellaneous smaller orders (33). In addition there are approximately 50,000 vials of specimens in alcohol and 23,000 microscope slides. Special strengths of the collection are: Miridae, Cicindelidae, Chrysomelidae, and Thysanoptera. Special collections: V. M. Kirk collection of Coleoptera [**Kirk, Vernon M.**, VMKC]; P. Slaybaugh collection of Cicindelidae, the R. L. Post collection of Thysanoptera, and M. Hebard collection of determined Orthoptera from North Dakota. Publications sponsored: the Shaffer-Post "North Dakota Insects." [1992]

OHIO

Bowling Green

NATIONAL *DROSOPHILA* SPECIES RESOURCE CENTER, DEPARTMENT OF BIOLOGICAL SCIENCES, BOWLING GREEN STATE UNIVERSITY, BOWLING GREEN, OH 43403. [NDSR]

Director: Dr. Jong S. Yoon. Phone: (419) 372-2742; 372-2096. Professional staff: Kay Yoon, Linda Treeger, and Sharon Lockhart. The Center is governed by a council appointed by the American Society of Naturalists and its purpose is to provide cultures of a wide variety of species to researchers, teachers, and students. Approximately 350 species representing 8 genera and 34 species groups of the family Drosophilidae are maintained. This is the largest collection of living eucaryotic organisms ever assembled whose evolutionary relationships and genetic biology have been extensively studied. Stocks of many species include strains having visable mutants, electromorphs, and chromosomal rearrangements. A list of cultures and details on their maintenance is available. The Center intends to expand its holdings of species, especially those with mutants and chromosomal rearrangements and solicits information concerning them. New cultures will be added at the discretion of the Director. Limited facilities are available for visiting scientists, including the large reference collection of pinned specimens assembled and provided by Professor Emeritus Marshall R. Wheeler of the University of Texas at Austin, as well as a nearly complete reprint file on *Drosophila* systematics. Modern equipment for research on the genetics and biology of

Drosophila is also available to qualified persons through the cooperation of the members of the genetics faculty in the Department of Biological Sciences. Cultures are available for purchase. [1986]

Cincinnati

ARTHROPOD COLLECTION, CINCINNATI MUSEUM OF NATURAL HISTORY, 1720 GILBERT AVE., CINCINNATI, OH 45202. [CNHM]

Director: Dr. DeVere Burt. Phone: (513) 621-3890. Professional staff: Dr. Robert Kennedy and Dr. Gene Kritsky. The non-marine arthropod collection numbers nearly 100,000 specimens, about 90% of these are pinned, 75% labelled and identified. There are 75,000 Coleoptera, 5,000 Lepidoptera, 6,000 Diptera, and 2,500 Araneae among others. The majority of the specimens were collected and identified by Charles Dury (1847-1931). He began collecting insects around 1877, and most of the specimens are dated from that time until around 1920. Other collectors of the time contributed specimens, including: W. S. Blatchley, T. L. Casey, H. C. Fall, J. L. LeConte, C. W. Leng, H. Ulke, E. C. Van Dyke, and J. B. Wallis. Other special collections that are housed in the Museum include the Ralph Kellog insect collection, a microlepidoptera collection of Annette Braun, the spider collection of Charles Oehler, and several other small collections of butterflies and beetles. Specimens are from all over the U.S.A., with the eastern portion of the country particularly well represented, especially Indiana, Ohio, and Kentucky. Some Canadian, European, South American, and Australian insect specimens, particularly Coleoptera, are also depoisited here. The collection is housed in 250 drawers and 385 Schmitt boxes. Material is not available for exchange; study at the Museum is encouraged; loans may be made for scientific study. Publications sponsored: "Journal of the Cincinnati Society of Natural History." [1992]

Cleveland

CLEVELAND MUSEUM OF NATURAL HISTORY, 1 WADE OVAL DRIVE, UNIVERSITY CIRCLE, CLEVELAND, OH 44106. [CEMU]

Director: Dr. Sonja Teraguchi. Phone: (614) 231-4600. The collection contains about 200,000 specimens, mostly from Northeastern U.S.A., including 100,000 moths from Northeastern Ohio, 30,000 Coleoptera, world-wide, and 1,500 Odonata from Ohio. Special collections include Zahrobsky collection of world Coleoptera, and Welling collection of Ohio Lepidoptera. These collections are housed in 696 drawers in unit trays. [1992]

Columbus

OHIO STATE UNIVERSITY COLLECTION OF INSECTS AND SPIDERS, 1735 NEIL AVE., COLUMBUS, OH 43210. [OSUC]

Director: Dr. Norman Johnson. Phone: (614) 292-6839. Professional staff: Dr. Charles A. Triplehorn, Dr. Norman Johnson, Dr. D. E. Johnston, and Mr. J. A. Wilcox. The collection contains about 4 million specimens. The basic collection was initiated by Prof. Josef N. Knull (1891-1975). Both Prof. Knull and his wife, Dorothy were outstanding collectors and spent over 40 years collecting, much of which was added to the collection. Many outstanding private collections were added over the years, including H. W. Wenzel, Coleoptera; Herbert Osborn, Homoptera and Hemiptera; James S. Hine, Diptera; R. A. Leussler and W. N. Tallant, Lepidoptera; C. H. Kennedy, ants; D. J. Knull, leafhoppers, and William M. Barrows, spiders. Increased acquisition of exotic specimens has occurred through the opportunistic efforts of Triplehorn (Brazil, Panama, Mexico), Paul H. Freytag (Mexico, Guatemala, Honduras), H. J. Harlan (South Vietnam, Panama), D. M. DeLong (Mexico, Chile, Argentina, Bolivia, Peru), L. E. Watrous (Southeast Asia), F. W. Fisk (Mexico, Costa Rica), and B. D. Valentine (Caribbean Islands). Although the primary emphasis is still on the North American fauna, we have extensive holdings in most groups from many parts of the world. The single most important collection obtained recently was the D. M. DeLong Homoptera collection. This collection, combined with those of Herbert Osborn and D. J. Knull, is probably the largest leafhopper collection in the Western Hemisphere and is especially rich in type specimens. Other significant collections acquired over the past 20 years are the Alvah Peterson collection of immature insects, the C. R. Cutright aphids, the Robert M. Geist Mallophaga, and the Homer Price and D. J. Borror Odonata. The extensive insect collection of the Ohio Historical Society was recently transferred to this collection on permanent loan. The outstanding feature of that collection is the Orthoptera, assembled by Dr. Edward S. Thomas; it is also strong in Asilidae, Tabanidae, Cicadidae, Mecoptera, and several other groups including Ohio Noctuidae. The collection occupies 3,800 drawers in cabinets. Unidentified specimens, most of which are sorted to family, are contained in about 2,000 Schmitt boxes in cabinets. The extensive larval collection is preserved in about 13,000 vials housed racks in 5 steel cabinets. Spiders, scorpions, and other non-insectan arthropods (except mites), are preserved in 6,000 alcohol vials and occupy 2 steel cabinets. Material is available for loan to qualified researchers. Publications sponsored: The J. N. Knull series of the "Ohio Biological Survey." [1992]

Affiliated Collections:

Aalbu, Rolf L., 11 Rue Des Orangers, 2080 Ariana, Tunis, Tunisia. [ALRC] This collection contains about 263,000 adult and immature specimens: one half worldwide Tenebrionidae, over 1,900 species; one quarter general Coleoptera; one quarter general insects, including numerous arachnids. The general collection is mainly from southwestern U.S.A., Baja California, Mexico, and Cameroon, West Africa. The collection is housed in 125 drawers, 50 Schmitt boxes, and numerous alcohol vials and slides. No holotypes are kept in the collection. Material may be borrowed; usual loan rules apply [1986]. (Registered with OSUC.)

Berry, Richard Lee, 899 Bricker Blvd., Columbus, OH 43221 USA. [RLBC] Phone: (614) 421-1078, ext. 70; 459-3198. This is a collection of 74 families of eastern U.S.A. beetles, esp. Tenebrionidae, Scarabaeidae, Curculionidae, Cerambycidae, Staphylinidae, and Elateridae, housed in drawers, 40 Schmitt boxes, and cardboard boxes. Specimens may be borrowed for study; usual loan rules apply [1986]. (Registered with OSUC.)

Graves, Dr. Robert C., Department of Biological Sciences, Bowling Green State University, Bowling Green, OH 43403 USA. [RCGC] The collection is primarily North American Coleoptera. Nearly all North American families are represented, with specialization in Cicindelidae and Carabidae. Areas most collected are the Great Lakes region, and southwestern U.S.A., but specimens from most of the U.S.A. and parts of Canada are present [1986]. (Registered with OSUC.)

Schultz, Dr. William T., 616 Sharon Woods Blvd., Columbus, OH 43229 USA. [WTSC] Phone: (614) 466-1500. This is a collection of over 6,000 specimens of all orders and families from central U.S.A., with about half of the collection Chrysomelidae (Coleoptera), particularly Eumolpinae. It includes several paratypes. It is housed in 12 drawers and 16 Schmitt boxes [1986]. (Registered with OSUC.)

OHIO STATE UNIVERSITY ACAROLOGY LABORATORY, MUSEUM OF BIOLOGICAL DIVERSITY, 1315 KINNEAR ROAD, COLUMBUS, OH 43212-1192; FAX (614) 292-7774. [OSAL]

Director: Dr. Donald E. Johnston. Phone: (614) 422-7180. Professional staff: Dr. Glen R. Needham, and Dr. W. Calvin Welbourn. Current estimates of specimen holdings are 125,000 determined specimens (preserved as slide preparations or in alcohol and including many paratypes) and over 1.5 million undetermined specimens (mostly preserved in alcohol and including many Berlese residues). This is a world collection, but the representations of soil Acari from North America, the Neotropics, and Europe (the European collections are the most extensive in the New World) are outstanding. Other strengths are the parasitengone Prostigmate (including chigger and water mites), parasites of mammals (Mesostigmata, Prostigmata, Astigmata) and the ticks (Ixodida). Publications sponsored: "Directory of Acarologists," Offers an annual three-week summer program in acarology. [1992]

OHIO HISTORICAL SOCIETY INSECT COLLECTION, 1982 VELMA AVE., COLUMBUS, OH 43211. [OHSC]

Director: Dr. William T. Schultz. Phone: (614) 466-1500. Professional staff: Robert Glotzhober, and David Dyer. The collection contains about 30,000 identified specimens and 28,000 unidentified arthropods, mostly insects. It is primarily a collection of Ohio insects for reference, display, and teaching, and is strong in Coleoptera, Lepidoptera, and Diptera. The main part of the collection is on permanent loan to the Ohio State University Insect Collection. [1986]

Dayton

DAYTON MUSEUM OF NATURAL HISTORY, 2629 RIDGE AVE., DAYTON, OH 45414. [DMNH]

Director: Mr. E. J. Koestner. Phone: (513) 275-7432. Professional staff: Gary A. Coovert. The collection is stored in 400 drawers and 50 Schmitt boxes. It contains about 81,000 pinned specimens and 1,000 vials of alcohol preserved material. Although emphasis of the collection is on species from southwestern Ohio, the collection is fairly strong for all of eastern North America. All orders are represented but the collection is especially strong in Diptera. Particularly well represented is the family

Syrphidae with about 12,000 specimens from North America north of Mexico. Over half the collection has been collected in the past 10 years and is accompanied by field notes so data is particularly good. A large amount of material has been added recently from Malaise trap collections. Older collections include the Pilate and Klages collection and recently acquired is about 6,000 specimens from the Harold Morrison collection. Material is available for exchange and student loan; the usual loan rules apply. The spider collection consists of about 2,000 vials of specimens, mostly from s.w. Ohio. Publication sponsored: "Isotelus," an irregularly issued scientific journal covering all areas of natural history research. [1986]

OKLAHOMA

Norman

OKLAHOMA MUSEUM OF NATURAL HISTORY, UNIVERSITY OF OKLAHOMA, NORMAN, OK 73069. [OMNO]

Director: Dr. Michael Mares. Phone: (405) 325-4312. Professional staff: Dr. Harley Brown and Dr. Cluff Hopla. From the standpoint of research, the most significant portion of the collection is a specialized collection of over 150,000 specimens of aquatic dryopoid beetles amassed by Harley Brown. Most are from the U.S.A., Brazil, and Mexico, but virtually all parts of the western hemisphere are represented and there are significant holdings from Europe and Asia. Expeditions have been made to Mexico, Central America, the West Indies, western Europe, and most countries of South America. The bulk of this collection is in alcohol, but many are pointed and stored in drawers in metal cabinets. Aside from numerous light-trap collections of mosquitoes from Alaska and Oklahoma, most of the remaining insects are from Oklahoma and are housed in drawers. The spider collection, also from Oklahoma, is in alcohol. The Oklahoma insects number about 130,000. The special collections are as follows: H. P. Brown dryopid beetle collection, 150,000 specimens; Ralph Bird odonate collection, about 3,400; and the Ortenburger-Banks spider collection, about 600. [1992]

Stillwater

K. C. EMERSON MUSEUM, DEPARTMENT OF ENTOMOLOGY, OKLAHOMA STATE UNIVERSITY, STILLWATER, OK 74078. [OSEC]

Director: Dr. Don C. Arnold. Phone: (405) 624-5530. Professional staff: Dr. K. C. Emerson (Sanibel, FL) and R. D. Price (Ft. Smith, AR) are adjunct faculty associated with the louse collection. This is a general collection of insects, spiders, mites, and other arthropods. Most of the material is from Oklahoma, with small holdings from other states and countries. The pinned collection is stored in about 1,300 drawers of which 250 drawers are unidentified material. Immature insects, spiders, and other arthropods are stored in alcohol. There are about 4,000 vials of spiders. Special collections include (1) a small but representative collection of North American mosquitoes donated by S. J. Carpenter, and (2) the K. C. Emerson collection of approximately 23,000 slides of Mallophaga, Anoplura, and Siphonaptera (including 1,800 slides of paratypes of Mallophaga). The Anoplura and Mallophaga are collections of world-wide in scope. The museum also holds reprint collections donated by Drs. Emerson and Carpenter. Material is available for exchange and student study; usual loan rules apply. [1992]

Affiliated Collections:

Emerson, Dr. K. C., 560 Boulder Drive, Sanibel, FL 33957 USA. [KCEC] Phone: (813) 472-9156. [1992]

Sulphur

CHICKASAW NATIONAL RECREATIONAL AREA (formerly PLATT NATIONAL PARK), P. O. BOX 201, SULPHUR, OK 73086. [PNPC]

Director: Ms. Chris Czazasty. Phone: (405) 622-3165. The collection consists of 19 drawers with approximately 1,200 specimens. The specimens are made up of a variety of species collected in and around the area. Specimens are all insects, no arachnids. [1992]

OREGON

Crater Lake

MUSEUM COLLECTIONS OF CRATER LAKE NATIONAL PARK, P. O. BOX 7, CRATER LAKE, OR 97604. [CLNP]

Director: Mr. Kent Taylor, Chief of Interpretation. Phone: (503) 594-221. The collection is restricted to Crater Lake National Park, Oregon Caves National Monument, and their immediate areas. Local families are represented. There are about 1,800 specimens housed in 21 drawers. Material is not available for exchange or loan. Those wishing to view the collection may do so by contacting a staff member. [1992]

Corvallis

DEPARTMENT OF ENTOMOLOGY COLLECTION, OREGON STATE UNIVERSITY, CORVALLIS, OR 97331. [OSUO]

[*This department has a collection, but numerous requests for information have not brought replies. The Hatch collection of Coleoptera is deposited here. Some additional information may be found in the 1968 edition.*]

Eugene

MUSEUM OF NATURAL HISTORY, UNIVERSITY OF OREGON, EUGENE, OR 97403. [UOIC]

Director: Dr. Don Dumond (Anthropology). Phone: (503) 686-5101. The collection consists of one case of Lepidoptera and a few other insects, but the collection is not active at present. [1986]

Salem

ENTOMOLOGY MUSEUM, PLANT DIVISION, OREGON DEPARTMENT OF AGRICULTURE, 635 CAPITOL N.E., SALEM, OR 97310-0110. [ODAC]

Curator: Mr. Richard L. Westcott. Phone: (503) 378-6458. Professional staff: Mr. Richard L. Westcott (Coleoptera: Buprestidae), Dr. Dan J. Hilburn and Mr. Alan D. Mudge; Mr. Gary L. Peters, Technician. This is a reference collection restricted almost entirely to Oregon material and containing about 90,000 pinned specimens representing 5,500 determined species housed in 18 cabinets. There are about 1,500 vials of specimens in alcohol and 3,000 slide mounts. In addition, there is a small collection of economic species which are either exotic or otherwise not recorded from Oregon. Coleoptera, Diptera, and Lepidoptera comprise 70% of the collection. Orthoptera make up another 10%, largely consisting of an especially well prepared and comprehensive collection of Acrididae. Other families particularly well represented are: Diptera: Asilidae (80% of the species known in Oregon), Dolichopodidae, Syrphidae, Tabanidae, Tephritidae; Hymenoptera: Tenthredinidae; Lepidoptera: Noctuidae, Geometridae; Coleoptera: Buprestidae, Carabidae, Cerambycidae, Curculionidae, Elateridae, Scarabaeidae; Homoptera: Cicadellidae. [1992]

PENNSYLVANIA

Harrisburg

PENNSYLVANIA DEPARTMENT OF AGRICULTURE ARTHROPOD COLLECTION, BUREAU OF PLANT INDUSTRY, PENNSYLVANIA DEPARTMENT OF AGRICULTURE, 2301 NORTH CAMERON ST., HARRISBURG, PA 17110. [PADA]

Director: Dr. Karl R. Valley. Phone: (717) 787-5609. Professional staff: Rayanne D. Lehman, James F. Stimmel, Dr. A. G. Wheeler, Jr. This collection represents one of the largest maintained by the 50 state departments of agriculture, ranking about fifth in size. It contains about 105,000 pinned specimens, 2,000 slides of mites, 500 slides of aphids and thrips, and 1,500 slides of scale insects and mealybugs, all identified to species. The pinned specimens are housed in 825 drawers in 33 cabinets. The alcohol preserved collection of immature stages and soft bodied adult insects contains more than 2,500 bottles and vials. The approximate number of identified specimens of the well represented orders include 43,000 Coleoptera, 15,000 Diptera, 15,000 Hemiptera, 10,000 Hymenoptera, 10,000 Lepidoptera, 3,000 Homoptera, 900 Orthoptera, and 800 Odonata. The Acarina collection is strongest in Tetranychidae and Phytoseiidae, and it contains representatives of 43 families, 98 genera, and about 125 species. The collection of immature stages is represented mainly by Coleoptera, Diptera, Hemiptera, Hymenoptera, and Lepidoptera. In addition to the material described here, 745 pinned specimens of exotic Lepidoptera and 500 of exotic Coleoptera are present. A few of the groups well represented in the collection include the Chrysomelidae and

Cerambycidae (Coleoptera), Ichneumonidae (Hymenoptera), Noctuidae (Lepidoptera), and Syrphidae and Tabanidae (Diptera). The most impressive lot represented is the plant bug family Miridae (Hemiptera). Holdings of this group include about 10,000 adults and 1,200 vials of alcohol preserved nymphs, representing about 450 species. Many well known entomologists, including D. M. DeLong, W. S. Fisher, J. N. Knull, J. G. Sanders, and W. R. Walton, contributed to the early development and growth of the collection. Fisher and Knull and local entomologists A. B. Champlain and H. B. Kirk were mainly responsible for the large number of Coleoptera represented (nearly 50% of the holdings). The collection mainly includes arthropods from Pennsylvania and other northeastern states and contains many species found on ornamental plants. The Acarina collection is especially strong in mites associated with conifers in Pennsylvania. Other material includes the collections of V. A. E. Daecke and C. S. Anderson, both former employees, and specimens from the western and southern U.S.A. Emphasis on survey work on ornamental plants and on detection of insects new to Pennsylvania and the northeastern U.S.A. has been responsible for the growth of the collection in recent years. Much of the recent material has host and other ecological data. Accompanying the collection is an accumulation of literature on insects and mites, including subscriptions to about 20 entomological publications. The specimens are available for loan; usual loan rules apply. [1992]

Philadelphia

DEPARTMENT OF ENTOMOLOGY, ACADEMY OF NATURAL SCIENCES, 19TH AND THE PARKWAY, PHILADELPHIA, PA 19103. [ANSP]

Director: Daniel Otte. Phone: (215) 299-1189. Professional staff: Donald Azuma, James Newlin, Ruth Griffith, Skip Glenn. The entire collection contains well over 3 million specimens with a major portion housed in about 7,500 drawers in cabinets. The remainder is preserved in alcohol and on microscope slides. There are significant holdings in all the major orders but the collection's strength is in the Orthoptera which consists of over 1 million specimens, and is one of the 3 largest in the world, equalling or surpassing in size those of the British Museum (Natural History) and the Paris Museum. The Academy is the major repository of New World Orthoptera and also contains substantial material from Africa, Europe, and Asia. This collection was started by James A. G. Rehn and Morgan Hebard. Their collections and the collections of many other great orthopterists have been deposited at the Academy. The J. L. Hancock collection of Tetrigidae (largest collection of the group in existence), the Lawrence Bruner collection (one of the major collections from the American tropics) and the Samuel H. Scudder-A. P. Morse collection is deposited here. The Hebard collection by itself consists of about 250,000 specimens with 1,389 holotypes and 2,000 paratypes. Altogether, the Orthoptera collection contains close to 3,000 primary types. There are substantial holdings in non-orthopterous groups also. The collection of Cresson (Hymenoptera, Diptera), Fox, Norton, Pate, Viereck, Bassett,

Cockerell, and Blake (Hymenoptera), Braun (Microlepidoptera), Darlington, Clemens, and Grote (Lepidoptera), Poey (Cuban fauna), Stone (northeastern U.S.A. insects), and Calvert (Odonata) are some of the more notable components of the collection. These collections include about 2,000 Hymenoptera types, 800 Lepidoptera types, 360 Coleoptera types, 750 Diptera types, and 200 types from other insect groups. The slide collection is largely the dipterous family Chironomidae, representing 30 years of research on this family by the late Dr. Selwyn Roback and over 30 years of limnological surveys over most of the continental U.S.A. by the Division of Environmontal Research. The Chironomidae collection is one of the largest and most important collections in North America. The geographic coverage of this collection is North American but there are important segments from the Neotropical Region and Europe. Other important collections are: other Diptera (Ceratopogonidae, Simuliidae, Culicidae), Helwig (orthopterous genetic slide material), Odonata (Calvert collection), Collembola (Scott collection) and representatives of most of the other insect orders, plus the Arachnida and Acarina. The alcohol preserved collection is large and diverse. The insect component represents over 30 years of limnological surveys plus older material of the aquatic insects in the Academy's collection. It covers North America and contains most of the genera of aquatic insects. There is, in addition, a good collection of Neotropical aquatics. The non-insect arthropods also represent an important collection. There are large holdings, including many the Araneae, Diplopoda, Chilopoda, Scorpionida, and Solpugida. The spider collection, which includes the Dietz collection, consists of about 50 primary types and 3,750 vials of mostly North American material. The tape recording collection is growing rapidly and now contains several hundred tape reels with the recordings of about 170 grasshopper species (mainly from the U.S.A. and Mexico) and about 400 cricket species from the Hawaiian Islands, the southwestern U.S.A., Mexico, Australia, the south Pacific, and Africa. Sound recordings have become a necessary part of the systematic and biogeographical research on Orthoptera. The sounds produced by crickets, tettigoniids, and grasshoppers can provide crucial taxonomic characters in groups of species that are almost indistinguishable morphologically. Publications sponsored: "Proceedings of the Academy of Natural Sciences, Philadelphia," "Notulae Naturae," "Special Publica tions of the Academy of Natural Sciences, Philadelphia," and "Monographs of the Academy of Natural Sciences, Philadelphia." [1986]

Affiliated Collection:

Boyd, Howard P., 232 Oak Shade Road, Tabernacle Twp., Vincentown (PO), NJ 08088 USA. [HPBC] Phone: (609) 268-1734. This is a collection of 25,000 specimens stored in unit trays in 50 drawers. Coleoptera of North America comprise the major portion (20,000 specimens) of the collection, of which 15,000 are Cicindelidae, and the balance is a small synoptic collection of most other orders. Specimens may be borrowed for study; usual rules apply if borrowers are known to the lender; exchanges are limited [1986]. (Registered with ANSP.)

MOTHS OF COSTA RICA [MOCR], INSECTS OF SANTA ROSA NATIONAL PARK COSTA RICA [SRNP], DEPARTMENT OF

BIOLOGY, UNIVERSITY OF PENNSYLVANIA, PHILADELPHIA, PA 19104.

Director: Dr. Daniel H. Janzen. Phone: (215) 898-5636. Professional staff: D. H. Janzen and Winifred Hallwachs. MOCR: This is a collection supporting a thorough survey of moths (macro and micro) of Costa Rica. It was begun in 1979, and we anticipate it will take at least 20 years before approximating completion. As of now there are about 200,000 spread and labelled specimens, and increasing at 20,000 per year. Eventually it will be housed at the U. S. National Museum of Natural History, British Museum (Natural History), and Museo Nacional de Costa Rica. The organizers of this collection are not taxonomists and therefore welcome inquiries that will aid in identification of specimens. SRNP: This is a collection of all insects (ongoing) from Santa Rosa National Park in the northwestern coastal lowlands of Costa Rica (0-350 m elevation); geographic area is 10,500 ha. It was begun in 1978, and we anticipate many tens of years before completion. It now contains about 100,000 pinned and labelled specimens. Eventually it too will be placed in the U. S. National Museum of Natural History, Museum Natural History of London, and Museo Nacional de Costa Rica. The eventual product is family level field guides for Santa Rosa. Both collections are housed in drawers and various types of hinged insect boxes. The larvae are stored in alcohol vials. [1986]

Pittsburgh

SECTION OF INSECTS AND SPIDERS, CARNEGIE MUSEUM OF NATURAL HISTORY, 900 FORBES AVE., PITTSBURG, PA 15213. [CMNH] [=ICCM]

Director: Dr. John E. Rawlins. Phone: (412) 622-3259. Professional staff: Dr. Chen W. Young, Mr. Robert Davidson, Mr. Richard T. Satterwhite. The collection was formed in 1895 from earlier separate collections. Former curators and important staff included: W. H. Holland (founder), H. Kahl, W. Sweadner, G. Wallace, H. Klages, A. Avinoff, F. Marloff, R. Fox, and H. Clench. The collection contains about 5.5 million prepared specimens with over 1 million specimens available for preparation in fluids or papered. There are over 7,000 types, primary and secondary, covering all orders. However, the collection is strongest in Coleoptera, Odonata, and Lepidoptera; weakest in Orthopteroidea and Hemipteroidea. Geographically the collection contains material from throughout the world, with one of the largest Old World collections now in the New World. Abundant material from Ecuador, Bolivia, Venezuela, arctic North America, and tropical West Africa is present. It is weak in Australia, Indonesia, east African, and Central American specimens. The collection is especially strong in the following families: rhopaloceran families, Noctuidae, Arctiidae, Thyretidae, Notodontidae, Lymantriidae, Sphingidae, Cossidae, Pyralidae, Zygaenidae, Limacodidae, and Apatelodidae among the Lepidoptera; in Coleoptera: Carabidae, Cicindelidae, Erotylidae, Pselaphidae, Cerambycidae, Scarabaeidae, and Chrysomelidae. Odonata: all anisopteran families, world-wide. Among the special collections are: W. H. Edwards collection of North American Rhopalocera; Knyvette collection of Indian Rhopalocera; Coleoptera collections of J. Hamilton, H. Ulke, and L. Peña; Sphingidae collection of B. Preston Clark. There is also a large collection of eggs, larvae, pupae, hybrids, gynandromorphs, and anomalies of Lepidoptera. The Lepidoptera are housed in Holland drawers (equal to 1 1/2 the size of Cornell drawers) and other orders are in U.S.N.M. drawers in compactor storage. Publications sponsored: "Annals of the Carnegie Museum of Natural History," and "Bulletin of the Carnegie Museum of Natural History." [1986]

Affiliated Collections:

Acciavatti, Dr. Robert E., 2111 Cherry St., Marion Meadows, Morgantown, WV 26505 USA. [REAC] Phone: (304) 291-4133; 292-0036. This is a collection of Cicindelidae and Carabidae, about 10,000 specimens, primarily Nearctic, but Palearctic and Oriental Regions represented; the largest number are *Cicindela*. The collection is housed in 30 display cases and 35 Schmitt boxes. (Registered with ICCM.)

Bellamy, Dr. Charles L., 1651 S. Juniper St. #215, Escondido, CA 92025 U.S.A. [CLBC] Phone: (619) 739-9246. Buprestidae and Schizopodidae; buprestids of the World, over 35,000 specimens representing 75% of World genera; numerous secondary types; representation particularly strong from North America, southern South America, Australia, southern Africa, and Eurasia. Loans available with normal rules; exchanges are welcomed. [1992] (Registered with ICCM.)

Surdick, Robert W., 107 Santa Fe Drive, Bethel Park, PA 15102 USA. [RWSC] Phone: (412) 835-5389. All orders are represented in this collection of 94,000 specimens, with emphasis on North American Coleoptera, especially Cerambycidae. The collection contains 75,800 specimens pinned and labelled as follows: 15,000 Cerambycidae (11 paratypes); 37,000 Coleoptera of all families; 13,000 Lepidoptera, and 7,000 specimens of other orders. Duplicate papered specimens amount to 7,000 and there are 20,000 Coleoptera in alcohol waiting curating. Pinned Coleoptera are mounted with legs extended in a lifelike position and stored in drawers [1992]. (Registered with ICCM.)

Latrobe

SAINT VINCENT ARCHABBEY MUSEUM, LATROBE, PA 15650. [SVAM] COLLECTION DISBANDED. DEPOSITED AND ACCESSED BY CARNEGIE MUSEUM OF NATURAL HISTORY.

University Park

FROST ENTOMOLOGICAL MUSEUM, DEPARTMENT OF ENTOMOLOGY, PENNSYLVANIA STATE UNIVERSITY, UNIVERSITY PARK, PA 16802. [PSUC]

Curator: Dr. K. C. Kim. The major portion of the collection was originated and developed by Dr. Stuart W. Frost through a project, "The Ecological Insect Survey of Pennsylvania." After his retirement in 1957 Dr. Frost made excellent additions of Florida insects to the Museum by lighttrapping. The Museum has also acquired many other collections, *e.g.*, C. A. Thomas collection of Coleoptera, J. O. Pepper collection of Homoptera, and W. W. Long collection of Lepidoptera. During the period of the 1970's, Pennsylvanian insects were extensively added through a project entitled "Economic and Faunistic Survey of Pennsylvania." The primary objective is to have a collection of Pennsylvanian insects for research along with field data. The Museum now holds about 250,000 specimens, including pinned insects, microscope slides, and vials of alcohol preserved specimens. It is housed in 100 steel cabinets in drawers. It includes more than 45,000 specimens of beetles, 25,000 flies, 25,000 moths and butterflies, among others. The collection is strongest in groups that include pests, parasites, and predators affecting crops, forests, man, and domestic animals. The collections of particular groups are continuously expanded and improved by the special interests of systematists associated with the Museum. Active research projects utilizing the collection include taxonomic and ecological studies of many different insects, especially sucking lice, plant lice, and flies (Simuliidae, Sphaeroceridae), and ectoparasites of birds and mammals. Material is available for taxonomic studies on loan to qualified personnel; usual loan rules apply. [1992]

RHODE ISLAND

Kingston

UNIVERSITY OF RHODE ISLAND DEPARTMENT OF ZOOLOGY COLLECTION, UNIVERSITY OF RHODE ISLAND, KINGSTON, RI 02881. [URIC]

Manager: Dr. Kerwin E. Hyland. Phone: (401) 792-2650. This is a general collection of North American families with emphasis on Rhode Island fauna dating from the turn of the century. It contains about 425 drawers of identified and unidentified specimens. The slide collection of N. A. vertebrate ectoparasites is housed in 300 slide boxes, and several hundred vials. Special collection: Blanchard-Albro collection of exotic Lepidoptera and Coleoptera acquired in 1954, about 75 drawers. The entire collection is housed in unit trays in drawers in cabinets. [1992]

SOUTH CAROLINA

Clemson

DEPARTMENT OF ENTOMOLOGY COLLECTION, CLEMSON UNIVERSITY, CLEMSON, SC 29631. [CUCC]

Director: Dr. John C. Morse. Phone: (803) 656-3111. This is the largest collection of spiders and insects in South Carolina. The collection includes about 110,000 pinned, labelled, and identified specimens and another 40,000 vials of alcohol preserved specimens, and about 29,000 slides. Several hundred thousand specimens in alcohol await preparation. Thus there is a total of about 500,000 specimens available for study, primarily from South Carolina. No holotypes are kept in the collection, but over 200 paratypes and many voucher specimens from research by the South Carolina Agricultural Experiment Station are maintained here. The pinned specimens are housed in 660 drawers and 70 Schmitt boxes. Specimens in 75% EtOH are stored in vials in wooden racks. Slides are kept in cabinets. The library maintains a fair collection of taxonomic journal and books of use for systematics. Material is available for loan to qualified taxonomists; usual loan rules apply. [1986]

Affiliated Collections:

Manley, Prof. Donald G., Pee Dee Research and Education Center, Route 1, Box 531, Florence, SC 29501-9603. [DGMC] This is a special collection of 6,334 specimens of Mutillidae, representing all seven subfamilies and each tribe and subtribes, representing 77 genera. The bulk of the collection (over 85%) is concentrated in four genera, *Pseudomethoca* (866 specimens), *Dasymutilla* (3,639 specimens), *Timulla* (subgenus *Timulla*) (658 specimens), and *Ephuta* (337 specimens)[1992]. (Registered with CUCC.)

SOUTH CAROLINA DEPARTMENT OF HEALTH AND ENVIRONMENTAL CONTROL, 2600 BULL STREET, COLUMBIA, SC 29201. [SCDH]

Director: [Unknown.]. Phone: (803) 758-3944; 791-2901. This collection contains all aquatic and semiaquatic groups, including other aquatic invertebrate and vertebrate groups. About 180,000 specimens, including 12 paratypes are housed in vials. Specimens loaned; usual rules apply. [1986]

THE DOMINICK MOTH AND BUTTERFLY COLLECTION, INTERNATIONAL CENTER FOR PUBLIC HEALTH RESEARCH, P. O. BOX 699, MCCLELLANVILLE, SC 29458. [DMBC]

Director: Dr. Mac Tidwell. Phone: (803) 527-1371 or 527-1372. Professional staff: Dr. F. Lane Wallace, The Citadel, Charleston, SC 29408 (Curator). The collection contains 26,973 specimens of moths and butterflies representing 1,189 species in 688 genera, and 52 families, housed in drawers in cabinets. [1986]

SOUTH DAKOTA

Brookings

SEVERIN-McDANIEL INSECT COLLECTION, PLANT SCIENCE DEPARTMENT, S. D. STATE UNIVERSITY, BROOKINGS, SD 57007. [SDSU]

Head of the Department: Dr. Fred A. Cholick. Phone: (605) 688-4149; FAX: (605) 688-6065. The H. C. Severin Collection has been incorporated. Chrysomelidae, Dytiscidae, Hydrophilidae, and Meloidae are especially well represented. The Coleoptera collection is housed in 227 Cornell type drawers and 175 Schmitt boxes. [1992]

TENNESSEE

Gatlinburg

MUSEUM COLLECTION OF GREAT SMOKY MOUNTAINS NATIONAL PARK, GATLINBURG, TN 37738. [GSNP]

Director: Mr. Randall R. Pope. Phone: (615) 436-1201. Professional staff: Donald H. DeFoe, and Kathleen L. Manscill. The collection has fair coverage of families from the Park area, including Coleoptera (esp.

Carabidae, Cerambycidae, Chrysomelidae, Elateridae, and Scarabaeidae), in 24 drawers; Lepidoptera (27 drawers), and Hymenoptera (14 drawers). The entire collection is housed in 90 drawers. Material is available for study but is not loaned or exchanged. [1992]

Knoxville

DEPARTMENT OF ZOOLOGY AND ENTOMOLOGY COLLECTION, UNIVERSITY OF TENNESSEE, KNOXVILLE, TN 37916. [ECUT]

Director: Dr. M. L. Pan. Phone: (615) 974-2371. The collection contains 300 drawers of insects, primarily for teaching purposes. [1986]

Nashville

CUMBERLAND MUSEUM AND SCIENCE CENTER, 800 REDLEY AVE., NASHVILLE, TN 37203. [CMSC]

Director: Mr. William C. Bradshaw. Phone: (615) 259-6099. This is the Harry C. Monk collection of 760 specimens of insects, including 80 species of butterflies and skippers collected mainly in 1959-1963 in the Nashville area. It includes 390 specimens of Bell's Roadside Rambler, *Amblyscirtes belli*, collected July-Sept., 1960. These are stored in envelopes. [1992]

Affiliated Collections:

Treadway, Stephan R., 202 Ridgeway Dr., Nashville, TN 37214 USA. [SRTC] Phone: (615) 883-2231. This collection contains about over 40,000 specimens and includes many classes of arthropods, including 295 spiders, 80 millipeds and centipeds, 200 mites, but the majority of specimens are insects arranged in display cases [1992]. (Registered with CMSC.)

TEXAS

Alpine

INSECT COLLECTIONS, SUL ROSS STATE COLLEGE, ALPINE, TX 79830. [SRSC]

Director: Dr. Jim V. Richerson. Phone: (915) 837-8112. The collection consists of 13,000 specimens; 5,000 in alcohol. It is a reference collection for insects found on various desert plants and aquatic insects from desert streams. Some British Columbia forest insects. Two paratypes of *P. hyllophaga*. [1992]

Austin

TEXAS MEMORIAL MUSEUM, 2400 TRINITY ST., AUSTIN, TX 78712. [TMMC]

Director: Dr. William G. Reeder. Phone: (512) 472-2604. Professional staff: Dr. Christopher J. Durden (Phone: (512) 471-4823), Dr. James R. Reddell. The collection consists of 645 drawers of pinned insects (65% Lepidoptera), 1,434 boxes of papered and layered insects (57% Lepidoptera), 124 drawers of fossil insects (83% Palaeozoic), 1,862 jars of vials of arthropods in alcohol, insect microfossils, unsorted, arthropod field samples (unsorted Berlese and sifted); microslides, chiefly Diptera. There is a total of about 900,000 specimens. Material is from N.A., Africa, Europe, Asia, and South America. Special collections include fossil insects, arachnids, and cave fauna. The material is housed in unit pinning trays, or in boxes, jars, and in papers. Publications sponsored: "Bulletin of Texas Memorial Museum," "Monographs in Speleology." [1986]

Big Bend

MUSEUM COLLECTIONS OF BIG BEND NATIONAL PARK, BIG BEND NATIONAL PARK, TX 79834. [BBNP]

Curator: Mark Herberger. Phone: (915) 477-2251. Professional staff: Cherl Long. 1,500 specimens. [1992]

College Station

DEPARTMENT OF ENTOMOLOGY INSECT COLLECTION, DEPARTMENT OF ENTOMOLOGY, TEXAS A & M UNIVERSITY, COLLEGE STATION, TX 77843. [TAMU]

Director: Dr. Horace R. Burke. Phone: (409) 845-9712. Professional staff: Dr. Horace R. Burke (Curculionidae), Mr. Edward G. Riley (Chrysomelidae, Scarabaeidae), Dr. Joseph C. Schaffner (Miridae), Dr. Robert A. Wharton (Braconidae, Tephritidae), and Dr. James B. Woolley (Chalcidoidea). Others associated with the collection: Dr. Charles Cole (Thysanoptera), Mr. Allen Dean (Spiders), Dr. John A. Jackman (Buprestidae, Mordellidae), and Mr. Mike Rose (Aphelinidae). The collection dates back to 1902, but most of its growth has taken place since 1947 when the collection of the Texas Agricultural Experiment Station and that of the department were combined. Substantial growth has taken place during the last 10 years. The curated collection consists of just over 1 million pinned specimens, 9,000 vials, and 25,000 slides. Additionally, there is a large backlog of recently collected bulk samples. Some primary types are maintained in the collection, but the current policy is not to retain primary types. Pinned material is housed in 85 48-drawer cabinets containing approximately 4,080 Cornell-style drawers with foam bottom unit trays. The breakdown of pinned material by order is 40% Coleoptera, 25% Hymenoptera, 20% Heteroptera-Homoptera, 9% Diptera, 4% Lepidoptera, and 2% other orders. Geographically, the collection is strong in insects of Texas and surrounding areas, including Mexico. Material may be borrowed for study; usual loan rules apply. Computerized lists are available for types and for the determined species of some taxa in the general collection. [1992]

Affiliated collections:

Dean, D. Allen, 314 Tee Dr., Bryan, TX 77801 U.S.A. [DADC] Phone (409) 845-3412 (work); 823-0565 (home). This collection contains more than 4,000 vials of spiders comprising 400 species. It is Nearctic, but mostly Texas with more than 400 vials from Mexico. [1992] (Registered with TAMU.)

Neff, Dr. John L., 7307 Running Rope, Austin, TX 78731 U.S.A. [JLNC] Phone: (512) 345-7219. This is primarily a collection of New World bees, with

20,000 specimens, nearly all with floral records. Geographic emphasis is on arid areas of Texas, Arizona, California, northern Mexico, northern Chile, and northern Argentina. The collection is housed in 60 drawers and numerous Schmitt boxes. A small number of bee larae are kept in alcohol. Specimens are available for exchange. Specimens may be borrowed for study; usual loan rules apply. [1992] (Registered with TAMU.)

Riley, Edward G., 1409 Todd Trail, College Station, TX 77845 U.S.A. [EGRC] Phone: (409) 845-9711 (work); 693-4377 (home). This collection contains 100,000 leaf beetles (Chrysomelidae) stored in 200 Cornell drawers. The collection's primary purpose is to serve as a resource for research on chrysomelid systematics. It also serves as a reference collection for survey-inventory work on this family. Coverage is best for North and Central America. Several thousand beetles of other families are also present. Primary types are not retained. Material may be borrowed for study; the usual loan rules apply. [1992] (Registered with TAMU.)

Stidham, John A., 301 Pebblecreek Dr., Garland, TX 75050 U.S.A. Phone: (214) 495-6947. [JASO] This collection contains about 11,000 specimens of Orthoptera from the state of Texas. In addition about 4,000 insects from other parts of the U.S.A. are included. [1992] (Registered with TAMU.)

Thomas, Donald B., 1209 W. 3rd St., Weslaco, TX 78596 U.S.A. [DBTC] Phone: (512) 968-9545. This collection contains about 20,000 pinned specimens (95% identified to species) including 184 paratypes, stored in 60 museum drawers. Two-thirds of the collection is Pentatomidae with the remainder Scarabaeidae and Tenebrionidae. The coverage is primarily New World. [1992] (Registered with TAMU.)

Fort Davis

TAYLOR, TERRY W., COMBINED SCIENTIFIC SUPPLIES, P. O. BOX 1466, FT. DAVIS, TX 79734 USA. [TWTC]

This is a general collection containing specimens from 85 countries. South America, Mexico, and U.S.A. are well represented. The collection is housed in 960 drawers and over 6,000 jars. The owner specializes in *Plusiotis* and *Megasoma* (Coleoptera). All material is complete with ecological data and is available for exchange and loan; usual rules apply [1986].

Houston

HOUSTON MUSEUM OF NATURAL SCIENCE, ONE HERMAN CIRCLE DRIVE, HOUSTON, TX 77030 [HMNS]

Curator of Invertebrates: Dr. Raymond W. Neck. Phone: (713) 639-4678. Approximately 100,000 specimens of insects (a few non-insect arthropods), largely Coleoptera and Lepidoptera, especially strong in tropical World-wide. No type specimens. [1992]

Kingsville

DEPARTMENT OF BIOLOGY COLLECTION, TEXAS A & I UNIVERSITY, KINGSVILLE, TX 78363. [TAIU]

Director: Dr. James E. Gillaspy. Phone: (512) 595-3803. This collection contains 70,000 specimens of which 45,000 are pinned (strongest in

Lepidoptera, Hymenoptera, and Coleoptera); alcohol preserved (5,000) in vials, and 25,000 in Riker mounts. The collection is stored in 412 drawers in 39 cabinets, with 346 Riker mounts of butterflies and moths (world); and the P. A. Glick collection (25,000 specimens). Spiders are stored in 3 cabinets on 16 hardware cloth frames. Also there are about 100 Schmitt boxes of specimens. [1986]

Lubbock

TEXAS TECH UNIVERSITY ENTOMOLOGICAL COLLECTION, DEPARTMENT OF AGRONOMY, HORTICULTURE, AND ENTOMOLOGY, TEXAS TECH UNIVERSITY, LUBBOCK, TX 79409. [TTCC]

Director: Dr. Robert W. Sites. Phone: (806) 742-2828. The collection consists of 250,000 specimens in 300 drawers, 110,000 pinned insects, and 10,000 vials of insects and arachnids in alcohol. There are 8,000 vials of ants of western Texas, and 1,000 jars and vials of scorpions, world-wide. Publications sponsored: "Occasional Papers of the Museum," and "Special Publications of the Museum." [1986]

Nacogdoches

DEPARTMENT OF BIOLOGY INSECT COLLECTION, DEPARTMENT OF BIOLOGY, STEPHEN F. AUSTIN STATE COLLEGE, NACOGDOCHES, TX 75961. [SFAC]

Director: Dr. William W. Gibson. Phone: (409) 569-3601. About 70,000 pinned and labelled insects are in this collection, representing eastern Texas. The collection is heaviest on Coleoptera, primarily minor families, Scarabaeidae, Cerambycidae, and Cicindelidae. Several hundred larvae and such are stored in alcohol vials. Among the special collections are several hundred identified spiders from a survey of Nacogdoches County, Texas. Material is available for loan or exchange with usual loan rules applying. [1986]

Wichita Falls

DEPARTMENT OF BIOLOGY, MIDWESTERN STATE UNIVERSITY, 3410 TAFT, WICHITA FALLS, TX 76301 [MSSC]

Director: Dr. Norman Horner. Phone: (817) 689-4253. This collection contains over 10,000 specimens of insects mostly from North Central Texas, the majority identified only to family. The Arachnid collection contains: Opiliones, 20 species; scorpions, 25 species; amblypygids, 2 species; solpugids, 6 species, and Araneae, 350 species. The insects are stored in Schmitt boxes and the arachnids in alcohol vials. [1992]

UTAH

UTAH FIELD HOUSE OF NATURAL HISTORY. LOCATION OF COLLECTION UNKNOWN. This collection was listed in the 1969 edition as a collection of Coleoptera. [UFNH]

Logan

ENTOMOLOGICAL MUSEUM, DEPARTMENT OF BIOLOGY, UTAH STATE UNIVERSITY, LOGAN, UT 84332. [EMUS]

Director: Dr. Wilford J. Hanson. Phone: (801) 750-2554. Professional staff: Dr. Terry Griswold (USDA, Apoidea), Dr. George E. Bohart (Prof. Emeritus). The insect collection has nearly 2 million specimens. The Great Basin, Rocky Mountains, Southwestern U.S.A., and Costa Rica are the regions most completely covered. There are also significant collections from Mexico, Panama, Trinidad, South America, Spain, India, Indonesia, and Africa. Many other areas of the world are represented. The pinned specimens are stored in drawers in Cornell drawers in metal cabinets. Primary types are normally deposited in the U. S. National Museum of Natural History collection or other major museums. Specimens are available for loan to specialists for research purposes. [1992]

U. S. NATIONAL POLLINATING INSECT COLLECTION, BEE BIOLOGY AND SYSTEMATICS LABORATORY, UTAH STATE UNIVERSITY, LOGAN, UT 84322. [BLCU]

Director: John D. Vandenberg; Curator: Terry Griswold. Phone: (801) 750-2524. This collection consists of 500,000 specimens in unit trays in cabinets of aculeate Hymenoptera (except Formicidae), primarily from the Nearctic and Neotropical Regions,with emphasis on Southwestern North America and Costa Rica. Significant holdings also include the Palearctic, a trap nest reared collection, aculeate larvae. [1992]

Provo

ENTOMOLOGY SECTION, MONTE L. BEAN LIFE SCIENCE MUSEUM, BRIGHAM YOUNG UNIVERSITY, PROVO, UT 84602. [BYUC]

Head Curator: Dr. Richard W. Bauman. Phone: (801) 378-5492. Professional staff: Marek J. Kaliszewski, Curator of Arachnids and Merv. W. Nielson, Research Associate. Phone: (801) 378-6356; Dr. Stephen L. Wood, Curator Emeritus. Phone: (801) 378-2226. The collection includes 1,250,000 specimens of insects and 400,000 specimens of Arachnida. Areas of strength: Coleoptera, Diptera, Plecoptera, Trichoptera, Homoptera, Neuroptera, Siphonaptera, and Acarina. Primary types: over 100; secondary types, over 1,000. Periodicals sponsored: "The Great Basin Naturalist." [1992]

Affiliated collections:

Clark, Dr. Shawn M., 7 Guthrie Center, Charleston, WV 25312 U.S.A. [SMCI] Phone: (304) 342-5545. The collection consists of an estimated 25,000 pinned specimens of Chrysomelidae. No other families are included. It is strongest in Nearctic species, but also includes major holdings from the Neotropical Region and less extensive holdings from other areas. Only a few paratypes and no primary types are included. [1992] (Registered with BYUC.)

Mower, Robert C., 378 N. 650 E., Orem, UT 84057 U.S.A. [RCMC] This is a collection of Arctidae from North America shored in 24 drawers, more than 2,000 spread specimens and hundreds in papers, representing about 160 species. [1992] (Registered with BYUC.)

Springdale

MUSEUM COLLECTIONS OF ZION NATIONAL PARK, SPRINGDALE, UT 84767. [ZNPC]

Director: Harold L. Grafe. Phone: (801) 772-3256. This collection consists of two metal specimen cabinets containing 24 drawers of insects representing the following orders: Coleoptera, Collembola, Diptera, Ephemeroptera, Hemiptera, Hymenoptera, Isoptera, Lepidoptera, Mallophaga, Neuroptera, Odonata, Orthoptera, Plecoptera, Trichoptera, and Thysanura. Special collection: Alice Lindahl collection from Grapevine Springs (Great West Canyon of Zion National Park). [1986]

St. George

DIXIE COLLEGE, ST. GEORGE, UT 84770. [AHBC]

Director: Dr. Andrew H. Barnum. Phone: (801) 673-4811, ext. 336. A collection of approximately 100,000 pinned and wet specimens from the mid western and western regions of the U.S.A. The Orthoptera collection, all species identified, is most prominent, followed by Coleoptera (about half of the species identified). Most other orders are identified to family or genus only. Specimens of spiders and other arachnids are representative of the southwest, but are relatively few in number. These are primarily wet specimens. Since most of the specimens represent the director's field work, ecological data and behavioral information are available for most specimens. The pinned specimens are housed in drawers; wet specimens in jars. [1986]

VERMONT

Burlington

UNIVERSITY OF VERMONT INVERTEBRATE COLLECTION, DEPARTMENT OF ZOOLOGY COLLECTION, UNIVERSITY OF VERMONT, BURLINGTON, VT 05405-0086 [UVCC]

Director: Dr. Ross T. Bell. Phone: (802) 656-2922. This collection covers all groups of invertebrates, and is especially strong in Coleoptera. The Vermont fauna is strongly represented, but the collection also includes extensive material from other parts of the U.S.A., Canada, Mexico, West Indies, Central and South America, and Papua New Guinea. Among the special collections are: boreal and alpine insects from the Green Mountains; Carl Parsons collection of insects from southern Vermont; a synoptic collection of Vermont insects; Carabidae of Vermont. The collection is housed in 380 drawers; also extensive alcohol collections. Material is available on loan and for exchange; usual loan rules apply. [1992]

VIRGINIA

Blacksburg

VIRGINIA MUSEUM OF NATURAL HISTORY AT VIRGINIA TECH, VIRGINIA POLYTECHNIC INSTITUTE AND STATE UNIVERSITY, BLACKSBURG, VA 24061-0542. [VPIC]

Founding Director: Dr. Michael Kosztarab. Phone: (703) 231-6773. Professional staff: Dr. J. Reese Voshell, Jr., and Ms. Mary Rhoades. This collection is the oldest (established in 1888), and largest (875,000 specimens) in Virginia. Besides the unique herbarium of insect and mite damage, the collection contains a good representation of seven orders of insects of national and international importance. An especially notable resource is the collection of over 146,000 slide mounted specimens of scale insects, mites, and biting midges. The collection was expanded and curated from 1891 to 1925 by Ellison A. Smyth, whose large sphinx moth collection was donated to the Smithsonian Institution. After 1950 Richard L. Hoffman curated and reorganized part of the collection, including the orders Odonata, Hemiptera, Mecoptera, and some families of Coleoptera. Edgar M. Raffensperger (1959-1961) organized the orders Diptera and Hymenoptera, while Charles V. Covell managed the Lepidoptera. Dr. Covell continued his work through 1964. Dr. Kosztarab assumed duties as curator in 1962, assisted by Mary Rhoades and others. Among the special collections are: the fourth largest collection of scale insects; one of the larger collections of Nearctic biting midges; one of the largest and fastest growing aquatic insect collections in North America. Material is stored in 1,902 Cornell drawers. Specimens may be borrowed for taxonomic studies; usual loan regulations apply. Publications sponsored: "Studies on the Morphology and Systematics of Scale Insects," "Insects of Virginia," and Occasional Papers from the Department of Entomology." [1992]

VIRGINIA MUSEUM OF NATURAL HISTORY, 1001 DOUGLAS AVENUE, MARTINSVILLE, VA 24112. [VMNH]

Curator, Recent Invertebrates: Dr. Richard L. Hoffman. Phone (703) 666-8629. The main goal of the insect collection is to conduct intensive surveys of the biota of Virginia with the intention of issuing handbooks on the various major taxa. Virtually all specimens are from Virginia except Diplopoda (Worldwide). Emphasis is on taxa from fresh water and litter biotypes. Approximatey 13,000 processed specimens of which about 7,500 are Carabidae, 4,000 Heteroptera, and 1,500 other groups. Paratypes are kept of several Carabidae (undescribed), Cerambycidae, Plecoptera, with holotypes and paratypes of new species of Miridae. Also included are holotypes of about 300 species of Diplopoda and Chilopoda, and paratypes of about 300 additional Diplopoda species. Publications sponsored: Memoirs, "Jeffersoniana," "Myriapodologica," and "Insects of Virginia." Specimens may be borrowed under the usual loan rules. [1992]

Richmond

INSECT COLLECTION, VIRGINIA DEPARTMENT OF AGRICULTURE AND COMSUMER SERVICES, 1 NORTH 14th ST., RM. 254, RICHMOND, VA 23219. [VDAC]

Curator: Ms. Sylvia A. Shives. Phone: (804) 786-2415. About 1,000 general specimens. [1992]

WASHINGTON

Ashford

MUSEUM COLLECTIONS OF MOUNT RAINIER NATIONAL PARK, TACOMA WOODS, STAR ROUTE, ASHFORD, WA 98304. [MRNP]

Director: Mr. William F. Dengler. Phone: (206) 569-2211, ext. 46. The collection is neither large nor extensive. Material is restricted to the National Park. Specimens are kept in three Schmitt boxes and have little supporting ecological data. Material is not available for exchange or loan. The Chief Park Naturalist should be contacted for study at the collection. [1986]

Port Angeles

MUSEUM COLLECTIONS OF OLYMPIC NATIONAL PARK, 600 EAST PARK AVE., PORT ANGELES, WA 98362. [ONPC]

Director: Mr. Henry C. Warren. Phone: (206) 452-4501. Professional staff: Sydney Jacobs. This collection contains good coverage for families in the park and surrounding areas. Specimens are housed in 12 drawers. Supporting ecological data is limited to the labels. Material is not available for exchange, but institutional loans may be arranged. No specimens have been added to this collection for many years. [1986]

Pullman

JAMES ENTOMOLOGICAL COLLECTION, DEPARTMENT OF ENTOMOLOGY COLLECTION, WASHINGTON STATE UNIVERSITY, PULLMAN, WA 99163. [WSUC]

Director: Dr. Richard S. Zack. Phone: (509) 335-3394. The collection has over 1 million specimens and is well represented in all major insect orders. The collection is especially strong in Diptera, Coleoptera (except aquatics), Lepidoptera, and Hymenoptera, and aquatic Heteroptera. The majority of the material is representative of the Pacific Northwest, however, large holdings of material from South and Central America, western Canada, Indonesia, and the U.S.A. in general are contained in the collection. A separate type collection is maintained. Specimens are housed in USNM drawers in cabinets. Publication sponsored: "Friends of the James Entomological Collection Newsletter" and "Melanderia" which is a periodical published by the Washington State Entomological Society, not directly sponsored by the collection. [1992]

Affiliated Collection:

Schroeder, Dr. Paul C., SW 145 Arbor St., Pullman, WA 99163 USA. [PCSC] Phone: (509) 335-2766; 334-3446. This collection is restricted to Coleoptera, with special emphasis on Tenebrionidae, Cerambycidae, and Scarabaeidae of the World, although general Coleoptera of most families are represented. Best coverage is for western North America, but the collection includes specimens collected while resident for 15 months in Germany and southern France, and, most recently, a year in the Brisbane area of Australia. Smaller collections from central Mexico, the Peruvian Amazon and East Africa are also included. Most identifications have been made by myself, including work at museums in Munich and Brisbane as well as the California Academy of Sciences. The collection includes an estimated 30,000 specimens housed in 65 drawers and miscellaneous boxes, and about 150 vials of small specimens in alcohol; much unmounted material remains on hand in addition. Material is available for exchange or loan; usual loan rules apply [1992]. (Registered with WSUC.)

Seattle

BURKE MUSEUM, DB-10, UNIVERSITY OF WASHINGTON, SEATTLE, WA 98195. [UWBM]

Curator of Lepidoptera: Mr. Jonathan P. Pelham; Curator of Arachnids: Mr. Rodney L. Crawford. Phone: (206) 543-9853. Professional staff: Dr. Lars G. Crabo, Curatorial Associate, Lepidoptera, and one student assistant. The arthropod collections are segregated as follows: Lepidoptera, 42,000 specimens; Siphonaptera, 1,040 specimens; Spiders, 63,220 specimens; other arachnids 7,330 specimens; myriapods, 5,500 specimens; isopods, 10,000 specimens; cave invertebrates, 8,000 specimens. The museum has an official collection management policy (copy available on request) and does loan specimens for research purposes to borrowers with institutional affiliation. Series sponsored: "Burke Museum Controbitions in Anthropology and Natural History"; "Burke Museum Research Reports"; "Burke Museum Monographs." [1992]

WEST VIRGINIA

Morgantown

WEST VIRGINIA UNIVERSITY ARTHOPOD COLLECTION, RM. G176, AGRICULTURAL SCIENCES BLDG., WEST VIRGINIA UNIVERSITY, P. O. BOX 6108, MORGANTOWN, WV 26505. [WVUC]

Director: Dr. James W. Amrine, Jr. The collection consists of approximately 100,000 specimens housed in drawers, in vials, and on slides. The strong collections are primarily West Virginia material and include: Macrolepidoptera, 1,000 species, several thousand papered specimens, 150 drawers of pinned specimens; Diptera-Tabanidae, 75 species, 1,500 pinned specimens; Diptera-Simuliidae, 30 species in 2,000 vials, and 500 pinned specimens. The collection also contains important representatives of most orders of insects and arachnids, especially Acari and Araneae. A considerable number of pinned specimens of forest insects originally collected by A. D. Hopkins are in the collection. There is also a large series of the diopsid, *Sphyracephala brevicornis* Say, including eggs, larvae, pupae, and adults reared successfully for the first time at the University. [1986]

Affiliated Collections:

Adler, Peter H., Department of Entomology, Long Hall, Clemson University, Clemson, SC 29631 USA. [PHAC] Phone: (803) 656-3111; 654-8709. About 400 specimens of Lepidoptera (macrolepidoptera primarily) are contained in this collection. Most specimens are from near Charleston, WV. They are housed in drawers [1986]. (Registered with WVUC.)

Allen, Thomas J., Route 3, Box 468, Elkins, WV 26241 USA. [TJAC] Phone: (304) 636-1767; 636-0856. This collection includes butterflies from around the world, but mostly from U.S.A. Foreign specimens are dominantly Papilionidae and Morphidae, about 1,000 specimens. Fifty percent of the species of Morphidae are represented. About 1,200 specimens of U.S.A. butterflies, mostly from West Virgina, are included. The moths are mostly worldwide Saturnidae, and U.S.A. Sphingidae and Catocalas, about 800 specimens. Beetles are represented by about 200 specimens, mostly Lucanidae, Scarabaeidae, and Cicindelidae. The collection is housed in 12 large display cases, plus 28 drawers. Some specimens (about 1,000) remain in papers [1986]. (Registered with WVUC.)

Curtis, Daniel L., *Address and location of collection unknown.* [DLCC] This is a Lepidoptera collection of about 400 specimens representing 25 families, housed in insect boxes. All specimens are identified [1992]. (Registered with WVUC.)

Estep, Edward E., 72 Mound St., New Martinsville, WV 26155 USA. [EEEC] Phone: (304) 455-2200, ext. 3309; 455-1727. This is a general collection from Wetzel/Marshall counties, West Virginia, consisting of about 2,500 pinned specimens, including 1,400 identified Lepidoptera, and 1,100 unidentified specimens representing 9 orders, housed in 26 drawers in cabinets [1986]. (Registered with WVUC.)

Norris, Sam, P. O. Box 113, Norton, WV 26285 USA. [SNSC] Phone: (304) 636- 6251. This collection of about 1,500 spiders is mostly from Randolph County, West Virginia. All specimens are preserved in alcohol in vials stored in boxes [1986]. (Registered with WVUC.)

Charleston

PLANT INDUSTRIES DIVISION, WEST VIRGINIA DEPARTMENT OF AGRICULTURE, CHARLESTON, WV 25305. [PPCD]

Director: Dr. Charles C. Coffman. Phone: (304) 558-2212. Professional staff: Sherri C. Hutchinson, Shawn M. Clark, Thomas W. McCutcheon. The pinned collection consists of 55,000 identified specimens representing 6,000 species in all orders, and 50,000 unidentified specimens, with the major orders Heteroptera, Coleoptera, Lepidoptera, Diptera, and Hymenoptera. There is a considerable vial collection as well. The emphasis is on West Virginia, but other areas are also represented. Among the special collections are 26 species of Dermestidae donated by Dr. R. S. Beal. [1992]

WISCONSIN

Kenosha

DEPARTMENT OF ENTOMOLOGY COLLECTION, CARTHAGE

COLLEGE, KENOSHA, WI 53140. [CCCC]

[There is a collection here, but several letters have failed to provide information other than that published in 1968.]

Madison

INSECT RESEARCH COLLECTION, DEPARTMENT OF ENTOMOLOGY, 346 RUSSELL LABS., 1630 LINDEN DRIVE, UNIVERSITY OF WISCONSIN, MADISON, WI 53706. [IRCW] [=UWEM]

Curator: Acting Director: Dr. Daniel K. Young. Phone: (608) 262-2078. Academic Curator: Mr. Steven Krauth. Phone: (608) 262-0066. The collection contains about one million pinned insects stored in 152 U. S. National Museum cabinets with a large amount of additional material stored in alcohol. In the past, emphasis has been placed on the Lepidoptera (408 USNM drawers) and Coleoptera (367 USNM drawers). In 1959 Professor C. L. Fluke deposited his worldwide collection of 16,050 specimens of Syrphidae (Diptera) in the collection. His collection was the result of 40 years of work and is one the best of its kind in North America. An additional 156 USNM drawers house the remaining families of Diptera. More recently efforts have been concentrated on developing the microhymenoptera (70,413 specimens) section of the collection from Malaise trapped samples collected in conjunction with a program recover released parasitoids of the Gypsy Moth. Specimens of Chalcidoidea, Proctotrupoidea, Bethyloidea, Ceraphronoidea, and Cynipoidea have been prepared on card points and labeled. Identifications have been made to the level the current knowledge of the group will permit. The entire Hymenoptera section is housed in 312 USNM drawers.

Efforts are under way to develop a computerized database of insect collection data by recapturing the collection data from the specimens and adding it to the existing list of species present in the collection.

Loans are available to qualified specialists with institutional associations. Visitors are welcome and will be provided with space and a reasonable amount of equipment to examine specimens of interest. A literature collection is associated and is available for visitor use on site. Steenbock Agriculture library is next door and includes an extensive collection of Entomological literature. The collection is a member of ENT-LIST, a computer mail service coordinated by Mark O'Brien at Michigan State University. [1992]

Affiliated collection:

Schmude, Dr. Kurt L., Lake Superior Research Institute, University of Wisconsin-Superior, Superior, WI 54880 U.S.A. [KLSC] Phone: (715) 394-8525 (Lab.); 394-8158 (Office). A general aquatic insect collection, with specimens mainly collected from Wisconsin, is maintained and kept in alcohol. A North American collection consisting of adult and larval riffle beetles in the families Elmidae, Dryopidae, Psephenidae, and Lutrochidae, and numbering about 14,000 specimens, is actively maintained, with more than 50% housed in alcohol in one dram vials, and the remainder housed in Schmitt boxes. About 12,500 of the riffle beetles are in the genus *Stenelmis*, which includes paratypes of nearly all of the species found in North America [1992]. (Registered with UWEM.)

Milwaukee

INSECT COLLECTION, MILWAUKEE CITY PUBLIC MUSEUM, 800 W. WELLS ST., MILWAUKEE, WI 53233. [MCPM]

Director: Dr. Allen M. Young. Phone: (414) 278-2789. The collection contains about 400,000 insects and 8,000 spiders. The two best represented orders are Coleoptera and Lepidoptera, with about 130,000 and 175,000 repsectively. There are holotypes, lectotypes, or syntypes of 117 species of insects. The Lepidoptera collection includes the recent donation of specimens from James R. Neidhoefer and William E. Sieker. The approximately 95,000 specimens donated by Neidhoefer are particularly strong in Papilionidae, Nymphalidae, Heliconidae, Ithomiidae, Morphidae, Pieridae, Saturnidae, and Sphingidae. All major faunistic regions are represented, with particular strengths in the Neotropical and Indo-Australian Regions. The approximately 9,000 specimens donated by Sieker represent about 90% of the world's known species of Sphingidae and all known Wisconsin species. The Coleoptera collections comprise a synoptic representation of major North American groups; the best represented group of the order is the family Carabidae, with about 55,000 specimens, mostly from the New World, including ones from remote areas of the Andes. The spider collection includes representatives of North and Central American groups. Most of the insects are pinned, labelled and stored in metal cabinets with drawers. Some of the Lepidoptera are papered and housed in metal storage cabinets. Three cabinets contain soft bodied insects stored in alcohol. The spiders are stored in vials and jars of alcohol within metal cabinets. Dr. Allen M. Young and Ms. Susan S. Borkin curate the Lepidoptera and Odonata; Dr. Gerald R. Noonan cares for Coleoptera and other orders of insects. Ms. Joan P. Jass takes care of the spiders as part of her curation of invertebrates other than insects. Current areas of staff and collection emphasis include midwestern North America and Central America for insects and spiders in general. Dr. Noonan has research interest in the systematics, biogeography, and cladistics of the family Carabidae, and is building on those of tribe Harpalini. Dr. Young has research interest in the ecology and biology of Lepidoptera, Cicadidae, and selected Neotropical insects and is assembling collections of these, including that of Ceratopogonidae and Cecidomyiidae associated with the pollination of cocoa and related species of *Theobroma*. Loans are available to qualified specialists. Publication sponsored: "Milwaukee Public Museum, Contributions in Biology and Geology." Large papers are published as separate publications. [1992]

WYOMING

Laramie

ROCKY MOUNTAIN SYSTEMATIC ENTOMOLOGY LABORATORY, DEPARTMENT OF PLANT, SOIL, AND INSECT SCIENCES, BOX 3354, UNIVERSITY OF WYOMING, LARAMIE, WY 82071. [ESUW]

Curator: Dr. Scott R. Shaw. Phone: (307) 766-5338. The collection contains 130,000 pinned specimens, 400 slides, and 4,000 vials. The collection is rich in rangeland insects, especially from southeastern Wyoming and northeastern Colorado with some material from Australia, Brazil, China, Costa Rica, Puerto Rico, Somalia, and Mexico. There are about 300 paratypes in the collection. Among the special collections are: Pawnee National Grassland insect collection; Yellowstone National Park collection, and Costa Rican Braconidae (over 20,000 specimens). Specimens are housed in 500 drawers in cabinets. Material is available for exchange. Student loans may be arranged by major professor; usual rules apply. [1992]

Yellowstone Park

YELLOWSTONE PARK MUSEUMS, YELLOWSTONE NATIONAL PARK, P. O. BOX 168 YELLOWSTONE PARK, WY 82190. [YPMC]

Acting curator: Ms. Lyn Riley. Phone: (307) 344-7381, ext. 2319. This is essentially a research collection and not on display to the general public. There is available a complete list of the taxa in the collection, identified by various specialists. [1992]

(United States of America, Miscellaneous Pacific Islands, includes Kingman Reef, Howland, Jarvis, Midway, and Wake islands (3 sq. mi., population 300), Johnston Atoll, and Guam, **Size:** 209 sq. mi.; **population:** 144,000.]

AGRICULTURAL EXPERIMENT STATION, UNIVERSITY OF GUAM, MANGILAO 96913. [ESUG]

[Contact: Dr. Donald Nafus. Phone: (671) 734-3113. Staff includes Dr. Ilse Schreiner. This is the principal collection of Guam. It is expanding and includes specimens from various parts of the Micronesian area. Eds. 1992]

(Upper Volta, see Burkina Faso.)

195. URUGUAY, Oriental Republic of

[Neotropical. Montevideo. **Population:** 2,976,138. **Size:** 68,037 sq. mi.]

ENTOMOLOGIA, INSTITUTO DE BIOLOGÍA, FACULTAD DE TRISTAN NARVAJA 1674, C. C. 10.773, MONTEVIDEO. [UYIC]

Director: Dr. Carlos E. Casini. Phone: 48-74-19. Professional staff: Lic. Fernando Pérez Miles, Enrique Morelli, Alba Bentos Pereira, Carmen Viera, Loreley Amari, Miguel Simó, Ana Verdi, Estrellita Lorier. This collection is the most important in Uruguay and includes about 150,000 specimens from Uruguay, Argentina, Brazil, Chile, Paraguay, and Bolivia. The pinned specimens are stored in boxes in cabinets; some material is in alcohol, and Crustacea in formalin, stored in special metal boxes. Special collections include spiders from Prof. Roberto Capocasale, as well as scorpions, Opiliones, Myriopoda, and Crustacea. [1992]

COLECCION ENTOMOLOGICA DEL MUSEO NACIONAL DE HISTORIA NATURAL, CASILLA DE CORREO 399, MONTEVIDEO. [URMU]

Director: Dr. Héctor S. Osorio. Phone: 96-09-08. Professional staff: Lucrecia Covelo de Zolessi, and Roberto Capocasale. The collection consists of about 500 species, in general, Uruagayan material, especially Coleoptera, Lepidoptera, and some other orders. This collection has historical value as well as scientific. It is stored in boxes in cabinets. There is type material from Deyrolle. The spider collection contains about 5,000 specimens in vials of alcohol. About 80% of the Uruguayan species are represented. They are identified to order, and families (Araneae, Opiliones, Scorpiones) and 70% are determined to species (approximately 250 spider, 30 Opiliones, and 10 scorpion species). The collection contains holotypes and paratypes. Publications sponsored: "Boletin del Museo Nacional de Historia Natural," "Anales del Museo Nacional de Historia Natural," and "Comunicaciones Zoologicas del Museo Nacional de Historia Natural." [1992]

196. VANUATU

[=New Hebrides Islands. Oceanian (Melanesia). Port Vila. **Population:** 154,691. **Size:** 4,707 sq. mi. *No known insect collection.*]

197. VATICAN CITY, State of

[Palearctic. Vatican City. **Population:** 752. **Size:** 0.17 sq. mi. *No known insect collection.*]

(Venda, see R. South Africa.)

198. VENEZUELA, Republic of

[Neotropical. Caracas. **Population:** 18,775,780. **Size:** 352,144 sq. mi.]

DECANATO DE AGRONOMÍA, UNIVERSIDAD CENTRO OCCIDENTAL, APDO. POSTAL 400, BARQUISIMETO, LARA 3002. [MJMO] [=UCOB]

Director: Mr. Rafael Gonzalez, Ing. Agr. Phone: (051) 62014; 62012. Professional staff: Dr. Hugo A. Chávez T., Dr. Francisco A. Díaz B.; Ing. Agr. M.Sc., José Morales S., Ing. Agr. M.Sc. Alvaro Chávez T., Ing. Agr. M.Sc. Franklin Gutierrez, Ing. Agr. M.Sc. Carlos Pereira N. The collection consists of 50,000 pinned specimens in a special collection, 25,000 Ichneumonoidea. [1992]

INSTITUTO DE ZOOLOGIA AGRICOLA, FACULTAD DE AGRONOMIA, UNIVERSIDAD CENTRAL DE VENEZUELA, APT. 4579, 2010A MARACAY, ARAGUA. [IZAV]

This is effectively the national collection of Venezuela and contains the collection of the late Dr. Francisco Fernández Yépez. [*No reply, no further information available.* Eds. 1992]

FACULTAD DE AGRONOMIA, UNIVERSIDAD DEL ZULIA, APTDO. 526, MARACAIBO, ZULIA. [UZMC] [*No reply.*]

LABORATORIA NACIONAL DE PRODUCTOS FORESTALES, UNIVERSIDAD DE LOS ANDES, AVE. CHORROS DE MILLA, APTDO 220, MERIDA, MERIDA. [UAMM] [*No reply.*]

PRODUCCION AGRICOLA, UNIVERSIDAD NACIONAL EXPERIMENTAL DE LOS LLANOS OCCIDENTAL, CARRERA 3, NO. 16-40, GUANARE, PORTUGUERA. [UNLO] [*No reply.*]

199. VIETNAM, Socialist Republic of (Unified)

[Indomalayan. Hanoi. **Population:** 65,185,278. **Size:** 127,246 sq. mi.].

INSECT COLLECTION, PLANT PROTECTION DEPARTMENT, HANOI. [VICH] [*No reply.*]

200. VIRGIN ISLANDS (BRITISH)

[Neotropical. Road Town. **Population:** 12,075. **Size:** 59 sq. mi. *No known insect collection.*]

201. VIRGIN ISLANDS OF THE U.S.A.

[Neotropical. Charlotte Amalie. **Population:** 112,636. **Size:** 137 sq. mi. *No known insect collection.*]

(Volcano Islands, see Japan.)

(Wake Island, see U.S.A. Miscellaneous Pacific Islands.)

(Wales, see United Kingdom.)

202. WALLIS AND FUTUNA ISLANDS, Overseas Territory of

[French territory, includes Iles de Horne, Ile Uves, and Ile Alofi. Oceanian (Melanesia). Mata-Utu. **Population:** 14,254. **Size:** 106 sq. mi. *No known insect collection.*]

(West Germany, see Germany.)

(West Indies, see specific islands.)

(West Irian, see Irian Jaya.)

203. WESTERN SAHARA

[Formerly Spanish Sahara. Palaearctic. El Aaium. **Population:** 181,411. **Size:** 102,703 sq. mi. *No known insect collection.*]

204. WESTERN SAMOA

[Oceanian (Polynesia). Apia. **Population:** 178,045. **Size:** 1,093 sq. mi. *No known insect collection.*]

(Windward Islands, see Dominica, Grenada, St. Lucia, St. Vincent.)

(Wrangel Island (Vrangelya Ostrov), island territory of Russia.)

205. YEMEN, People's Democratic Republic of South)

[=Aden, formerly Southern Yemen, and North Yeman. Palearctic. San'a and Aden. **Population:** 9,158,040. **Size:** 207,286 sq. mi. *No known insect collection.*]

206. YUGOSLAVIA, Socialists Federal Republic

[This country is in the process of dividing into five countries, each treated separately in this work. Yugoslavia now consists of Montenegro and Serbia. Palearctic. Belgrade. *No known insect collection.*]

207. ZAIRE, Republic of

[=Belgium Congo. Afrotropical. Kinshasa. **Population:** 33,293,946. **Size:** 875,525 sq. mi.]

INSTITUTE DE RECHERCHE SCIENTIFIC, BP 3474, KINSHASA/GOMBE. [IRSC] [*No reply; mail service suspended.*]

208. ZAMBIA, Republic of

[=Northern Rhodesia. Afrotropical. Lusaka. **Population:** 7,546,177. **Size:** 285,994 sq. mi.]

LIVINGSTONE MUSEUM, MOSI-OA-TUNYA ROAD, P.O. BOX 498, LIVINGSTONE. [LMRZ] [*No reply.*]

209. ZIMBABWE

[=Southern Rhodesia. Afrotropical. Harare. **Population:** 9,728,547. **Size:** 149,293 sq. mi.]

INSECT COLLECTION, MUTARE MUSEUM, P. O. BOX 920, MUTARE. [MMMZ]

Director: Mr. J. Chipoka. Phone: 63630. Professional staff: Mr. L. Mutisi. This is a small reference collection of insects and spiders of the Manicaland Province. It is stored in cabinets. [1992]

INVERTEBRATE COLLECTION, NATIONAL MUSEUM. P.O. BOX 240, CENTENARY PARK, BULAWAYO. [NMBZ]

Head curator: Mrs. M. J. Fitspatrick. Phone: 60045. Professional staff: Mrs. R. Sithole. The collection encompasses the whole of Africa, but is especially rich in Central African species (Zimbabwe, Zambia, Malawi, Mozambique). Particularly well represented are Odonata and Lepidoptera. About 4,000 colony samples of Zimbabwean Isoptera are housed here, resulting from a National Survey of Termites, 1972-1975. Over a thousand holotypes (many of Odonata) and numerous paratypes are held here. Excluding the termites, the collection contains about 1 million specimens, a large portion of which are identified. The collection is housed in 3,000 drawers and about 100 storage boxes, 13,000 bottles of alcohol preserved arachnids, including 20 types, and 4,000 colony samples in alcohol bottles. Publications sponsored: "Arnoldia, Zimbabwe," and "Smithersia," formerly "Occasional Papers," series B. [1992]

ENTOMOLOGY COLLECTION, PLANT PROTECTION RESEARCH INSTITUTE, P. O. BOX 8100, CAUSEWAY, HARARE. [PPRZ]

Director: Dr. S. Mlambo. Phone: 704531. This collection is mostly agricultural pests, but with reasonably good collection of Diptera, all from Zimbabwe. It is housed in 25 20-drawer insect cabinets. [1986]

UNKNOWN AND UNREGISTERED COLLECTCIONS

The following collections are unregistered. Some have once been included in our directory, but the owners cannot be found. Although we believe that the registration system is useful, not everyone agrees. Some will not register their collections. So that those who have already been listed are not lost entirely, we have included them in the following list.

Austin, George T., c/o Nevada State Museum and Historical Society, 700 Twin Lakes Drive, Las Vegas, NV 89107 USA. [GTAC] This collection contains about 100,000 Lepidoptera, principally North American (some from Mexico, Ecuador, Panama), primarily from Nevada, housed in 250 drawers and over 150 Schmitt boxes. (Unregistered.) [1992]

Beer, Frank M. [FMBC]. Now in the U. ID., W. F. Barr Collection. [1992]

Bejsák, Vratislav Richard, 53 Lamrock Ave., Bondi Beach, N.S.W. 2026 [VRBC] Phone: (02) 612-365-5253. This is a collection of beetles housed in 246 Schmitt boxes: 1500 (mostly Palearctic), esp. Tenebrionidae, Zopheridae, Lagriidae, and Alleculidae); 1000 specimens of unusual beetles (monsters, *etc.* of all families), and other insects. There are approximately 1,200 genera, 5,000 species, 15,000 specimens of beetles from Central Europe, and over 6,000 specimens of beetles from Australia, including 10 paratypes (Kaszab, Reitter, *etc.*). Specimens may be borrowed; usual loan rules apply. [1992] (This collection is not registered.) [1992]

Cerda-G., Dr. Miguel, Juan Moya 374, Nunoa, Santiago de Chile, Chile. [MCGC] Phone: 227-1036. This collection consists of 35,000 specimens of Chilean insects, especially Cerambycidae and other Coleoptera families; Hymenoptera, Apoidea, with 9,000 specimens identified by Prof. Haroldo Taro; Hemiptera, Homoptera, Neuroptera, and Odonata, stored in boxes. (This collection is not registered.) [1992]

Covell, Dr. Charles Van Orden, Jr., University of Louisville, Louisville, KY 40292 USA. [CVCJ] Phone: (502) 588-5942. This is a collection of about 200,000 specimens of Lepidoptera from North America and Neotropical Region, but primarily from Kentucky. It also includes a bee collection (types are in the USNM), and the Burt Monroe collection of Coleoptera. (Unregistered.) [1992]

The BCCC (**Bruce Cutler**) collection listed in the Coleoptera collection directory has been given to Ronald Huber [RHCC]. [There is also a spider collection but details are unknown.]

Ford, Everett J., Route 2, Box 302, Woodbury, TN 37190-9638 USA. [EJFC] This is a collection of Coleoptera of North America of about 27,000 specimens representing 3,534 species, housed in 48 drawers in cabinets, and 12 Schmitt boxes. It includes 1,840 species from the state of Maryland. Few paratypes are included. (This collection is not registered.) [1992]

Ferrington, Dr. Leonard C., Jr., 2045 Constant Ave., Campus West, University of Kansas, Lawrence, KS 66044 USA. [LFJR] The collection consists primarily of larval and pupal Chironomidae collected from eastern deciduous forest streams in western Pennsylvania and from alpine lakes and streams of the Beartooth Mountains of Wyoming and Montana. Over 30,000 specimens are slide mounted and an additional 5-6,000 specimens are in alcohol. In addition, approximately 15,000 adult specimens are included in the collection; however, they are not sorted or identified to date. (This collection is not registered.)

Francoeur, Dr. Andre, Departement des Sciences Fondamentales, Universite du Quebec a Chicoutimi, Chicoutimi, PQ G7H 2BI, Canada. [CAFQ]. Phone:

(418) 545-5430; 549-0064. This is a specialized collection of ants, mainly from the Holarctic Region, including over 1,000 species, over 100 primary types, represented by over 100,000 specimens pinned or in alcohol vials. Associated with the collection is a large data file of over 15,000 entries, and an ant library with over 5,000 papers and books. The collection is stored in 160 drawers and includes 2,000 slides of mouthparts and male genitalia. [This collection is not registered.]

Gilbert, Dr. Lawrence E., Department of Zoology, University of Texas, Austin, TX 78712 USA. [LGUT] Phone: (512) 471-4705. About 10,000 spread specimens of Mexican and Costa Rican butterflies are housed in 125 drawers. There is included a special collection of about 3,000 bred species of *Heliconius* representing various crosses. [This collections is not registered.]

Huber, Ronald L. 4637 W. 69th Terrace, Prairie Village, KS 66208 USA. [RHIC] Phone: (913) 236-4043. Worldwide collection of Cicindelidae, housed in 120 Cornell drawers and 60 Schmitt boxes, 45,000 specimens which includes 32 of the 35 classic genera, about 1,000 species. Types include 2 holotypes, 3 allotypes, and 427 paratypes of 103 taxa in 10 genera. All faunal realms are represented, but riches in New World taxa (with emphasis on topotypical specimens). The collection has incorporated the collections of Grant Gaumer, Harold L. Willis, Jens Knudson, Bruce Cutler, Charles Wolfe, James Lawton, Rev. Bernard Rotger, F. Martin Brown, Don Stallings, Don Frechin, and Don Shaw. A research collection of Nearctic butterflies is housed in 50 Cornell drawers and 10 Schmitt boxes, about 5,000 specimens, including six paratypes of 3 species; also a synoptic collection of Neotrophical Morphidae. In addition, specimens of Nearctic moths, families Sphingidae, Saturniidae, Sesiidae, Noctuidae, genus *Schinia*, comprising about 1,000 specimens. About 100 specimens of miscellaneous Holocene fossils (McKittrick and LaBrea faunas), Miocene (Dominican amber), and Eocene (Florissant) insects specimens are also housed. A general collection of about 50 Schmitt boxes of Coleoptera and other orders of insects, as well as centipedes, millipedes, scorpions, tarantulas, *etc.*, is maintained as exchange specimens. [RLHI] [1992]

Karasjov, V. P., Academicheskaya ul. 23-28, 220012 Minsk, RUSSIA. [VPKC] This is a collection of about 1,000 species of Curculionidae from the USSR and Mongolia. (Registered with Academy of Science, BSSR, Institute of Bioorgan, Chemistry, Minsk, a collection which is not listed in this directory because we do not have an address or details.)

Kistner, Dr. David H., Department of Biology, California State University at Chico, Chico, CA 95926 USA. [DHKC] Phone: (916) 895-5116. The collection is restricted primarily to Staphylinidae and other myrmecophilous and termitophilous Insecta, with about 250,000 specimens housed in drawers or preserved on slides and in alcohol. All specimens have supporting ecological data. The termite host collection is extensive and is probably the fourth largest in the world. Doryline ants are also well represented. The myrmecophile and termitophiles are more than twenty times the total specimens in all the collections of the world put together, if donations by me to collections are ignored. [The collection is not registered.]

Krahmer, Ernesto, Casilla 546, Valdivia, Chile. [EKIC] Phone: 3981. Mainly insects from local regions in Chile, the collection now consists of about 3,500 specimens. [The collection not registered.]

Lawson, C. S., 6633 Mountainwood Lane, Las Vegas, NV 89103 USA. [CSLC] This is a collection of about 5,000 butterflies, mostly from Nevada, housed in 50 boxes. [This collection is not registered.]

McCleve, Scott, 2210 13th St., Douglas, AZ 85607 USA. [SMCC] This collection of beetles contains over 50,000 adult specimens, mostly from Arizo na, New Mexico, and Mexico. About 8,000 specimens are of the scarab genus *Diplotaxis*. The collection is stored in Schmitt type boxes. [The collection is not registered.]

Mignot, Dr. Edward C., [ECMC], location of collection unknown.

Pakaluk, Dr. James, National Museum of Natural History, Smithsonian

Institution, Washington, DC 20560. [JPCC] World-wide collection of Clavicornia (Coleoptera), especially New World species. About 10,000 specimens (1988) stored in 9 x 5 Mason boxes, some immatures, disarticulated specimens, and slide-mounted material, including a few secondary types. (Registered with SEMC.)

Peters, Gary L., 1445 NW Menlo Drive, Corvallis, OR 97330 USA. [GLPC] Phone: (503) 754-6461. This is a general collection of over 30,000 specimens of Coleoptera representing nearly all families known to occur in U.S.A. Material is primarily from California and the Pacific Northwest, with some from Arizona. Foreign specimens include those from Mexico, Costa Rica, and southeast Asia (Laos). The collection is housed in 60 drawers and over 30 Schmitt boxes. Specimens are available for exchange. (Collection not registered.)

Rodriguez, Tomas Moore, [*ddress unknown.*] [TPMR] This is a collection of 3,000 Buprestidae (Coleoptera) from Chile, housed in 21 boxes. (Not registered. [1992]

Stibick, Dr. J. N. L., USDA, APHIS, PPQ, Federal Center Bldg., Hyattsville, MD 20782 USA. [JNLS] Phone: (301) 436-6464. This is a general insect collection, but predominately Coleoptera, housed in 20 drawers, 12 of which are Elateridae [*No reply in 1992*]. (This collection is not registered.)

Ziff, Seymour, 1441 Beverwil Dr., Los Angeles, CA 90035 USA. [SZIC] Phone: (213) 643-6165; 553-4011; 553-4243. This is a collection of 3,000 specimens of Coleoptera. Specimens may be borrowed for study; usual loan rules apply [*No reply in 1992*]. (No longer registered with LACM.)

CODENS FOR INSECT AND SPIDER COLLECTIONS OF THE WORLD

The following list of codens (we use this term in preference to acronyms, as explained in the introduction) is complete for this directory, and includes variations that have appeared in the literature if known to us. We doubt that the list is complete and we will welcome additions.

Please note that public collections are listed by country, state/province, and in some areas, by city, while private collections are listed under the name of the collector. Where they appear in the directory can be determined by using the index.

A name followed by a question mark (?) indicates that the person did not respond to the questionnaire and, therefore, we are not sure of the location or status of the collection.

AAAG Alan and Anita Gillogly
AAPI Canada, Alberta, Edmonton, Alberta Agriculture
ABSC USA, Florida, Lake Placid, Archbold Biological Station
ACBC Canada, British Columbia, Summerland, Agriculture Canada Research Station
ACBV Canada, British Columbia, Vancouver, The Aphids of British Columbia
ACNB Canada, New Brunswick, Fredericton, Agriculture Canada Research Station
ACRM Canada, Manitoba, Winnipeg, Criddle Collection
ACSK Canada, Nova Scotia, Kentville, Agriculture Canada Research Station
AEIC USA, Florida, Gainesville, American Entomological Institute
AESB India, Bogb Agriculture Experiment Station
AFBC Andrew F. Beck
AFLC Alvin F. Ludtke
AGRL Canada, Alberta, Lethbridge, Lethbridge Research Station
AHBC USA, Utah, St. George, Dixie College
AHBS Allen H. Benton
AHCC see ARHC
AJKC Alfred J. Kistler ?
ALBH Arno L. van Berge Henegoueven
ALRC Rolf L. Aalbu
AMCL Brazil, Amapa, Macapa, Museu Territorial de Historia Natural
AME see FSMC
AMG see AMGS
AMGS South Africa, Grahamstown, Albany Museum
AMIC Antonio Martinez
AMMM see MMKZ
AMNH USA, New York, New York, American Museum of Natural History
AMNZ New Zealand, Auckland, Auckland Institute and Museum
AMS see AMSA
AMSA Australia, Sydney, Australian Museum

AMUZ India, Uttar Pradesh, Aligarh Muslini University
ANCB Bolivia, La Paz, Coleccion Entomological
ANDC Annie Dozier
ANIC Australia, Canberra City, Australian National Insect Collection
ANLW Austria, Wien, AMT Der Niederösterreichischen
ANSP USA, Pennsylvania, Philadelphia, Academy of Natural Sciences
APEI Canada, Prince Edward Island, Charlottetown, Agriculture Canada Research Station
APHI USA, Maryland, Baltimore, U.S. Department of Agriculture, Animal and Plant Health Inspection Service
ARCC Anthony Ross ?
ARCM Canada, New Brunswick, St.Andrews, Atlantic Reference Centre
ARHC Alan R. Hardy
ARMC Andrew R. Moldenke ?
ARSC Arthur R. Strong
ASAY USSR, Armenia, Institute of Zoology
ASUA Egypt, Cairo, Ain Shams University
ASUT USA, Arizona, Tempe, Arizona State University
AUA see AUEM
AUBL Lebanon, Beirut, Museum of Natural History
AUCE Egypt, Cairo, El Azhar University
AUEM USA, Alabama, Auburn, Auburn University
AUTC Turkey, Erzurum, Ataturk Universitesi
AVEC Arthur V. Evans

BAAC Algeria, Beni Abbes, Musee de Beni Abbes
BAFC Benjamin A. Foote
BAU see BAUC
BAUC China, Beijing, Beijing Agricultural University
BBNP USA, Texas, Big Bend, Big Bend National Park
BCCC Bruce Cutler, see Ronald L. Huber RLHI
BCPM Canada, British Columbia, Victoria, British Columbia Provincial Museum
BCRC Brett C. Ratcliffe
BCWL Italy, Roma, Biological Control of Weeds Laboratory
BDLU Canada, ON, Sudbury, Laurentian University
BDMU Canada, Ontario, Hamilton, McMaster University
BDUC Canada, Alberta, Calgary, University of Calgary
BDUW Canada, Ontario, Waterloo, University of Waterloo
BDVC Barry D. Valentine
BDWC Canada, Ontario, Windsor, University of Windsor
BDWL Canada, Ontario, Waterloo, Wilfred Laurier University
BENH United Kingdom, London, British Entomological and Natural History Society
BESM Malawi, Limbe, Bvumbwe Experiment Station
BFIC France, Brunoy, Museum National d'Histoire Naturelle
BHLC Benjamin Landing
BHMH Brazil, Minas Gerais, Belo Horizonte, Museu de Historia Natural
BHPC Brian H. Patrick
BIDA USA, Idaho, Boise, Boise State University

BIPB Barbara I. P. Barratt
BKDC Byrd K. Dozier
BLCU USA, Utah, Logan, Utah State University
BLGA Austria, Eisenstadt, Burgenlandischer Landesmuseum
BLPC Brian Lyford
BMB see BMBN
BMBN United Kingdon, Brighton, Booth Museum of Natural History
BMDC Bastiaan M. Drees
BMGB Barbados, St. Ann's Garrison, Barbados Museum and Historical Society
BMHP Bermuda, Hamilton, Department of Agriculture and Fisheries
BMIC Bryant Mather
BMKB Brunei, Kota Baru, Brunei Museum
BMNH United Kingdom, London, The Natural History Museum
BMSA South Africa, Bloemfontein, National Museum Bloemfontein
BMSC USA, New York, Buffalo, Buffalo Museum of Science
BMUK United Kingdom, Bolton, Bolton Museum
BMUW see UWBM
BNHD India, Darjeeling, Bengal Natural History Museum
BNHS India, Bombay, Bombay Natural History Society
BPBM USA, Hawaii, Honolulu, Bernice P. Bishop Museum
BSTC Barney Streit
BYU see BYUC
BYUC USA, Utah, Provo, Brigham Young University

CAES USA, Connecticut, New Haven, Agriculture Experiment Station
CAFQ Andre Francoeur
CARD Barbados, St. Thomas, Caribbean Agricultural Research Institute
CARE Trinidad and Tobago, Port of Spain
CARS Surinam, Paramaribo, University of Surinam
CAS see CASC
CASC USA, California, San Francisco, California Academy of Sciences
CASM USA, Illinois, Chicago, Chicago Academy of Sciences
CBFC Bolivia, La Paz, Coleccion Boliviana de Bolivia
CCAC Brazil, Ceara, Fortaleza, Centro Ciencias Agrarias
CCCC USA, Wisconsin, Kenosha, Carthage College
CCCZ Malawi, Zomba, University of Malawi
CCFL Chad, Fort Lamy, Chad National Museum
CCNP USA, New Mexico, Carlsbad, Carlsbad Caverns National Park
CCPC Charles C. Porter
CCUF Brazil, Alagoas, Maceio, Departamento de Biologia
CDAE USA, California, Sacramento, Department of Food and Agriculture
CDRS Ecuador, Galapagos Islands, Charles Darwin Research Station
CEAM Mexico, Chapingo, Colegio de Postgraduados
CEEF Honduras, Siguatepeque, Escuela Nacional de Ciencias Forestales
CELM Colombia, Bogota, Coleccion Entomologica "Luis Maria Murillo"
CEMU USA, Ohio, Cleveland, Cleveland Museum of Natural History
CENA Nicaragua, Managua
CENG Nicaragua, Nueva Guinea

CEST Trinidad and Tobago, Centeno, Central Experiment Station
CEWC Charles E. White [deceased] see FSCA
CFUA Chile, Valdina, Universidad Austral de Chile
CFRB China, Beijing, Forest Research Institute
CFZC Charles F. Zieger
CGEC China, Guangzhou, China Entomological Research Institute
CGNZ C. J. Green
CHAH Henry A. Hespenheide
CHIC Christopher Hair
CIAN Mexico, Ciudad Obregon, Centro de Investigaciones Agricolas Nortoeste
CIBC Trinidad and Tobago, Curepe, Commonwealth Institute of Biological Control
CICA see IDEA
CIDA USA, Idaho, Caldwell, College of Idaho
CIJC Romania, Judetul Covasna, Muzeului Judetean Covasna
CISC see EMEC
CISM USA, Michigan, Bloomfield Hills, Cranbrook Institute of Science
CJSC James S. Cope
CKSC Christopher K. Starr
CKSF Charles Kristenren
CLBC Charles L. Bellamy
CLCC Canada, Alberta, Camrose, Camrose Lutheran College
CLNP USA, Oregon, Crater Lake, Crater Lake National Park
CLSJ C. L. Staines, Jr.
CMB see CMBK
CMBK United Kingdom, Bristol, The City Museum
CMC see CMNZ
CMEI USA, Florida, Leesburg, Clements' Museum of Exotic Insects
CMNH USA, Pennsylvania, Pittsburgh, Carnegie Museum of Natural History
CMNC Canada, Ottawa, Canada Museum of Nature
CMNS China, Shanghai, Museum of Natural History
CMNZ New Zealand, Christchurch, Canterbury Museum
CMP see CMNH
CMSC USA, Tennessee, Nashville, Cumberland Museum and Science Center
CNC see CNCI
CNCI Canada, Ontario, Ottawa, Canadian National Collection of Insects
CNHM USA, Ohio, Cincinnati, Cincinnati Museum of Natural History
CNHP China, Beijing, Beijing Natural History Museum
CNHS United Kingdom, South Croydon, Croydon Natural History and Scientific Society
CNM see CNMS
CNMC USA, Colorado, Fruita, Colorado National Monument
CNMS Sri Lanka, Colombo, National Museum
CNPS Brazil, Parana, Centro Nacional de Pesquisas da Soja
CNSM see CEMU
CPAC Brazil, Distrito Federal, Plantina, Centro Pesquisas Agropecuarias do Curado

CPAP Brazil, Para, Belem, Centro de Pesquisas Agropecuarias do Tropico Unido
CPDC Brazil, Bahia, Itabuna, Centro de Pesquisas do Cacau
CPMM Mozambique, Lourenco Marques, Dr. Alvaro de Castro Provincial Museum
CPQC see ULQC
CPUP USA, California, California Polytechnic University
CSCC USA, Nebraska, Chadron, Laboratory of Arthropod Diversity
CSDA see CDAE
CSDS USA, California, Baker, Desert Studies Center
CSLB USA, California, Long Beach, California State College
CSIR see ANIC
CSLC C. S. Lawson
CSUC USA, Colorado, Fort Collins, Colorado State University
CTAM USA, Hawaii, Honolulu, University of Hawaii, Manoa
CTPC Carl T. Parsons [deceased] see UVCC
CU see CUIC
CUCC USA, South Carolina, Clemson, Clemson University
CUGE Egypt, Cairo, Cairo University
CUIC USA, New York, Ithaca, Cornell University
CUMZ United Kingdom, Cambridge, University Museum
CVCJ Charles Van Orden Covell, Jr.
CWOB Charles W. O'Brien
CWRC Carl W. Rettenmeyer
CZAA Argentina, Mendoza, Caledra de Zoologica Agricola

DACL Canada, Ontario, London, Research Centre Annex
DADC D. Allen Dean
DAFB see BMHP
DAFH Hong Kong, Kowloon, Department of Agriculture and Fisheries
DAWC Donald A. Wilson
DBAI Brazil, Distrito Federal, Brasilia, Departmento de Biologia Animal
DBAU Brazil, Rio de Janeiro, Rio de Janeiro, Departamento de Biologia Animal
DBRC David B. Richman
DBSE Brazil, Sergipe, Aracaju, Departamento de Biologia
DBTC Donald B. Thomas
DBUM see QMOR
DCBU Brazil, Sao Paulo, Sao Carlos, Universidade Federal de Sao Carlos
DCLG D. C. L. Gosling ?
DCMB Brazil, Amazonas, Manaus, Universidade do Amazonas
DCMC David C. Miller [deceased] ?
DCMP Brazil, Parana, Curitiba, Universidade Federal do Parana
DCTC David C. Taylor
DDFF Brazil, São Paulo, Jaboticabal, Departamento de Defensa Fitossanitarista
DEBU Canada, Ontario, Guelph, University of Guelph
DEES Brazil, São Paulo, Piracicaba, Universidade de Sao Paulo
DEFS Brazil, São Paulo, São Paulo, Universidade de Sao Paulo

DEI see DEIC
DEIC Germany, Eberswalde Finow, Institut fur Pflanzenschutz-forschung
DENH USA, New Hampshire, Durham, University of New Hampshire
DERC Donald E. Rich ?
DEUN= UNSM
DFCZ Malawi, Zomba, Forest Research Institute
DFEC USA, New York, Syracuse, State University of New York
DFLC Brazil, Minas Gerais, Lavras, Ecole Superior de Agricultura
DFRU Canada, New Brunswick, Fredericton, University of New Brunswick
DGKC David G. Kissinger ?
DGMC Donald G. Manley
DHGC Donald H. Gudehus
DHKC David H. Kistner
DJBC Denis J. Brothers
DKYC Daniel K. Young
DLBC Donald L. Baumgartner
DLCC Daniel L. Curtis
DLDC D. L. Deonier
DLPC David L. Pearson ?
DMBC USA, SC, McClellanville, Dominick Moth and Butterfly Collection
DMDC Cameroon, Douala, Douala Museum
DMNH USA, Ohio, Dayton, Dayton Museum of Natural History
DMSA South Africa, Durban, Durban Museum
DNHC USA, Colorado, Denver, Museum of Natural History
DORC United Kingdom, Dorchester, Dorset County Museum
DPBA Argentina, Buenos Aires, Departemento de Pathologia
DPIC Brazil, Minas Gerais, Belo Horizonte, Departamento de Parasitologia
DPNC USA, Connecticut, Mystic, Denison Pequotsepos Nature Center
DPPC Taiwan, Nau-Tan, Department of Agriculture
DPUP Brazil, Parana, Maringa, Universidade Federal de Maringa
DPWC David P. Wooldridge ?
DRMC David Robert Maddison
DRPC Dennis R. Paulson
DSEC Brazil, Paraiba, Joao Pessoa, Universidade Federal da Paraiba
DSIR see NZAC
DSVC David S. Verity
DTIC Brazil, Minas Gerais, Departamento Parasitologia
DVCC USA, California, Pleasant Hill, Diablo Valley College
DVNM USA, California, Death Valley, Death Valley National Monument
DWJC Dale W. and Joanne F. Jenkins
DZCU India, Calcutta, Calcutta University
DZEC see MTEC
DZIB Brazil, São Paulo, Campinas, Departamento de Zoologia Zoologia
DZUP Brazil, Parana, Curitiba, Departamento de Zoologia

EAPC see EAPZ
EAPZ Honduras, Tegucigalpa, Escula Agricola Panamericana

EBCC Mexico, Jalisco, San Patricio, Estación de Biología "Chamela"
EBDS Spain, Sevilla, Estacion Biologicas de Donana
ECMC Edward C. Mignot ?
ECOL Belgium, Louvain-la-Neuve, Collection du Laborataire d'Ecologie
ECUT USA, Tennessee, Knoxville, University of Tennessee
EDNC USA, North Carolina, Raleigh, N. C. Department of Agriculture
EDUM Canada, Manitoba, Winnipeg, University of Manitoba
EDWC Edward D. Weidert ?
EEEC Edward E. Estep
EELM Peru, Lima, Estacion Experimental Agricola de la Molina
EGCC Edward V. Gage?
EGFC Edward G. Franworth?
EGNP USA, Florida, Everglades National Park
EGRC Edward G. Riley
EGSC Eric Gowling-Scopes
EHSC Eric H. Smith ?
EIHU Japan, Sapporo, Hokkaido University
EINS Ecuador, Quito, Ecuadorian Institute of Natural Sciences
EISC China, Shanxi, Taigu
EJFC Everett J. Ford
EJGC see FSCA
EJKC Eric J. Kiteley
EKIC Ernesto Krahmer
ELMF USA, Maine, Augusta, Maine Forest Service
ELSC E. L. Sleeper?
EMAU Germany, Greifswald, Ernst-Moritz-Arndt-Universitat Greifswald
EMB see EMBT
EMBT Thailand, Bangkok, Department of Agriculture
EMEC, USA, California, Berkeley, Essig Museum
EMUS USA, Utah, Logan, Utah State University
ENIH Japan, Tokyo, National Institute of Health
EPNZ Eric D. Pricthard
EPRL USA, Puerto Rico, Mayaguez, University of Puerto Rico
ESRC Canada, Nova Scotia, Belmont, N.S. Dept. Nat. Resources
ESRN Brazil, Rio Grande do Norte, Mossoro, Ecole Superior de Agricultura
ESUG USA, Guam
ESUW USA, Wyomong, Laramie, University of Wyoming
ETHZ Switzerland, Zurich, Eidgenossische Technische Hoch-schule-Zentrum
EUMJ Japan, Matsuyama, Ehime University
EVGC Edward V. Gage
EYCC Eric Yensen ?

FACS China, Shaxian, Fujian Agricultural College
FAMU USA, Florida, Tallahassee, Florida A & M University
FAUN Colombia, Pasta, Universidad de Narino
FAVU Brazil, Rio Grande do Sul, Porto Alegre, Faculdade Agronomia e Veterenaria

FBUB Germany, Bielefeld, Universitat Bielefeld
FBWA Austria, Wien, Forstliche Bundsversuchsanstalt
FCAP Brazil, Para, Belem, Faculdade Ciencias Agrarias
FCDA USA, California, Fresno, Fresno County Department of Agriculture
FCNI Australia, Sydney, Forest Commission
FCNS see FCNI
FCNZ Francis Dudley Chambers
FCTH Australia, Hobart, Forestry Commission of Tasmania
FDA see FSCA
FDUC USA, New Jersey, Rutherford, Fairleigh Dickinson University [collection transferred to FSCA].
FEM see PSUC
FFCL Brazil, Sao Paulo, Itu, Faculdade de Filosofia Ciencias e Letras
FGAC Fred G. Andrews?
FGIC François Génier
FHKS USA, Kansas, Fort Hays, Fort Hays Kansas State College
FHUB F. H. U. Baker
FICB Papua New Guinea, Lae, Forest Research Centre
FIDS Canada, Ontario, Sault Ste. Marie, Forest Insect and Disease Survey
FIEC Canada, Manitoba, Winnipeg, Fisheries and Aquatic Science
FIJI Fiji, Suva, University of the South Pacific
FIOC Brazil, Rio de Janeiro, Rio de Janeiro, Fundacao Instituto Oswaldo Cruz
FMBC Frank M. Beer ?
FML see IMLA
FMNH USA, Illinois, Chicago, Field Museum of Natural History
FMSS El Salvador, San Salvador, Natural History Museum
FNML Netherlands, Leeuwarden, Fries Natuuhistorisch Museum
FNYC Frank N. Young
FRCS Malaysia, Forest Research Centre, Sandakan
FRIM Malaysia, Kula Lumpur, Forest Research Institute
FRLC Canada, New Brunswick, Fredericton, Forest Insect and Disease Survey Reference Collection
FRNZ New Zealand, Rotorua, Forest Research Institute
FSAC Forrest St. Aubin ?
FSAG Belgium, Gembloux, Zoologie Generale et Faunistique
FSCA USA, Florida, Gainesville, Florida State Collection of Arthropods
FSMC USA, Florida, Gainesville, Florida State Museum
FSSF Germany, Fors. Senchenberg
FTHC Frank T. Hovore
FWJP Floyd W. and June D. Preston
FWMC Frank W. Mead

GADC Gary A. Dunn
GAGC Glenn A. Gorelick
GASC G. Allan Samuelson, see BPBM
GBFC G. B. Fairchild
GBFM Panama, Panama, Universidad de Panama

GCGC Grant C. Gaumer ?
GCNP USA, Arizona, Grand Canyon, Grand Canyon National Park
GCSC George C. Steyskal
GCTP Singapore, Singapore, Global Colosseum
GCWC George C. Walters
GDCB Greg Daniels
GEIC China, Guangdong, Guangdong Entomology Institute
GEFC Gerard E. Flory ?
GFHC Gary F. Hevel, see USNM
GHBC George H. Bick
GHNC G. H. Nelson
GIES Edmund F. Giesbert
GLCC Gilbert L. Challet ?
GLFR Canada, Ontario, Sault Ste. Marie, Great Lakes Forest Research Centre
GLIC Graeme Lowe
GLNP USA, Montana, West Glacier, Glacier National Park
GLPC Gary L. Peters
GNHM Greece, Kifissia, Goulandris Natural History Museum
GNME Sweden, Goteborg, Naturhistoriska Museet
GPPT Russia, Georgia, Tbilisi
GPTA Virendra Gupta
GRNC Gerald R. Noonan see MCPM
GRSW Namibia, Waluis Bay, Desert Ecological Research Unit
GSAT USA, Alabama, Tuscaloosa
GSIC Gordon Small
GSNP USA, Tennessee, Gatlinburg, Great Smoky Mountains National Park
GTAC George T. Austin
GZHC G. van der Zanden

HAHC Henry and Anne Howden
HAUH India, Hissar, Haryana Agricultural University
HCCA USA, Nebraska, Hastings, Hastings College
HCOE see OXUM
HCTR USA, Georgia, Statesboro, Hoogstraal Center for Tick Research
HDBC Howard David Baggett
HDEC Dodge Engleman
HDOA USA, Hawaii, Honolulu, Hawaii Department of Agriculture
HEMS United Kingdom, Haslemere, Haslemere Educational Museum
HENA Honduras, Catacamas, Escuela Nacional de Agricultura
HHCC Hilary Hacker ?
HISC Harvey I. Scudder ?
HLBC Hugh L. Burns, Jr. ?
HLD see HLDH
HLDC Herbert L. Dozier
HLDH Germany, Darmstadt, Hessisches Landesmuseum Darmstadt
HLWC Harold L. Willis, see RLHC
HMNS USA, Texas, Houston Museum of Natural History
HMOX see OXUM

HMUG United Kingdom, Glasgow, Glasgow University
HNHB see HNHM
HNHH China, Heilongiian, Natural History Museum
HNHM Hungary, Budapest, Hungarian Natural History Museum
HNSA Austria, Salzburg, Haus der Natur
HPBC Howard P. Boyd
HPSC H. P. Stockwell
HRSC Harrison R. Steeves, Jr. ?
HSIC Solomon Islands, Honiara, Ministry of Natural Resources
HSUE Ethiopia, Addis Ababa, Natural History Museum
HVNP USA, Hawaii, Honolulu, Hawaii Volcanoes National Park
HWIC Henk Wolda
HZMZ Crotia, Zagreb, Hrvatski Narodni Zooloski Muzej

IAAA Brazil, São Paulo, Araras, Instituto do Azucar e do Alcool
IACC Brazil, São Paulo, Campinas, Instituto Agronomico de Campinas
IANZ I. C. Andrew
IARC New Zealand, Mosgiel, Invermay Research Centre
IBSP Brazil, São Paulo, São Paulo, Instituto Biologico
IBUP Poland, Torun, Uniwersytet Mikolaja Koperniki
IBUS Brazil, Rio de Janeiro, Seropedica, Instituto de Biologia
IBUT Brazil, Sao Paulo, Sao Paulo, Instituto Butarita
ICBU Canada, Quebec, Lennoxville, Natural History Museum
ICCM see CMNH
ICIS USA, Idaho, Pocatello, Idaho State University
ICPR India, Bangalore, Biological Control Research Institute
ICRC USA, Maryland, Baltimore, Insect Control and Research
ICRG China, Canton, Institute of Entomology
ICRI China, Guangzhou, Research Institute of Entomology
ICUI USA, Iowa, Iowa City, University of Iowa
IDEA Chile, Arica, Instituto de Agronomia
IDMC I. D. McLellan
IEAS China, Shanghai, Institute of Entomology
IECA Czechoslovakia, Budejovice, Czechoslovakia Academy of Science
IEEM see MNMS
IEGG Italy, Bologna, Instituto de Entomologia
IEME Russia, Moscow, Institute for Evolution, Morphology, and Ecology of Animals
IEMM Mexico, Mexico City, Instituto de Ecologia
IEUS Italy, Instituto di Entomologia Dell'Universita degi Studi
IEVB Belgium, Liege, Institut Ed. Van Beneden
IFPE see DEIC
IFRI India, Dehra Dun, Indian Forest Research Institute
IFSA Brazil, Sao Paulo, Sao Paulo, Instituto Florestal
IGAA I. G. Andrew
IGUS Germany, Saarbrucken, Universitat des Saarlandes
IIES Argentina, Salta, Instituto Investigaciones Entomologicas Salta
IICT Portugal, I. I. C. T., Lisboa
IJSM Jamaica, Kingston, Science Museum
ILRC Ira LaRivers [Deceased] ?

IMCC Ian Moore ?
IMI see IPSM
IML see IMLA
IMLA Argentina, Tucuman, Instituto Miguel Lillo
INBC Costa Rica, Santo Domingo, Instituto Nacional de Biodiversidad
INBP Paraguay, Asuncion, Museo Nacional de Historia Natural del Paraguay
INER Italy, Roma, Instituto Nazionale di Entomologia
INHM Iraq, Baghdad, Iraq Natural History Museum
INHS USA, Illinois, Urbana, Illinois Natural History Survey
INIA Mexico, Chapingo, Institut Nacional de Investigaciones Agricultura
INIR Mexico, Xalapa, Coleccion de Termites Mexicanas
INLA Chile, La Cruz, INIA Subestacion Experimental
INPA Brazil, Amazonas, Manaus, Coleccio Sistematica da Entomologia
INPC India, Hayara National PUSA Collection
IOC see FIOC
IPCN Argentina, San Martin de los Andes, Instituto Patagonia de Ciencias Naturales
IPSM United Kingdom, Ipswich, Ipswich Museum
IPTB Brazil, São Paulo, São Paulo, Instituto de Pesquisas Tecnologica
IPZE see DEIC
IRAG Guadeloupe, Petit-Bourg, Institut National de la Recherche Agronomique de Antilles et Guyane
IRCW USA, Wisconsin, Madison, University of Wisconsin
IREC Guadeloupe, Pointe-a-Pitre, Institut de Recherches Entomologique de la Caribe
IRRI Philippines, Manila, International Rice Research Institute
IRSB see ISNB
IRSC Zaire, Kinshasa, Institute de Recherche Scientific
IRSN see ISNB
ISAR Byelorussia, Minsk
ISAS China, Kunming, Institute of Zoology
ISCM Morocco, Rabat, Institut Scientifique Cheripen
ISMC USA, Indiana, Indianapolis, Indiana State Museum
ISMS USA, Illinois, Springfield, Illinois State Museum
ISNB Belgium, Brussels, Institut Royal des Sciences Naturelles de Belgique
ISNS see FSCA
ISU see ISUC
ISUC USA, Illinois, Normal, Illinois State University
ISUI USA, Iowa, Ames, Iowa State University
ISZK see ISZP
ISZP Poland, Krakow, Polish Academy of Sciences
ITLJ Japan, Tsukuba, Lab. of Insect Science
ITMM Mexico, Nuevo Leon, Monterrey, Instituto Technica de Monterrey
IUIC USA, Indiana, Bloomington, Indiana University
IVCB Israel, Volcani Center
IZAC Cuba, Havana, Academia de Ciencias de Cuba
IZAS China, Beijing, Institute of Zoology
IZAV Venezuela, Aragua, Maracay, Instituto de Zoologia Agricola

IZBE Estonia, Tartu, Institute of Zoology and Botany
IZP see IZPC
IZPC Portugal, Porto, Universidade do Porto
IZS see IZPC
IZSI Italy, Siena, Instituto di Zoologia
IZUC see UCCC
IZUE Germany, Erlangen, Universitat-Erlangen-Nurnberg
IZUI Austria, Innsbruck, Institut fur Zoologie

JARC J. A. Ramos?
JASC James A. Slater
JASO John A. Stidman
JATH Hungary, Szeged, Josef Attila Tudomanyegyeten
JBHC John B. Heppner
JBJC James B. Johnson
JBKC Jay B. Karren ?
JBVC John B. Vernon
JBWC John Bowden
JCAC Jean Charles Aube ?
JCCC James C. Cokendolpher
JCSC Jack C. Schuster
JCVB Jack C. Von Bloeker ?

JELC James E. Lloyd
JEWC James E. Wappes
JFBC J. F. Brimley ?
JFCC J. F. Cornell ?
JGEC J. Gordon Edwards
JGFC John G. Franclemont
JHBC James H. Baker ?
JHFC J. Howard Frank
JHRC John R. Robinson ?
JHSC Jorge Hendrichs ?
JJDC Jerrell James Daigle
JJMD Jean and Jerry M. Davidson ?
JKBC John K. Bouseman ?
JLNC John L. Neff
JMCC Joseph M. Cicero
JMCI John M. Coffman
JMFC Julius Ganev
JMPC John M. Plomley
JMSC John M. Snider
JNKC Josef N. Knull [Deceased] see FMNH
JNLS Jeffrey N. L. Stibick
JOIC J. Oppewall
JPCC James Pakaluk
JPMP Hungary, Pecas, Janus Pannonius Museum
JRCC James Robertson
JRIC Jonathan Reiskind
JRPC Jack R. Powers ?

JSCC Joe Schuh [Deceased] see AMNH
JSNC John S. Nordin
JSTC John Stamatov ?
JTGC James T. Goodwin
JTPC John T. Polhemus
JTNM USA, California, Twentynine Palms, Joshua Tree National Monument
JVMJ John V. Matthews, Jr.
JWJC James W. Johnson
JWTC J. W. Tilden [Deceased] see CASC

KARI Kenya, Narobi, Kenya Agricultural Research Institute
KARS Uganda, Kampala, Kawanda Agricultural Research Station
KCEC K. C. Emerson
KDIC Keith Dobry
KEIU Korea, Seoul, Korea University
KFNZ Kenneth J. Fox, *Deceased*
KFRI India, Kerala, Kerala Forest Research Institute
KHSC Karl H. Stephan
KKUK Korea, Seoul, Kon-Kuk University
KLSC Kurt L. Schmude
KMFC Kenneth M. Fender ?
KMVC Czechoslovakia, Hradec Kralove, Krajske Muzeum Vychodnich Cech
KNHM Kuwait, Kuwait, Kuwait Natural History Museum
KNUC Korea, Kangweon, Kangweon National University
KPMC Michigan, Kalamazoo, Kalamazoo Public Museum
KSBS USA, Kansas, Lawrence, State Biological Survey of Kansas
KSUC USA, Manhattan, Kansas State University
KUEC Japan, Fukuoka, Kyushu University
KUIC Japan, Kagoshima, Kagoshima University
KUKI India, Kurukshetra, Kurukshetra University
KWBC Kirby W. Brown
KWKC Kenneth W. Knopf
KWVC Kenneth W. Vick

LACM USA, California, Los Angeles, Los Angeles County Museum of Natural History
LBIC Lewis Berner
LBIT France, Toulouse, Laboratoire de Biologie des Insectes
LBOB Lois B. O'Brien
LCAC L. Clair Armin [Deceased] ?
LCDC Linwood C. Dow
LCM see LCMI
LCMI India, Madras, Loyola College
LCNZ New Zealand, Lincoln, Lincoln University College
LEM see LEMQ
LEMG Louis E. McGee ?
LEMQ Canada, Quebec, Ste. Anne de Bellevue, Lyman Entomological Museum

LEPG Luis E. Peña
LFGC Lawrence F. Gall
LFJR Leonard C. Ferrington, Jr.
LGCC Louis G. Gentner [Deceased] ?
LGUT Lawrence E. Gilbert
LGWC Laurel G. Woodley ?
LLLJ Lester L. Lampert, Jr.
LMAD Germany, Düsseldorf, Löbbecke Museum und Aquazoo
LMRZ Zambia, Livingstone, Livingstone Museum
LMZG Guinea, N'Zerekore, Local Museum
LNK see LNKD, SMNK
LNKD see SMNK
LNM see LNMD
LNMD Germany, Munster, Landessammlungen fur Naturkunde
LPBC Lincoln P. Brower
LPSP USA, Delaware, Kirkwood, Whale Wallow Nature Center
LRGC Lorin R. Gillogly ?
LS see LSUK
LSUC USA, Lousiana, Baton Rouge, Louisiana State University
LSUK United Kingdom, London, Linnean Society
LVNP USA, California, Mineral, Lassen Volcanic National Park
LZLP Portugal, Lisboa, Faculdade de Ciencias

MACA Macau, Government
MACB M. Alvarenga ?
MACN Argentina, Buenos Aires, Museo Agentina de Ciencias Naturales
MAFC Manfredo Fritz
MAGB Botswana, Gaberones, National Museum and Art Gallery
MAGD Australia, Darwin, Museum and Art Galleries of N. T.
MAIC Michael A. Ivie
MAKD see ZFMK
MAMU Australia, Sydney, Macleay Museum
MATC M. A. Tidwell
MATH Bryant
MBBJ Indonesia, Bogor, Museum Zoologicum Bogoriense
MBCG Italy, Berganio, Museo di Scienze Naturali
MBR see MACN
MBSR Romania, Sibiu, Muzeul Brukenthal
MBUC Venezuela, Caracas, Universidad Central de Venezuela
MCCM India, Tambaraw, Madras Christian College
MCGC Miguel Cerda-G.
MCM see MCMC
MCMC Mexico, Mexico, Museo de Historia Natural
MCMF Marc C. Minno
MCN see MCNZ
MCNV Italy, Venice, Museo Civico di Storia Naturale
MCNZ Brazil, Rio Grande do Sul, Porto Alegre, Museu de Ciencias Naturais
MCPM USA, Wisconsin, Milwaukee, Milwaukee City Public Museum
MCPU Brazil, Rio Grande do Sul, Porto Alegre, Pontificia Universidade

MCSC USA, Colorado, Colorado Springs, May Natural History Museum
MCSN Italy, Genoa, Museo Civico de Storia Naturale "Giacomo Doria"
MCTC M. C. Thomas
MCZ see MCZC
MCZC USA, Massachusetts, Cambridge, Museum of Comparative Zoology
MDGC Manfred Döberl
MDLA Angola, Luanda, Museu do Dundo
MECN Ecuador, Quito, Museo Ecuadoriano de Ciencias Naturales
MEMU USA, Mississippi, Mississippi, Mississippi State University
MEUC Chile, Santiago, Universidad de Chile
MEUP Panama, Panama, Universidad de Panama, Museo de Entomologia
MFAP Madagascar, Tananarive, Museum of Folklore
MFNB see ZMHB
MGA see MGAP
MGAB Romania, Bucharest, Muzeul de Istoria Naturala "Grigore Antipa"
MGAP Brazil, Rio Grande do Sul, Porto Alegre, Museu Anchieta
MGBC see EBDS
MGDL Luxembourg, Luxembourg, Museum d'Histoire Naturalle
MGF see MGFT
MGFT Germany, Tutzing, Museum G. Frey
MHMC Martin H. Muma, *deceased*, see FSCA
MHNA France, Museum d'Histoire Naturalle D'Autun
MHNC Switzerland, La Chaux-de-Fonds, Musée d'Histoire Naturalle
MHND Dominican Republic, Santo Domingo, Museo Nacional de Historia Natural
MHNG Switzerland, Geneva, Museum d'Histoire Naturelle
MHNL France, Lyon, Musee Guimet d'Histoire Naturelle de Lyon
MHNM see URMU
MHNS see MNNC
MIMM Mauritius, Port Louis, Mauritius Institute
MINC Spain, Madrid, Universidad Politecnica
MINI Romania, Iasi, Muzeul de Istoria Naturala
MIUP Panama, Panama, Museo de Invertebrates
MIZT Italy, Torino, Universita di Torino
MJLC M. J. Laliberte ?
MJMO Venezuela, Decanato de Agronomía
MJPL see MUSM
MLP see MLPA
MLPA Argentina, La Plata, Universidad Nacional de La Plata
MLPC Manuel L. Pescador
MLUH Germany, Halle, Wissenschaftsbereich Zoologie
MMB see MMBC
MMBC Czech, Brno, Moravian Museum
MMCM Malawi, Blantyre, Museum of Malawi
MMKZ South Africa, Kimberley
MMMN Canada, Manitoba, Winnipeg, Manitoba Museum of Man and Nature

MMMZ Zimbabwe, Mutare, Mutare Museum
MMUE United Kingdom, Manchester, The University
MNCE Brazil, Parana, Curitiba, Museu de Historia Natural Capao da Embuia
MNCR Costa Rica, San Jose, Museo Nacional de Costa Rica
MND see MNNW
MNDG El Salvador, San Salvador, Museo Nacional "David J. Guzman"
MNF see MNFD
MNFD Germany, Breisgau, Museum fur Naturkunde
MNGC Guatemala, Guatemala City, Museo Nacional de Historia Natural
MNHC Cuba, Habana, Museo Nacional
MNHN France, Paris, Museum National d'Histoire Naturelle
MNHP USA, New Jersey, Princeton, Princeton University
MNHP see also MNHN
MNM see MNMS
MNMS Spain, Madrid, Museo Nacional de Ciencias Naturales
MNNC Chile, Santiago, Museo Nacional de Historia Natural
MNNW Germany, Dortmund, Museum fur Naturkunde
MNRJ see QBUM
MNSL Germany, Leipzig, Museum of Natural Sciences
MNV see MSNV
MOCR see SRNP
MONA Monaco, Musée
MONZ New Zealand, Wellington, Museum
MOSG Romania, Sf. Gheorghe, Muzeul Orasului Sf. Gheorghe
MPEG Brazil, Para, Belem, Museu Paraense Emilio Goeldi
MPGB Guinea-Bissau, Bissau, Museum of Portuguese Guinea
MPM see MCPM
MPMP Philippines, Manila, National Museum of the Philippines
MPRL Marshall Islands, Enewetak Atoll, Mid-Pacific Research Laboratory
MRAC Belgium, Tervuren, Musee Royal de l'Afrique Centrale
MRKC Marc Roger Kutash
MRNP USA, Washington, Ashford, Mount Ranier National Park
MRSN Italy, Torino, Museo Regionale Scienze Naturali
MSIE China, Shanghai, Museum of Shanghai
MSIR Mauritius, Maritius Sugar Industry
MSJC India, Tiruchirapalli, St. Joseph's College
MSNG see MCSN
MSNM Italy, Milan, Museo Civico di Storia Naturale
MSNV Italy, Verona, Museo Civico di Storia Naturale
MSSC USA, Texas, Wichita Falls, Midwestern State University
MSU see MEMU
MSUC USA, Michigan, East Lansing, Michigan State University
MSUE see MEMU
MSUT Albania, Tirane, Museum of Natural History
MTEC USA, Montana, Bozeman, Montana State University
MUCP see UNCP
MUCR Costa Rica, Universidad

MUDH Netherlands, The Hague, Museon
MUIC= MEMU
MUNC Canada, Newfoundland, St.John's, Memorial University
MUNZ New Zealand, Palmerston North, Massey University
MUSM Peru, Lima, Universidad Nacional
MUT see ZLMU
MVEN Netherlands, Enschede, Naturhistorisch Museum
MVNP USA, Colorado, Mesa Verde National Park [No longer listed; no collection.]
MVMA Australia, Victoria, Abbotsville, Museum of Victoria
MWNH Germany, Wiesbaden, Museum Wiesbaden
MZBS Spain, Barcelona, Museo Zoologia
MZC see MZCP
MZCP Portugal, Coimbra, Universidade de Coimbra
MZCR Costa Rica, Ciudad Universitaria, Universidad de Costa Rica
MZF see MZRF
MZHF Finland, Helsinki, Zoological Museum
MZLS Switzerland, Luzern, Musee Zoologique
MZLU Sweden, Lund, Lund University
MZRF Italy, Forli, Museo Zangheri di Storia Naturale della Romagna
MZRO Italy, Verona, Museo P. Zangheri
MZS see MZSF
MZSF France, Strasbourg, University of Strasbourg
MZSP Brazil, São Paulo, São Paulo, Museu de Zoologia da Universidade de São Paulo
MZUC Italy, Cagliari, Universita di Cagliari
MZUF Italy, Florence, Museo Zoologico "La Specola"

NABF North American Beetle Fauna Project, see FSCA
NARI, Guyana, Demerara
NARL Kenya, National Agricultural Research Laboratories
NAUF USA, Arizona, Flagstaff, Northern Arizona University
NAUJ China, Nanjing Agricultural University, Jiangsu
NBM see NBMB
NBMB Canada, New Brunswick, St. John's, New Brunswick Museum
NBME United Kingdom, Bramber, National Butterfly Museum
NCAW China, Shensi, North West College of Agriculture
NCDA see EDNC
NCHU Taiwan, Taichung, National Chung Hsing University
NCM see NCMK
NCMA USA, North Carolina, Raleigh, Vector Control Branch
NCMK United Kingdom, Norwich, Norwich Castle Museum
NCSR see NCSU
NCSU USA, Raleigh, North Carolina, North Carolina State University
NDAT Tunisia, Tunis, Department of Agriculture
NDFC Canada, Newfoundland, Corner Brook, Forest Protection Division
NDSR USA, Ohio, Bowling Green, Bowling Green State University
NDSU USA, North Dakota, Fargo, North Dakota State University
NFRC Canada, Alberta, Edmonton, Northern Forest Research Centre
NFRN Canada, Newfoundland, St. John's, Insect Reference Collection

NHMA Denmark, Jutland, Natural History Museum
NHMB Switzerland, Basel, Naturhistorisches Museum
NHMC Burma, Rangoon, Natural History Museum
NHME Netherlands, Maastricht, Natural History Museum
NHMK Austria, Klagenfurt, Landesmuseum fur Karnten
NHML Libya, Tripoli, Natural History Museum
NHMM Germany, Mainz, Naturhistorisches Museum
NHMN United Kingdom, Nottingham, Natural History Museum
NHMR Netherlands, Rotterdam, Natuurhistorisch Museum
NHMS see SOFM
NHMW Austria, Wien, Naturhistorisches Museum
NHNC Switzerland, La Chaux-de-Fons
NHRI Iceland, Reykjavik, Islandic Museum of Natural History
NHRS Sweden, Stockholm, Naturhistoriska Riksmusset
NHSP Reunion, Saint-Denis, Museum of Natural History
NICC Belize, Cayo, National Insect Collection
NICD India, Delhi, Malaria Research Center
NINF Canada, Newfoundland, Newfoundland Insectarium
NJSM USA, New Jersey, Trenton, New Jersey State Museum
NKMG see NMPG
NKUM China, Tianjin, Nankai University
NLEC Neal L. Evenhuis
NLHD Germany, Hanover, Niedersachsisches Landesmuseum
NLH see NLHD
NLRC Norman L. Rumpp
NMB see NMBZ
NMBA Austria, Admont, Naturhistorisches Museum
NMBS Switzerland, Bern, Naturistorische Museum
NMBZ Zimbabwe, Bulawayo, National Museum
NMCE see CMNC
NMCL Germany, Coburg, Natur-Museums der Coburger Landesstiftung
NMDC N. M. Downie
NMID Ireland, Dublin, National Museum of Ireland
NMK see NMKE
NMKE Kenya, Nairobi, National Museum of Kenya
NMKL Malaysia, Kuala Lumpur, National Museum
NMLS Switzerland, Luzern, Natur-Museum Luzern
NMND India, New Delhi, National Museum of Natural History
NMNH see USNM
NMNK Nepal, Chlauni, National Museum of Nepal
NMNS Taiwan, National Museum of Natural Science, Taichung
NMNZ see MONZ
NMP see NMPC
NMPC Czech, Praha, National Museum (Natural History)
NMPG Germany, DDR, Gotha, Museum der Natur-Gotha
NMPI USA, New Mexico, Las Cruces, Division of Plant Industry
NMS see NMSC
NMSA South Africa, Pietermaritzberg, Natal Museum
NMSC Singapore, National University of Singapore
NMSU USA, New Mexico, Las Cruces, New Mexico State University

NMTC see NMNS
NMTT Trinidad and Tobago, Port-of-Spain, National Museum and Art Gallery
NMVM see MVMA
NMWC United Kingdom, Cardiff, National Museum of Wales
NNKN Netherlands, Tilburg, Noordbrabants Natuurmuseum
NRBC N. Rae Brown and Sandra L. Brown
NREA see NHRS
NRNZ New Zealand, Whangarei, Northland Regional Museum
NRS see NHRS
NRSS see NHRS
NREA Sweden, Stockholm, Naturhistoriska Riksmuseum
NSMC Canada, Nova Scotia, Halifax, Nova Scotia Museum
NSMC see also NVMC
NSMH see NSMC
NSMK Korea, Seoul, National Science Museum
NSMT Japan, Tokyo, National Science Museum (Natural History)
NSPM Canada, Nova Scotia, Halifax, Nova Scotia Museum
NSWA Australia, New South Wales, Rydalmere, Department of Agriculture
NTU see NTUC
NTUC China (Taiwan), Taipei, National Taiwan University
NVDA USA, Nevada, Reno, Nevada State Department of Agriculture
NVMC USA, Nevada, Carson City, Nevada State Museum
NWAU China, Department of Plant Protection, North-West Agricultural University, Shaanxi
NYSM USA, New York, Albany, New York State Museum
NZAC New Zealand, Auckland, DSIR
NZCS Suriname, Paramaribo, University
NZSI India, Calcutta, National Zoological Survey of India

OAMB Benin, Parakou, Open Air Museum
OCOA Canada, Alberta, Olds, Olds College
ODAC USA, Oregon, Salem, Oregon Department of Agriculture
OGAS Gary Shook
OHBR Canada, Ontario, Toronto, Ontario Hydro
OHSC USA, Ohio, Columbus, Ohio Historical Society
OKS see OSEC
OLAN Honduras, Juticalpa, Ministerio de Recursos Naturales
OMNO USA, Oklahoma, Normal, Museum of Natural History
OMNH Japan, Osaka, Osaka Museum of Natural History
OMNZ New Zealand, Dunedin, Otago Museum
ONNC New Caledonia, Noumea, Office de la Recherche Scientifique et Tecnique Outre-Mer
ONPC USA, Washington, Port Angeles, Olympic National Park
ORSC French Cayenne, Office de la Recherche Scientifique et Technique d'Outre-Mer
ORST France, Bondy, ORSTOM
OSAL USA, Ohio, Columbus, Ohio State University Acarology Laboratory

OSEC USA, Oklahoma, Stillwater, Oklahoma State University
OSMH see OHSC
OSU see OSUC
OSUC USA, Ohio, Columbus, Ohio State University
OSUO USA, Oregon, Corvallis, Oregon State University
OTSC Costa Rica, San Jose, Organization for Tropical Studies
OXUM United Kingdom, Oxford, University Museum

PADA USA, Pennsylvania, Harrisburg, Pennsylvania Department of Agriculture
PANZ see PPCC
PAUP India, Ladhiane, Punjab Agricultural University
PBZT Madagascar, Antanamarivo, Parc Botanique et Zoologique de Tsimbazaza
PCNZ New Zealand, Lincoln, Ministry of Agriculture and Fisheries
PCSC Paul C. Schroeder
PEBC Paul E. Blom
PEJS Pablo E. Jordan-Soto
PESC Paul E. Slabaugh ?
PFRA Canada, Saskatchewan, Indian Head, Tree Nursery
PFRS Canada, British Columbia, Victoria, Pacific Forest Research Station
PHAC Peter H. Adler
PHCC Paul H. Carlson
PIKN Fiji, Nausori, Koronivia Research Station
PIME see OSUC
PINN USA, California, Paicines, Pinnacles National Monument
PJLC Peter J. Landolt
PLFV Liechtenstein
PMAE Canada, Alberta, Edmonton, Provincial Museum of Alberta
PMAG United Kingdom, Perth, Perth Museum and Art Gallery
PMAU United Kingdom, Priestgate, Peterborough Museum and Art Gallery
PMFL Slovenia, Ljubljana, Slovenian National Museum
PMFP French Polynesia, Tahiti, Papeete, Papeete Museum
PMIG Germany, Jena, Phyletisches Museum
PMNH USA, Connecticut, New Haven, Peabody Museum of Natural History
PMSL Slovenia, Ljubljana, Slovenian Natural History Museum
PMV see BCPM
PMY see PMNH
PNPC USA, Oklahoma, Sulphur, Platt National Park
PPCC New Zealand, Auckland, Lynfield Agricultural
PPCD USA, West Virginia, Charlestown, West Virginia Department of Agriculture
PPDD Egypt, Cairo, Ministry of Agriculture
PPNP Canada, Ontario, Leamington, Point Pelee National Park
PPRI South Africa, Pretoria, Plant Protection Research Institute
PPRZ Zimbabwe, Causeway, Plant Protection Research Institute
PRHC Parker R. Henry

PSAE Canada, Alberta, Vegreville, Alberta Environmental Centre
PSUC USA, Pennsylvania, University Park, Pennsylvania State University
PUCP India, Chandigarh, Punjab University
PUL see PURC
PURC USA, Indiana, West Lafayette, Purdue University
PUSA see INPC

QBUM Brazil, Rio de Janeiro, Rio de Janeiro, Quinta da Boa Vista
QCAZ Ecuador, Quito, Quito Catholic Zoology Museum
QDPC Australia, Brisbane, Queensland Dept. Primary Industries
QDPI see QDPC
QFMQ South Africa, Queenstown, Queenstown and Frontier Museum
QIBX China, Xining, Quihai Institute of Biology
QMB see QMBA
QMBA Australia, Brisbane, Queensland Museum
QMOR Canada, Quebec, Montreal, Collection entomologique Ouellet-Robert
QPIM Australia, Mareeba, Department of Primary Industries

RABC Robert A. Belmont
RARS Canada, Saskatchewan, Regina, Regina Research Station
RBDC Robert Dressler
RBSC Richard B. Selander, see FSCA
RCDM R. C. de Mordaigle
RCGC Robert C. Graves
RCMC Robert C. Mower
RCWC Richard C. Wilkerson
RDGC Robert D. Gordon ?
RDIC Robert Dirig
RDLC Robert D. Lehman
RDWA Robert D. Ward ?
RDWC Robert D. Ward
REAC Robert E. Acciavatti
RELC R. E. Lewis
RESC Ronald E. Somerby ?
REWC Robert E. Woodruff
RFAC Canada, Riveredge Foundation, see PMAE
RFCC Richard Freitag
RFSC Rebecca F. Surdick
RGBC Robert G. Beard [deceased] see CUIC
RGIC Richard Griffin
RHMP Ronald H. McPeak ?
RKIC Ronald Konopka
RLAC Rolf L. Aalbu
RLBC Richard Lee Berry
RLCA Rudolph Lenczy [Deceased] see USNM
RLHC Roger L. Heitzman
RLHI Ronald L. Huber
RLIC Ron Leuschner

RLIE Rudolph Lenezy, *deceased*, see USNM
RLJC Richard L. Jacques, Jr. *deceased* see FSCA
RLPC Richard L. Penrose
RLWC Richard L. Watson ?
RLWE Richard L. Westcott
RMBB R. M. Baranowski
RMBC R. Michael Brattain
RMHK see KMVC
RMNH Netherlands, Leiden, Nationaal Natuurhistorische Museum
RMSC USA, Colorado, Fort Collins, Rocky Mountain Forest and Range Experiment Station
RMSE see ESUW
RMYC Ronald M. Young ?
ROKC Roy O. Kendall
ROME Canada, Ontario, Toronto, Royal Ontario Museum
RPSP Brazil, Sao Paulo, Ribeirao Preto, Departamento de Biologia
RRHC Ronald R. Hooper
RSAN Robert S. Anderson
RSCC Robert Silberglied [Deceased] Collection?
RSME United Kingdom, Edinburgh, Royal Scottish Museum
RTAC Richard T. Arbogast
RTAS R. T. A. Schouten
RUDZ South Africa, Grahamstown, Rhodes University
RUIC USA, New Jersey, New Brunswick, Rutgers, the State University
RUNB see RUIC
RWBC Richard W. Boscoe
RWCJ Robert William Cavanaugh, Jr.
RWEC Rosser W. Garrison
RWFC R. Willis Flowers
RWGC Rosser W. Garrison
RWHC Robert W. Hamilton
RWHI R. W. Hornabrook
RWSC Robert W. Surdick

SACA China, Chongqing, Southwestern Agricultural College
SACC USA, Iowa, Davenport, St. Ambrose College
SACS China, Shenyang, Shenyang Agricultural College
SAFC Sergio Augusto Fragoso
SAFM Bulgaria, Sofia, Insect Collection
SAIM South Africa, Johannesburg, South African Institute for Medical Research
SAMA Australia, Adelaide, South Australian Museum
SAMC South Africa, Cape Town, South African Museum
SAMR see SAIM
SANC South Africa, Pretoria, South African National Collection of Insects
SBKA Austria, Kremsmunster, Stiftssammlungen des Benediktinerstifts
SBMN USA, California, Santa Barbara, Santa Barbara Museum of Natural History
SCAC China, Guangdong, South China Agric. College

SCDH USA, South Carolina, Columbia, South Carolina Department of Health and Environmental Control
SCSC USA, Minnesota, St. Cloud, Saint Cloud State College
SCUF Brazil, Santa Catarina, Florianopolis, Universidade Federal de Santa Catarina
SCUT see MRSN
SDMC USA, California, San Diego, San Diego Natural History Museum
SDSU USA, South Dakota, Brookings, South Dakota University
SEAC Mexico, Chihuahua, Estacion Experimental de Agricolas de la Campara
SEAN Nicaragua, Léon
SECM Mali, Malé, Science Center
SEMC USA, Kansas, Lawrence, University of Kansas
SEMM Central African Republic, Station Experimental de la Maboke
SFAC USA, Texas, Nacogdoches, Stephen F. Austin State College
SFVS USA, California, Northridge, San Fernando Valley State College
SGLC Sean G. Larsson
SGWC Stanley G. Wellso ?
SIAC China, Chengdu, Sichuan Inst. of Agriculture
SIIS USA, New York, Staten Island, Staten Island Institute of Arts and Sciences
SIUC USA, Illinois, Carbondale, Southern Illinois University
SJAC USA, California, Stockton, San Joaquin County Agriculture Commissioner
SJCA India, Agra, St. John's College
SJSC USA, California, San Jose, San Jose State University
SLJG Austria, Graz, Steirmarkisches Landesmuseum Joanneum
SLSC USA, Missouri, St. Louis, St. Louis Science Center
SLUB Philippines, Baguio City, St. Louis University Museum
SLWC Stephen L. Wood ?
SMCC Scott McCleve
SMCI Sharon M. Clark
SMDV Canada, British Columbia, Vancouver, University of British Columbia
SMEK see SMEC
SMF see SMFD
SMFD Germany, Frankfurt, Forschungsinstitut und Natur-museum Senckenberg
SMLJ see LJSM
SMJM Malaysia, Sabah Museum
SMKM Malaysia, Kuala Lumpar, Selangor Museum
SMMC China, Shanghai, Second Military Medical College
SMNG Germany, Gorlitz, Staatliches Museum fur Naturkunde
SMNH Canada, Saskatchewan, Regina, Museum of Natural History
SMNK Germany, Karlsruhe, Staatliches Museum
SMNS Germany, Ludwigsburg, Staatliches Museum fur Naturkunde
SMO see SMOC
SMOC Czech, Opava, Slezke Muzeum
SMPM USA, Minnesota, St. Paul, Science Museum of Minnesota
SMS see SMSM

SMSH see OMNH
SMSM Malaysia, Sarawak, Sarawak Museum of Natural History
SMTD Germany, Dresden, Staatliches Museum fur Tier-kunde
SMVM Seychelles, Union Vale, National Archives and Museum Division
SMW see MWNH
SMWN Namibia, Windhoek, State Museum of Windhoek
SNHB see SNMB
SNMB Germany, Braunschweig, Staatliches Naturhistorisches Museum
SNMB see also SNMC
SNMC Czech, Bratislava, Slovenske Narodne Muzeum
SNSC Sam Norris
SOFM Bulgaria, Sofia, National Natural History Museum
SOIC Oman, Muscat, National Insect Collection
SPIC Steven Passoa
SRNP USA, Pennsylvania, Philadelphia, Insects of Santa Rosa National Park
SRSC USA, Texas, Alpine, Sul Ross State College
SRTC Stephan R. Treadway
SSMS Suriname, Paramaribo, Suriname State Museum
SSNR Italy, Societa per GL Studi Naturalistica della Romagna
STMC India, Calcutta, School of Tropical Medicine
STRI Panama (via USA, Florida, Miami), Smithsonian Tropical Research Institute
SUEL Hungary, Budapest, Science University "Eotvos Lorand"
SVAM USA, Pennsylvania, Latrobe, St. Vincent Archabbey Museum [transferred to several collections, see especially CMNH].
SWDC Sidney W. Dunkle
SWGC Scott W. Gross
SWRS USA, Arizona, Portal, Southwestern Research Station
SZIC Seymour Ziff

TAIU USA, Texas, Kingsville, Texas A & I University
TAMU USA, Texas, College Station, Texas A & M University
TARI China (Taiwan), Taichung, Taiwan Agricultural Research Institute
TAUI Israel, Tel Aviv, Tel Aviv University
TCBC Thomas C. Barr, Jr. ?
TCDU Uganda, Kampala, Ministry of Animal Industry and Fisheries
TCMC Ted C. MacRae
TDAH Australia, Hobart, Tasmanian Dept. of Agriculture
TDMP Bhutan, Paro, Ta-Dzong Museum
TEDC Thomas E. Dimock
TFRI China (Taiwan), Taipei, Taiwan Forestry Research Institute
TGZC Thomas G. Zoebisch
THAC Thomas H. Atkinson
THDC Thomas H. Davies
TJAC Thomas J. Allen
TLEC Terry L. Erwin see USNM
TLMF Austria, Innsbruck, Tiroler Landsmuseum Ferdinandeum
TLMI see TLMF
TMAG Australia, Hobart, Tasmanian Museum & Art Gallery

TMB see SUEL
TMDU Japan, Tokyo, Tokyo Medical and Dental University
TMMC USA, Texas, Austin, Texas Memorial Museum
TMNH China, Tianjin, Tianjin Museum of Natural History
TMNJ Thomas N. Neal, Jr.
TMP see TMTC
TMSA South Africa, Pretoria, Transvaal Museum
TMT see TMTC
TMTC China (Taiwan), Taipei, Taiwan Provincial Museum
TNAU India, Coimbatore, Tamil Nadu Agricultural University
TOTC T. O. Thatcher ?
TPMR Thomas Moore Rodriguez
TPNG Papua New Guinea, Boroko, Department of Primary Industry
TTCC USA, Texas, Lubbock, Texas Tech University
TTRS USA, Florida, Tallahassee, Tall Timbers Research Station
TTUL see TTCC
TULE Japan, Tokyo, Tokyo University of Agriculture
TUTC Taiwan, Taichung, Tunghai University
TWDC Thomas W. Donnelly
TWTC Terry W. Taylor

UABD see UANH
UADE USA, Arkansas, Fayetteville, University of Arkansas
UAE see UASM
UAF see UADE
UAIC USA, Arizona, Tucson, University of Arizona
UAMM Venezuela, Merida, Universidad de los Andes
UANH USA, Alabama, University, University of Alabama
UASB India, Bangalore, University of Agricultural Sciences
UASC Bolivia, Santa Cruz de la Sierra, Museo de Historia Natural
UASK Ukrania, Kiev, Ukrainian Academy of Science
UASM Canada, Alberta, Edmonton, University of Alberta
UAT see UAIC
UCB see EMEC
UCCC Chile, Concepcion, Universidad de Concepcion
UCD see UCDC
UCDC USA, California, Davis, University of California at Davis
UCEC see UCMC
UCFC USA, Florida, Orlando, University of Central Florida
UCM see UCMC
UCMC USA, Colorado, Boulder, University of Colorado Museum
UCMS USA, Connecticut, Storrs, University of Connecticut
UCNZ New Zealand, Christchurch, University of Canterbury
UCOB see MJMO
UCPC Colombia, Popayan, Universidad del Cauca
UCR see UCRC
UCRC USA, California, Riverside, University of California
UCS see UCMS
UCSE see UCMS
UCV see UCMS

UCVC Chile, Valparaiso, Universidad Catolica de Valparaiso
UCVM see IZAV
UDCC USA, Delaware, Newark, University of Delaware
UDN see UDCC
UDSM Tanzania, Dar es Salaam, University of Dar es Salaam
UES see UESS
UESS El Salvador, San Salvador, Universidad de El Salvador
UFBI Italy, Universita di Firenze
UFMI Brazil, Mato Grosso, Instituto de Biociencias
UFNH USA, Utah, Utah Field House Natural History [address unknown]
UFPC see DCMP
UFRG Brazil, Rio Grande do Sul, Porto Alegre, Departamente de Zoologia
UFVB Brazil, Minas Gerias, Viçosa, Museum of Entomology
UGA see UGCA
UGCA USA, Georgia, Athens, University of Georgia
UGGE Ecuador, Guayaquil, Universidad de Guayaquil
UGCA see UGCA
UGG see UGGG
UGGG Guyana, Georgetown, University of Guyana
UICM see WFBM
UIM see WFBM
UKKY China, Kunming, University of Kunming
UKMS Sudan, Khartoum, Sudan Natural History Museum
ULCI Spain, Universidad de la Laguna, Canary Islands
ULKY USA, Kentucky, Louisville, University of Louisville
ULQC Canada, Quebec, Quebec, University of Laval
UMA see UMEC
UMAA see UMMZ
UMB see UMBB
UMBB Germany, Bremen, Ubersee-Museum
UMC see UMRM
UMCP see UMDC
UMDC USA, Maryland, College Park, University of Maryland
UMDE USA, Maine, Orono, University of Maine
UMEC USA, Massachusetts, Amherst, University of Massachusetts
UMED Kenya, University of Moi
UMIC USA, Mississippi, University, University of Mississippi
UMKU Uganda, Kampala, Uganda Museum
UMMZ USA, Michigan, Ann Arbor, University of Michigan
UMO see OXUM
UMRM USA, Missouri, Columbia, University of Missouri
UMSA Bolivia, La Paz, Instituto de Ecologia
UMSP USA, Minnesota, St. Paul, University of Minnesota
UNAC Nicaragua, Manogua, Museo Entomológico
UNAD Peru, Lima, Universidad Nacional Agraria
UNAM Mexico, Mexico, Universidad Nacional Autonoma
UNAN Nicaragua, Léon
UNCB Colombia, Bogotá, Muséo de Historia Natural

UNCC Colombia, Manizales, Museo de Historia Natural
UNCM Colombia, Medellin, Museo de Entomologia "Francisco Luis Gallego"
UNCP Colombia, Palmira, Universidad Nacional de Colombia
UND see NDSU
UNH see DENH
UNL see UNSM
UNLO Venezuela, Guanare, Universidad Nacional Experimental de los Llanos Occidental
UNMC USA, New Mexico, Albuquerque, University of New Mexico
UNSA South Africa, Pietermaritzburg, University of Natal
UNSA see also UNSP
UNSM USA, Nebraska, Lincoln, University of Nebraska
UOKN see OMNH
UOIC USA, Oregon, Eugene, University of Oregon
UOP see UOPJ
UOPJ Japan, Osaka, University of Osaka Prefecture
UPNG Papua New Guinea, University, University of Papua New Guinea
UPP see UPPC
UPPC Philippines, Laguna, University of the Philippines
UPRG Peru, Lambayique, Universidad Nacional "Pedro Ruiz Gallo"
UPRM see EPRL
UPSA South Africa, Pretoria, University of Pretoria
UQBA see UQIC
UQIC Australia, Queensland, University of Queensland
URI see URIC
URIC USA, Rhode Island, Kingston, University of Rhode Island
URM see URMU
URMU Uruguay, Montevideo, Museo Nacional de Historia Natural
USAC Australia, W. A., Perth, University of Western Australia
USCC USA, Colorado, Pueblo, University of Southern Colorado
USCP Philippines, Cebu City, University of San Carlos
USNM USA, D.C., Washington, Smithsonian Institution
USS see MAMU
USTF Suriname, Paramaribo, University of Suriname
USTK Gahana, Kumasi, Museum of Natural History
USUC see CSUC
USUL see EMUS
USWL see LSUC
UTKI Iran, Karadj, Universite de Teheran
UVCC USA, Vermont, Burlington, University of Vermont
UVG see UVGC
UVGC Guatemala, Guatemala City, Universidad del Valle de Guatemala
UWBM USA, Washington, Seatle, Burke Museum
UWCP Poland, Wroclaw, University of Wroclaw
UWEM see IRCW
UWI see UWIJ
UWIC Trinadad and Tobago, St. Augustine, University of the West Indies
UWIJ Jamaica, Kingston, University of the West Indies

UWOC Canada, Ontario, London, University of Western Ontario
UWEM see IRCW
UWM see IRCW
UYIC Uruguay, Montevideo, Facultad de Humanidades y Ciencias
UZIL see MZLU
UZIU Sweden, Uppsala, Universitets Zoologiska Institut
UZM see UZMC
UZMC Venezuela, Maracaibo, Universidad del Zulia
UZMC see also MZLU
UZMD see MZLU
UZMH see MZHF

VABL Vernon A. Brou
VDAC USA, Virginia, Richmond, Virginia Department of Agriculture and Commerce
VDAM Australia, Melbourne, Victoria Dept. of Agriculture
VDRC Vincent D. Roth
VICH Vietnam, Hanoi, Plant Protection Department
VMKC Vernon M. Kirk
VMNH USA, Virginia, Martinsville, Virginia Museum of Natural History
VNGA Austria, Dornbirn, Vorarlberger Naturschau
VOBC Vitor O. Becker
VPI see VPIC
VPIC USA, Virginia, Blacksburg, Virginia Polytechnic Institute
VPKC V. P. Karasjov
VRBC Vratislav Richard Bejsak
VSCA Philippines, Baybay, Visayas State College of Agriculture
VUWE New Zealand, Wellington, Victoria University

WACC Walter A. Connell?
WAMP Australia, Perth, Western Australian Museum
WARI Australia, South Australia, Duncan Swan Insect Collection
WARS Kenya, Nairobi, Wildlife Advisory and Research Service
WBMC William B. Muchmore
WEIC Papua New Guinea, Wau, Wau Ecology Institute
WETA Francis Dudley Chambers
WFBC William F. Barr
WFBM USA, Idaho, Moscow, University of Idaho
WHCC William H. Clark
WHNC Willard H. Nutting [Deceased]
WHTC W. H. Tyson
WIBG St. Lucia, Castries
WIUC USA, Illinois, Macomb, Western Illinois University
WPEC William Peters
WPNH No collection, deleted
WRCC William C. Rosenberg [deceased] see FSCA and USNM
WRSC Walter R. Suter ?
WSCC Wilfred S. Craig ?
WSU see WSUC
WSUC USA, Washington, Pullman, Washington State University

WTSC William T. Schultz
WVUC USA, West Virginia, Morgantown, West Virginia University
WWSP USA, North Carolina, Southern Pines, Weymouth Woods Sandhills Nature Preserve
WYWC W. Y. Watson ?

YEUX P. E. Hallett
YMUK United Kingdom, York, The Yorkshire Museum
YPMC USA, Wyoming, Yellowstone Park, Yellowstone National Park

ZAUC China, Hangzhou, Zhejian Agric. University
ZBMM India, Ernakulam, Maharaja's College
ZDUA see UASM
ZFMK Germany, Bonn, Zoologische Forschungsinstitut und Museum "Alexander Koenig"
ZGLC Cyprus, Limassol, Natural History Museum
ZIL see ZMAS
ZILS see MZLU
ZISB Bulgaria, Sofia, Institute of Zoology
ZIUS Austria, Salzburg, Zoologisches Institute der Universitat
ZLMU Japan, Nagoya, Meijo University
ZMAN Netherlands, Amsterdam, Zoologisch Museum
ZMAS Russia, Leningrad, Universitetskaya
ZMB see ZMHB
ZMBJ Indonesia, Bandung, Zoological Museum
ZMH see MZHF
ZMHB Germany, Berlin, Museum fur Naturkunde der Humboldt Universitat
ZMHU see ZMHB
ZMK see ZMKU
ZMKD see ZMUC
ZMKU Russia, Kiev, Zoological Museum
ZMLP Pakistan, Lahore, University of Punjab
ZMOF see ZMUO
ZMPA Poland, Warszawa, Polish Academy of Science
ZMS see ZMSZ
ZMSZ Bosnia-Herzegovina, Sarajevo, Zemaljski Muzej
ZMUA see ZMAN
ZMUB Norway, Bergen, University of Bergen
ZMUC Denmark, Zoologisk Museum, Copenhaven University
ZMUD Bangledesh, Dhaka, Zoology Museum
ZMUH Germany, Hamburg, Universitat von Hamburg
ZMUL see MZLU
ZMUM Russia, Moscow, University of Moscow
ZMUN Norway, Oslo, University of Oslo
ZMUO Finland, Oulu, Universitets Oulu
ZMUO see also ZMUN
ZNPC USA, Utah, Springdale, Zion National Park
ZSBS see ZSMC
ZSCI see NZSI

ZSMC Germany, Munchen, Zoologische Sammlung des Bayerischen Staates
ZUAC Madagascar, Antananarivo, Laboratoire de Zoologie

REFERENCES

Note: In addition to these publications, many societies publish membership lists which provide names of institutions with collections.

Arnett, Ross H. Jr., and Samuelson, G. Allan. 1969. Directory of Coleoptera collections of North America (Canada through Panama). Lafayette: Purdue University. 123 p. [Out of print.]

Arnett, Ross H. Arnett, Jr., G. Allan Samuelson, *et al.* 1986. The Insect and Spider Collections of the World. Gainesville, FL: Flora & Fauna Publications. 220 p. [Out of print.]

Hancock, E. G. and Morgan, P. J. (Eds.). 1980. A survey of zoological and botanical material in museums and other institutions of Great Britain. Report no. 1. Biology Curators Group. 32 p.

Heppner, John B., and Lamas, Gerardo. 1982. Acronyms for world museum collections of insects, with an emphasis on Neotropical Lepidoptera. Bull. Ent. Soc. America. 28:305-316.

Horn, W., and Kahle, I. 1935. Uber entomologische Sammlungen Entomologen & Entomo-Museologie. Berlin-Dahlem, Entomo-logische Beihefte, 536 p., 38 pl.

Horn, W., and Kahle, I. 1935. Supplement to: Uber entomologische Sammlungen Entomologen & Entomo-Museologie. Berlin-Dahlem, Entomologische Beihefte, 12 p.

Hudson, Kenneth and Nichols, Ann (Eds.), 1975. The directory of world museums. New York: Columbia University Press. xviii + 864 p.

Jacques, Richard L., Jr., 1974. Directory of American Coleopterists, including descriptions of their research projects and a list of abbreviations of collection names for the world. Latham: Biological Research Institute of America, iii + 86 p.

Kingsolver, J. M. 1979. A catalog of the Coleoptera of America north of Mexico. Washington, D.C.: Agriculture Handbook No. 529-1, pp.vii-ix.

Moucha, Josif, *et al.* 1971. Conscription of entomological collections in Czechoslovak museums. Bratislava. 55 p., map.

Sachtleben, Hans. 1961. Second Supplement to: Uber entomologische Sammlungen Entomologen & Entomo-Museologie. Beitrage Ent., 11:481-540.

Secretariat, Entomological Society of Canada. 1978. Collections of Canadian insects and certain related groups. Entomological Society of Canada, 12 p.

Watt, J. C. 1979. Abbreviations for entomological collections. New Zealand J. Zool. 6:519-520.

Williams, R. N., chairman, special committee. 1978. Worldwide directory of institutions with entomologists, part I: Latin America. Bull. Ent. Soc. America, 24:179-193.

INDEX TO TAXA

(Included are some genera, families, superfamilies, and orders, exclusive of Coleoptpera, Hymenoptera, Lepidoptera, Diptera, and Arachnida. Almost every collection contains spiders and the four major insect orders. For our use here, Heteroptera=Hemiptera for uniformity.)

INDEX TO PERSONNEL

NOTE: "coll." = collection; refers to individual's collections held by public collection. Private collections are either listed under the public collection with which they are registered, or in the separate section, pages 244-246.

QUESTIONNAIRES

These two pages of instructions, and the following two pages of questionnaires, may be photocopied and used for the submission of data for future editions of this directory.

INSECT AND SPIDER COLLECTIONS OF THE WORLD

The purpose of this directory is to provide a standard, world-wide, list of collections of insects and spiders useful for the location of specimens and, with the assigned codens, a uniform way to cite the location of specimens in publications treating the species stored in these collections.

Each collection is described by giving the size of the collection, either by citing the number of drawers, boxes, slides, and vials, or by the number of specimens they contain. The names of the curators are listed, along with notes on special collections, the regions covered, and similar data. Each collection is assigned a coden (composed of four letters) unique to that collection. This coden is, of necessity, four letters in order for enough unique codens to be available for the number of collections now known to occur throughout the world. If some of the previously used three-letter codens are continued, large blocks of letters become unavailable for the existing collections. Four letter codens provide enough possible combinations to add new collections as needed. The present list of codens has been adopted by several publications, particularly those used by coleopterists. The compilers hope that this list will continue to be the "standard."

Recognizing as we do, that many public, and especially, private collections have been omitted, either because we do not know of their existence, or because we did not reach the proper curator, we ask our readers' help in getting the following form to the proper people for completion and filing in our computer data base for use in revised editions of the work. Meanwhile, we are continuing our attempts to reach the proper person and to complete (or delete, as the case may be), those collections listed here that have not responded. Curators of collections listed are invited to resubmit the form or photocopies of updated listings.

Please note that two now only one form is used, both for "public" collections, *i.e.*, those collections owned and maintained by organizations instead of individuals, and those for the "private" owners of collections, generally, individuals, as opposed to museums, foundations, government institutions, and other groups that are expected to outlive individuals.

We do not wish to include any private collection not registered with a public collection. The reason for this is apparent when you see the number of private collections that have been lost even though listed in previous literature. The data reported can no longer be confirmed and the vouchered specimens represented are lost. However, we have not excluded unregistered collections, but ask that you seriously consider doing this simple procedure, as explained below.

Registering private collections: Several museums have offered to register private collections which we include in this directory. To register, private collectors must contacts a curator at a public collection, submit a copy of the form describing the collection, and asked to have the collection can be registered. When accepted, the original copy is submitted to us for recording. This registration does not imply or require the donation of the collection to the registering institution now or in the future. It is only an agreement to keep track of the collection, record movements of the material, other changes, and final disposition through sale, donation, abandonment, or destruction. Thus vouchered specimens may be found even after the collection owner no longer has them.

QUESTIONNAIRE FOR PUBLIC AND PRIVATE COLLECTIONS OF INSECTS AND SPIDERS

Name or owner of the collection: ________________________________

Mail address: ____

____

________________________(Postal codes & country) ____________

________________________ Phone (with access codes)____________

Collection coden, see instructions (______). Director (or owner) of insect and spider collection, *i.e.*, head curator, etc., include Dr., Mr., Ms., etc.:

Professional staff: ____

____

(List additional staff on back or send separate list.)

Description of collection (include number of specimens in collection or number of storage containers and an estimate of no. of specimens; include information about primary and secondary types, and special collections, including separate collections of vouchered specimens):

Periodicals or series sponsored by institution:

If this collection is already listed in the book, attached below (or with this form) is a photocopy of the listing.

IF THIS COLLECTION is privately owned, give name of public collection with which it is registered. (Give name and address of a public collection and the name of the curator accepting the registration.):

The next edition is not scheduled for publication. The next edition may be produced for electronic publication, but there may also be a supplement to this printed edition. Please help us keep this edition up to date by sending data to Dr. G. Allan Samuelson, Bishop Museum, P. O. Box 19000A, Honolulu, HI 96817. [Phone: (808) 848-4197.]

ORDER FORM

[ ] Please send us ______ copy(ies) of the cloth bound, definitive edition, 6 x 9" 350 pp. list price, $25.00 plus shipping. If payment is received with this order, shipping is free.

[ ] Payment enclosed: $35.00.

[ ] Please send invoice (individuals, Pro-forma) for $35.00 + $3.50 for shipping.

[ ] Institutional purchase order attached, $35.00 plus $3.50 for shipping.

Ship to:

Send order form to: AMERICAN INSECT PROJECTS, 2406 N. W., 47th Terrace, Gainesville, FL 32606 USA